Forum for Interdisciplinary Mathematics

Editors-in-Chief

Viswanath Ramakrishna, University of Texas, Richardson, USA
Zhonghai Ding, University of Nevada, Las Vegas, USA

Editorial Board

Ashis SenGupta, Indian Statistical Institute, Kolkata, India
Balasubramaniam Jayaram, Indian Institute of Technology, Hyderabad, India
P. V. Subrahmanyam, Indian Institute of Technology Madras, Chennai, India
Ravindra B. Bapat, Indian Statistical Institute, New Delhi, India

The *Forum for Interdisciplinary Mathematics* is a Scopus-indexed book series. It publishes high-quality textbooks, monographs, contributed volumes and lecture notes in mathematics and interdisciplinary areas where mathematics plays a fundamental role, such as statistics, operations research, computer science, financial mathematics, industrial mathematics, and bio-mathematics. It reflects the increasing demand of researchers working at the interface between mathematics and other scientific disciplines.

Khosro Shahbazi

Finite Difference Methods for Compressible Two-Fluid Dynamics

Springer

Khosro Shahbazi
Department of Mechanical Engineering
South Dakota School of Mines
and Technology
Rapid City, SD, USA

ISSN 2364-6748　　　　　　　ISSN 2364-6756　(electronic)
Forum for Interdisciplinary Mathematics
ISBN 978-3-031-87340-9　　　ISBN 978-3-031-87341-6　(eBook)
https://doi.org/10.1007/978-3-031-87341-6

© The Editor(s) (if applicable) and The Author(s), under exclusive license to Springer Nature Switzerland AG 2025

This work is subject to copyright. All rights are solely and exclusively licensed by the Publisher, whether the whole or part of the material is concerned, specifically the rights of translation, reprinting, reuse of illustrations, recitation, broadcasting, reproduction on microfilms or in any other physical way, and transmission or information storage and retrieval, electronic adaptation, computer software, or by similar or dissimilar methodology now known or hereafter developed.
The use of general descriptive names, registered names, trademarks, service marks, etc. in this publication does not imply, even in the absence of a specific statement, that such names are exempt from the relevant protective laws and regulations and therefore free for general use.
The publisher, the authors and the editors are safe to assume that the advice and information in this book are believed to be true and accurate at the date of publication. Neither the publisher nor the authors or the editors give a warranty, expressed or implied, with respect to the material contained herein or for any errors or omissions that may have been made. The publisher remains neutral with regard to jurisdictional claims in published maps and institutional affiliations.

This Springer imprint is published by the registered company Springer Nature Switzerland AG
The registered company address is: Gewerbestrasse 11, 6330 Cham, Switzerland

If disposing of this product, please recycle the paper.

Preface

This book aims at providing the essentials of high-order numerical methods for compressible single-fluid and two-fluid transport phenomena. The presented schemes enable computations for broad applications, including shock-induced interfacial instability and turbulence, shock-bubble interactions, and detonation, to name a few. The enabled direct numerical simulations also help obtain accurate data for tuning the emerging physics-based neuromorphic algorithms.

The field of numerical methods for fluid dynamics is vast. It includes various approaches, from those tailored to complex geometries with unstructured grids, such as finite elements and finite volumes, and infusion of the two into discontinuous Galerkin methods to those exploiting the simplicity of Cartesian grids, such as finite difference methods. This book is about finite difference methods.

I have written the book with two principles in mind. First, for a broad reach and impact, the numerical scheme should, to the extent possible, satisfy the simultaneous requirements of simplicity, extendibility, and efficiency on serial and parallel computers. Second, the physics of the compressible two-fluid system must guide the design and analysis of the numerical methods.

Guided by the first principle, the book aims to provide, with a sound pedagogy, the design, and analysis of a class of nonlinear high-order finite difference spatial schemes suitable for compressible single- and two-fluid systems. The text gives a detailed account of the primitive variable weighted essentially non-oscillatory (WENO) reconstructions, as well as the recently developed variants of the WENO scheme, namely the targeted essentially non-oscillatory (TENO) schemes, which yield a higher level of accuracy. The book presents, for time integration, high-order Runge-Kutta (RK) and spectral deferred correction (SDC) schemes up to the 12-order accuracy. The diagonally implicit RK schemes are specialized for stiff problems, while the explicit-implicit RK schemes are suitable for problems with non-stiff and stiff effects. Alternatively, SDC schemes provide a general, arbitrarily high-order framework for a semi-implicit treatment of transport phenomena involving non-stiff and stiff effects.

The second principle requires the book to address two interrelated tasks: (1) the discussion of the physical characteristics and modeling of the compressible single-

and two-fluid transport phenomena, and (2) the design and analysis of the appropriate numerical schemes of the partial differential equations (PDEs) governing the compressible transport phenomena. The book seeks the infusion of physical insights into the design and analysis of the corresponding numerical schemes. Examples include the following:

- The wave structure of the dynamics of a compressible medium dictates the choice of linearly stable spatial schemes based on directionally-biased approximations for the interior of the physical domain and one-sided approximations near the boundary of the spatial domain.
- The converging characteristics resulting in shock waves and the parallel ones in the advection of fluid interfaces with large jumps lead to the design of high-order nonlinear schemes that are capable of non-oscillatory capturing of discontinuities with minimal artificial dissipation.
- The hyperbolicity nature of the compressible two-fluid or two-phase models requires nonlinearly stable numerical schemes for advective, diffusive, and source terms that are positivity preserving for the square of characteristic speeds.
- The multi-scale nature of the two-fluid or two-phase model arising from reaction and phase change and the wide spread of the eigenvalues from the diffusive effects demand the use of temporal schemes with desirable stability properties for highly stiff PDEs.

The book focuses on the latest developments in the numerical modeling of compressible two-fluid dynamics and is unique in several ways. It discusses the high-order schemes for compressible two-fluid systems involving shock waves and material interfaces with potentially large jumps across them. The text presents certain high-order WENO and TENO finite difference methods that are extendable to two-fluid models and compressible diffusive fluxes in the Navier-Stokes equation. Furthermore, it details general multidimensional viscous numerical models for boundary treatments and approximation near boundaries for Cartesian and general non-Cartesian domains, which are often ignored. The second uniqueness of the book is its presentation of the diagonally implicit Runge-Kutta and spectral deferred correction schemes for a semi-implicit temporal discretization of compressible flows involving stiff terms. Finally, another unique aspect of the book is the extensive discussion of the nonlinear stability of the numerical schemes in preserving hyperbolicity in compressible single-fluid and two-fluid systems with and without compressible diffusive effects. Without a positivity-preserving mechanism, the numerical solution scheme lacks robustness, and in the presence of strong shock waves and material interfaces, it may fail, leading to a computational overflow.

The book can serve as a first course on the numerical methods for transport phenomena or fluid dynamics for students in mechanical, aerospace, and chemical engineering, applied mathematics, and physics at the senior level of an undergraduate or graduate degree. It also provides foundations and algorithmic details for implementing the most recent numerical schemes for compressible flows and extending them to include other physics, such as elasticity, reaction, and magnetohydrodynamics.

Chapter 1 describes the characteristics of single-fluid compressible dynamics. Chapter 2 introduces the polynomial interpolation problem and the high-order nonlinear finite difference schemes, while Chap. 3 deals with the approximations on and near domain boundaries for Cartesian and non-Cartesian domains. Chapter 4 presents the high-order temporal schemes of Runge-Kutta and spectral deferred correction approaches. Chapter 5 focuses on the high-order methods for nonlinear conservation laws, and Chap. 6 extends the numerical schemes to a simple compressible two-fluid model. Chapter 7 material extends the high-order methods to the diffusive fluxes in the compressible single- and two-fluid Navier-Stokes equations. Chapter 8 provides the necessary thermodynamic property relations for compressible two-fluid and two-phase modeling. Finally, Chap. 9 introduces a four-equation, two-phase model with and without heat and mass transfer. This chapter provides introductory material on two-phase and multi-phase flow modelling as they are currently the subject of extensive research. At the end of each chapter, exercises reinforce the comprehension of the materials presented; they may involve the implementation of particular numerical schemes. For the book to be self-contained, it highlights the classical results on low-order temporal and spatial methods for model problems in Appendix A.

Materials related to the numerical schemes for the compressible two-fluid flows are adopted from my research, carried out with financial support from the Office of Naval Research. I express my gratitude to the Office of Naval Research for supporting my research and providing support during the majority of the time I was writing this text.

I want to thank my former students, Daniel Boe and Tristan van Nieda, for many interesting discussions during their research on algorithms for two-fluid flows. I further thank Daniel Boe for generating Fig. 7.1 in Chap. 7.

Rapid City, SD, USA   Khosro Shahbazi

Competing Interests The author has no competing interests to declare that are relevant to the content of this manuscript.

Contents

1 Introduction and Basic Dynamics of Compressible Medium 1
   1.1 Equations of Compressible Fluid Transport 1
      1.1.1 Integral Forms of Compressible Fluid Transport 2
      1.1.2 Differential Forms of Compressible Fluid Transport 4
      1.1.3 Euler System and Entropy Equation 5
   1.2 Characteristics of the Compressible Fluid Dynamics 6
      1.2.1 Characteristic Speeds and Curves 7
      1.2.2 Shock Wave in a Nonlinear Conservation Law 16
      1.2.3 Shock Wave and Diffusive Model 17
      1.2.4 Characteristics of Diffusion and the Stiffness 18
      1.2.5 Conservation and Shock Speed 19
   1.3 Exercises 22
   References 24

2 Interpolation and High-Order Nonlinear Finite Difference Schemes 25
   2.1 Linear Interpolation 26
      2.1.1 Error in Interpolation 27
      2.1.2 Interpolation Error for Smooth Function 28
      2.1.3 Interpolation for Discontinuous Functions 30
   2.2 High-Order Finite Difference Scheme 32
      2.2.1 Why Mid-Point Approximations? 32
      2.2.2 Mid-Point Approximation 35
      2.2.3 Derivative Approximation 37
      2.2.4 Conservative Form of High-Order Derivative Approximation 42
   2.3 High-Order Finite Difference Method for Discontinuous Solutions 43
      2.3.1 Smoothness Indicator 44
      2.3.2 Essentially Non-oscillatory Scheme 46
      2.3.3 Weighted Essentially Non-oscillatory Scheme 46

		2.3.4	WENO Algorithm	48
		2.3.5	Conservative Derivative Approximation and WENO Interpolation	54
	2.4	Targeted Essentially Non-oscillatory Schemes	55
		2.4.1	Incremental Stencil Interpolation	56
		2.4.2	Strong Scale Separation	58
		2.4.3	Stencil Selection	58
		2.4.4	Assembled High-Order Scheme	59
	2.5	Exercises	59
	References	61
3	**Approximation On and Near Boundary**	63
	3.1	General Characteristic Boundary Treatments for System of Equations	64
		3.1.1	Modified Governing Equations on the Boundary	67
		3.1.2	Explicit Modification Terms for Various Boundary Conditions	68
	3.2	Distretization of Various Boundary Conditions on Cartesian Domains	75
		3.2.1	Periodic Boundary Condition	76
		3.2.2	Reflecting Boundary Condition	76
		3.2.3	Dirichlet Boundary Condition	78
		3.2.4	Derivative Approximation Near the Boundary	81
		3.2.5	Derivative Approximation on the Boundary	82
		3.2.6	Neumann Boundary Condition	83
	3.3	Boundary Condition Approach for General Non-Cartesian Domains	84
		3.3.1	Computing the Boundary-Intercept Point and the Normal Direction	85
		3.3.2	Ghost Data Approximation: Fluxes or Primitive Variables?	86
		3.3.3	Hermit Polynomial Reconstruction	87
	3.4	Exercises	90
	References	91
4	**High-Order Time Integration Methods**	93
	4.1	Runge-Kutta Methods	95
		4.1.1	Construction of Runge-Kutta Methods	96
		4.1.2	Alternative Form of RK Schemes	101
		4.1.3	Error of Runge-Kutta Schemes	102
		4.1.4	Error Due to Boundary Condition Imposition	105
		4.1.5	Example of Implicit RK Schemes	106
		4.1.6	Diagonally Implicit Runge-Kutta Schemes	107
		4.1.7	Examples of Diagonally Implicit Runge-Kutta Schemes	108
		4.1.8	Examples of Explicit Runge-Kutta Schemes	108

	4.1.9	Removal of the Boundary Condition Error in the Third-Order Runge-Kutta Scheme	109
	4.1.10	Implicit-Explicit Runge-Kutta Schemes	111
4.2	Spectral Deferred Correction Scheme		112
	4.2.1	Defect Corrections	112
	4.2.2	From Defect Correction to Deferred Correction	113
	4.2.3	Arbitrarily High-Order SDC Method	117
	4.2.4	Multi-implicit SDC Schemes	118
4.3	Final Comments		121
4.4	Exercises		122
References			124

5 High-Order Nonlinear Schemes for Nonlinear Conservation Laws ... 127

5.1	First-Order Lax-Friedrichs Scheme		128
	5.1.1	Monotonicity and Maximum Principle-Preserving	130
	5.1.2	Dissipation and Stability of the Lax-Friedrichs Scheme	131
5.2	High-Order Lax-Friedrichs Scheme for Scalar Conservation Law		132
	5.2.1	Conservative Form of the Discretized Scalar Nonlinear Conservation Law	134
	5.2.2	Maximum-Principle Preservation of High-Order Schemes	135
5.3	Lax-Friedrichs Scheme for System of Conservation Laws		138
	5.3.1	Numerical Example of Comparison of Approximation Orders for Ultrasound Pulse Propagation	139
	5.3.2	Positivity of Pressure in First-Order Scheme	141
	5.3.3	Positivity Enforcement of Pressure in High-Order Scheme	143
5.4	Numerical Examples for Positivity Preservation		145
	5.4.1	Linear Advection with Continuous Initial Data	146
	5.4.2	Inviscid Burgers Equation	147
	5.4.3	Euler System Interacting Blast Waves	148
	5.4.4	Euler System Shock, Interface Problem	151
	5.4.5	2D Isentropic Vortex	154
5.5	Exercises		157
References			158

6 Schemes for Compressible Two-Fluid Model ... 159

6.1	Compressible Two-Fluid Model		160
	6.1.1	Equation of State	160
	6.1.2	Two-Fluid Model	162
	6.1.3	Simple Advection of an Interface	164

		6.1.4	Lack of Consistency of the Conservative Model with the Simple Advection	165
		6.1.5	Consistency of the Quasi-Conservative Model with the Simple Advection	167
	6.2	Consistent Numerical Method for the Two-Fluid System		169
	6.3	Numerical Examples on High-Order Scheme for Two-Fluid System		170
		6.3.1	1D Strong Shock Helium-Air Interface Interaction	170
		6.3.2	Two-Dimensional Air Shock Cylindrical Helium Bubble Interaction	173
	6.4	Positivity-Preserving Property of Numerical Scheme for the Two-Fluid System		177
		6.4.1	Positivity Preservation of the First-Order Scheme	177
		6.4.2	Positivity Preservation of the High-Order Scheme	179
	6.5	Numerical Examples on Positivity Preservation for Two-Fluid System		179
		6.5.1	One-Dimensional Water and Air Shock, Interface Problem	180
		6.5.2	Two-Dimensional High Mach Shock Air Bubble Interaction	184
		6.5.3	Impact on CPU Time	189
	6.6	Exercises		189
	References			190
7	**Schemes for Compressible Two-Fluid Navier-Stokes Equations**			**191**
	7.1	Multi-dimensional Compressible Navier-Stokes Equations		191
	7.2	Property Relations		192
		7.2.1	Example of Property Relations for the Stiffened-Gas Equation of State	193
	7.3	Diffusive Flux Discretization		194
		7.3.1	First-Order Scheme	194
		7.3.2	High-Order Scheme	196
		7.3.3	Example of High-Order Positivity Preserving Scheme for the Navier-Stokes Equations	196
	7.4	Two-Fluid Compressible Viscous Model		197
		7.4.1	Positivity of the Two-Fluid Viscous Model	199
		7.4.2	Computation of Thermal Conductivity and Viscosity Coefficient	200

		7.5 Exercises	201
		References	202
8	**Thermodynamic Property Relations for Two-Phase Modeling**		203
	8.1	Fundamental Relations	203
	8.2	Conditions of Equilibrium	205
	8.3	Euler's Thermodynamics Equation	207
	8.4	Alternative Forms of Fundamental Relations	208
	8.5	Examples of Fundamental Equations	210
	8.6	General Relations for du, dh, and ds	211
	8.7	Speed of Sound as a Function of T and P	213
	8.8	Example: Internal Energy for Stiffened-Gas EoS	214
	8.9	Exercises	215
		References	215
9	**Mixture Theory Modeling of Compressible Two-Phase Systems**		217
	9.1	Mixture Theory Model for Two-Fluid System	218
		9.1.1 Reduction Under the Instantaneous Equilibrium	220
		9.1.2 System of Two Ideal Gases	221
		9.1.3 System of Gas and Liquid	222
		9.1.4 Modeling Mass Transfer or Phase Change	223
		9.1.5 Modeling Heat Conduction in Gas-Liquid Mixture with Amagat Law	224
	9.2	Justification for Single Temperature, Pressure, and Velocity Model	225
	9.3	Two-Phase Model with Finite Phase Change Rate	227
		9.3.1 Numerical Scheme for Two-Phase Flow Model	228
	9.4	Exercises	228
		References	228

Appendix: Basic Schemes for Model Problems 231

Index 259

Chapter 1
Introduction and Basic Dynamics of Compressible Medium

Transport phenomena are broad, diverse subjects with many specific areas, including compressible single-fluid, two-fluid, and two-phase fluid dynamics with phase change, chemical reactions, and heat transfer. This chapter discusses the mathematical models of compressible single-fluid dynamics and develop insights into their solutions and the physics, which will, in turn, guide the design of high-fidelity numerical algorithms for such models, as will be discussed in later chapters. Two-fluid and two-phase flows share similar characteristics with the single-fluid flows but have their unique complexities, which will be discussed starting in Chap. 6 and continuing through Chap. 9.

1.1 Equations of Compressible Fluid Transport

The integral form of the governing equations for the compressible fluid transport for a finite fluid element is first presented. The integral form is valid even when the flow has discontinuous features such as shock waves or two-fluid interfaces. Under the assumption that fluid properties are smooth (flow has no discontinuous features), the differential form of the governing equations is introduced. The differential form reveals the wave characteristics and mechanisms of shock formations; these insights guide the development of appropriate numerical schemes.

1.1.1 Integral Forms of Compressible Fluid Transport

The mass, momentum, and energy of a finite fluid element with a time-dependent volume $V(t)$ are represented as

$$\text{mass} = \int_{V(t)} \rho \, dV \tag{1.1a}$$

$$\text{momentum} = \int_{V(t)} \rho \vec{V} \, dV \tag{1.1b}$$

$$\text{energy} = \int_{V(t)} \rho e_{\text{total}} \, dV \tag{1.1c}$$

where ρ, e_{total}, and $\vec{V}$ are density, specific total energy, and velocity vector. The mass conservation, linear momentum equation (in an inertial reference frame), and the energy conservation for a fluid element moving about are thus

$$\frac{d}{dt} \int_{V(t)} \rho \, dV = 0 \tag{1.2a}$$

$$\frac{d}{dt} \int_{V(t)} \rho \vec{V} \, dV = \vec{F} \tag{1.2b}$$

$$\frac{d}{dt} \int_{V(t)} \rho e_{\text{total}} \, dV = \dot{E}_{\text{in,net}} \tag{1.2c}$$

where $\vec{F}$ denotes the sum of all forces acting on the fluid element, and $\dot{E}_{\text{in,net}}$ the net energy rate entering the fluid element. The force may act on the surface or volume of the fluid element. The pressure or viscous forces are examples of surface forces, while gravitational and electromagnetic forces are volume forces. Similarly, energy can enter a fluid volume via surface effects or volume mechanisms. Examples of surface energy exchange are heat conduction, the work done by pressure, and viscous forces on the boundary of the fluid elements. The work done to overcome the gravitational force and the radiation heat transfer to a fluid body are examples of volume energy exchange.

Alternatively, we write the governing equations for a control volume that is instantaneously occupied by a fluid element. The control volume is stationary, moving, or deforming, but makes the analysis simpler compared to the approach of following fluid elements in a flow (Fig. 1.1). The time rate of change of mass, momentum, and energy of a fluid element, the left side of (1.2) relate to the time rate of change of the corresponding properties of a control volume via the Reynolds Transport Theorem (RTT). Applying RRT to Eq. (1.2) results in

1.1 Equations of Compressible Fluid Transport

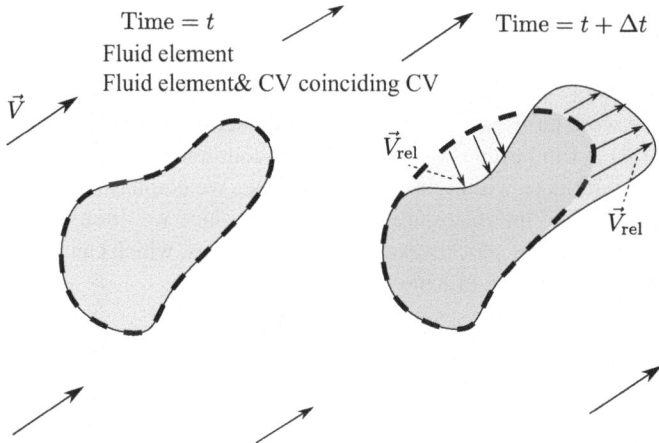

Fig. 1.1 A fluid element occupying a control volume (whose surface shown by dashed line) at time t moves and deforms while the control volume may move and deforms but with a different rate resulting in relative fluid velocity $\vec{V}_{\text{vel}}$ at the control surface and fluid flux in or out of the control volume

$$\frac{d}{dt}\int_{CV} \rho dV + \int_{CS} \rho \vec{V}_{\text{rel}} \cdot \vec{n} dA = 0 \tag{1.3a}$$

$$\frac{d}{dt}\int_{CV} \rho \vec{V} dV + \int_{CS} \rho \vec{V} \vec{V}_{\text{rel}} \cdot \vec{n} dA = \vec{F} \tag{1.3b}$$

$$\frac{d}{dt}\int_{CV} \rho e_{\text{total}} dV + \int_{CS} \rho e_{\text{total}} \vec{V}_{\text{rel}} \cdot \vec{n} dA = \dot{E}_{\text{in,net}} \tag{1.3c}$$

The second term on the left side of Eq. (1.3) are mass, momentum, and energy fluxes, i.e., the net rate of mass, momentum, and energy exiting the control volume from its surface due to the nonzero fluid velocity relative to the control surface velocity ($\vec{V}_{\text{rel}} \neq 0$). The vector $\vec{n}$ is the local unit outward normal vector to the surface, and $\vec{V}_{\text{rel}}$ is the velocity of the fluid relative to the control surface.

If the inertial reference frame coincides with a constant (or zero) velocity control volume, all velocities in the governing equations are relative to the control volume; hence, the subscript "rel" can be dropped from $\vec{V}_{\text{rel}}$, and d/dt replaced with $\partial/\partial t$. Also, after separating the pressure force and work from other mechanisms, we express the governing equations (1.3) in the form

$$\frac{\partial}{\partial t}\int_{CV} \rho dV + \int_{CS} \rho \vec{V} \cdot \vec{n} dA = 0 \tag{1.4a}$$

$$\frac{\partial}{\partial t}\int_{CV} \rho \vec{V} dV + \int_{CS} \rho \vec{V} \vec{V} \cdot \vec{n} dA = -\int_{CS} p\vec{n} dA + \vec{F}_{\text{other}} \tag{1.4b}$$

$$\frac{\partial}{\partial t}\int_{CV} \rho e_{\text{total}} dV + \int_{CS} \rho e_{\text{total}} \vec{V} \cdot \vec{n} dA = -\int_{CS} p\vec{V} \cdot \vec{n} dA + \dot{E}_{\text{in,net, other}} \tag{1.4c}$$

where p is the pressure of the fluid. The "other" mechanisms of force action denoted by $\vec{F}_{\text{other}}$ can represent viscous, elastic, and electromagnetic forces, while $\dot{E}_{\text{in,net, other}}$ can represent energy exchange rate via heat conduction and radiation, viscous and elastic surface work terms, and gravitational and electromagnetic forces.

System (1.4) is valid even if one considers discontinuous features such as shock waves or fluid interfaces within the control volume. We emphasize that we assume shock waves and fluid interfaces are discontinuous since we often intend to avoid resolving the thickness of shock waves or fluid interfaces, which can be of the order of micrometers or a fraction of nano-meters.

1.1.2 Differential Forms of Compressible Fluid Transport

In the absence of discontinuous features and the assumption of the smoothness of fluid properties, the control volume can be chosen as infinitesimally small. The application of the divergence theorem on the surface terms in Eq. (1.4) then yields the differential form of the governing equations as follows:

$$U_t + \nabla \cdot f = \nabla \cdot f^{d,e} + f^{\text{other}} \tag{1.5}$$

where $U = (\rho, \rho u, \rho v, \rho w, \rho e_{\text{total}})^t = (\rho, \rho \vec{V}, \rho e_{\text{total}})^t$, and $\vec{V} = (u, v, w)^t$. The vector $\nabla \cdot f$ represents the derivatives of advective fluxes and pressure force and work and its components are

$$\nabla \cdot f = \begin{pmatrix} \nabla \cdot \rho \vec{V} \\ [\nabla \cdot (\rho \vec{V} \otimes \vec{V} + p\mathbb{I})]^t \\ \nabla \cdot (e_{\text{total}} + p)\vec{V} \end{pmatrix} \tag{1.6}$$

In Eq. (1.6), $\mathbb{I}$ is the identity matrix, and

$$\vec{V} \otimes \vec{V} = \begin{pmatrix} u^2 & vu & wu \\ uv & v^2 & wv \\ uw & vw & w^2 \end{pmatrix} \tag{1.7}$$

Flux $f^{d,e}$ in Eq. (1.5) represents the sum of diffusive and elastic effects, while f^{other} is a density term from all volume forces.

In the context of numerical methods, it is customary to refer to vectors f and $f^{d,e}$ as fluxes, even though it is partially a misnomer.

1.1 Equations of Compressible Fluid Transport

1.1.3 Euler System and Entropy Equation

The system (1.5) with only pressure force and work terms is called the Euler system. The one dimension version of the Euler system with velocity v and the specific internal energy e is given as

$$\frac{\partial \rho}{\partial t} + \frac{\partial (\rho v)}{\partial x} = 0 \tag{1.8a}$$

$$\frac{\partial (\rho v)}{\partial t} + \frac{\partial (\rho v^2 + p)}{\partial x} = 0 \tag{1.8b}$$

$$\frac{\partial [\rho (e + \frac{v^2}{2})]}{\partial t} + \frac{\partial [(\rho e + \frac{\rho v^2}{2} + p)v]}{\partial x} = 0 \tag{1.8c}$$

This system is valid for negligible diffusive effects (viscous friction and heat conduction) when the flow involves small gradients. The contribution of the gravitational and other forces are also neglected. In the case of shock waves with large gradients, the model without diffusive effects is no longer applicable, as will be discussed later in this chapter. These equations demonstrate the conservation of the property variables

$$\boldsymbol{u} = [\rho, \rho v, \rho(e + \frac{v^2}{2})]^t \tag{1.9}$$

which is obtained from the corresponding integral form of system (1.8) for a control volume. Thus, the vector $\boldsymbol{u}$ is referred to as the conservative variable vector. The vector of fluxes is denoted as

$$\boldsymbol{f}(\boldsymbol{u}) = [\rho v, \rho v^2 + p, (\rho e + \frac{\rho v^2}{2} + p)v]^t \tag{1.10}$$

The Euler equations in the vector form can be written as

$$\frac{\partial \boldsymbol{u}}{\partial t} + \frac{\partial \boldsymbol{f}(\boldsymbol{u})}{\partial x} = 0 \tag{1.11}$$

An additional equation for the property entropy s can be derived using Eqs. (1.8) and the Gibbs property relation following a blob of fluid which has the form

$$T ds = de - \frac{p}{\rho^2} d\rho \tag{1.12}$$

The entropy equation can be derived as follows. First, the spatial and temporal derivatives in the Euler system are distributed, which is a valid operation under the assumption of the smoothness of the solution variables. Then, substituting the mass equation into momentum and energy equations gives the following system

$$\frac{\partial \rho}{\partial t} + v\frac{\partial \rho}{\partial x} + \rho\frac{\partial v}{\partial x} = 0 \tag{1.13a}$$

$$\frac{\partial v}{\partial t} + v\frac{\partial v}{\partial x} + \frac{1}{\rho}\frac{\partial p}{\partial x} = 0 \tag{1.13b}$$

$$\rho\left(\frac{\partial e}{\partial t} + v\frac{\partial e}{\partial x}\right) + \frac{\rho v}{2}\left(\frac{\partial v}{\partial t} + v\frac{\partial v}{\partial x}\right) + \frac{v}{2}\left(\frac{\partial(\rho v)}{\partial t} + \frac{\partial(\rho v^2)}{\partial x}\right) + \frac{\partial(pv)}{\partial x} = 0 \tag{1.13c}$$

Substituting Eq. (1.13b) into (1.13c) yields

$$\frac{d\rho}{dt} = -\rho\frac{\partial v}{\partial x} \tag{1.14a}$$

$$\frac{dv}{dt} = -\frac{1}{\rho}\frac{\partial p}{\partial x} \tag{1.14b}$$

$$\frac{de}{dt} = -\frac{p}{\rho}\frac{\partial v}{\partial x} \tag{1.14c}$$

where

$$\frac{d}{dt} = \frac{\partial}{\partial t} + v\frac{\partial}{\partial x} \tag{1.15}$$

Substituting Eq. (1.14) into the Gibbs relation yields

$$T\frac{ds}{dt} = \frac{de}{dt} - \frac{p}{\rho^2}\frac{d\rho}{dt} = -\frac{p}{\rho}\frac{\partial v}{\partial x} + \frac{p}{\rho}\frac{\partial v}{\partial x} = 0 \tag{1.16}$$

or

$$\frac{ds}{dt} = \frac{\partial s}{\partial t} + v\frac{\partial s}{\partial x} = 0 \tag{1.17}$$

As seen from the above equation, due to the lack of diffusive effects (viscous and heat conduction), the Euler system (1.8) with smooth solutions supports no destruction and generation of the entropy. As discussed later, diffusive effects are significant in the presence of large gradients such as shock waves. They must be included in the governing equations, which correspondingly yield an increase in entropy [1].

1.2 Characteristics of the Compressible Fluid Dynamics

As seen above, the Euler system and the entropy equation model the inviscid, non-heat conducting compressible single-fluid flows that are consistent with Newton's second law and the first and second laws of thermodynamics. However, the entropy equation is obtained from the other three equations. The energy equation can also be obtained using the mass, momentum, and entropy equation. To characterize the full

1.2 Characteristics of the Compressible Fluid Dynamics

field, we thus require solving either the Euler system or the Euler system in which the entropy equation replaces the energy equation.

This section first discusses the characteristics of compressible single-fluid dynamics with smooth flows without shock waves and diffusive effects. We then discuss how shock waves are generated and why diffusive effects are essential for shock waves' physical and numerical validity. The importance of conservation laws for shock waves and other discontinuities is also highlighted. The varying diffusion rates in the presence of multiple spatial scales are also discussed, leading to the concept of stiffness and the challenges it brings to temporal numerical schemes.

1.2.1 Characteristic Speeds and Curves

Let us first discuss the characteristics of the entropy equation in isolation before considering the system of Euler's equation.

1.2.1.1 Entropy Equation and Transport Equation

What does the entropy equation (1.17) mean? Equation (1.17) reveals that the entropy of a given fluid particle remains constant as it moves along the particle path, verified by taking the total derivative of the entropy of a fluid particle $s = s(t, x(t))$, with $x(t)$ being the particle path, as

$$\frac{d}{dt}[s(t, x(t))] = \frac{\partial s}{\partial t} + \underbrace{\frac{dx}{dt}}_{=v}\frac{\partial s}{\partial x} = \frac{\partial s}{\partial t} + v\frac{\partial s}{\partial x} = 0 \tag{1.18}$$

Equivalently, given an initial condition $s_0(x) = s(t = 0, x)$ and a known velocity field $v(t, x)$, the solution to the entropy equation is determined as

$$\left.\begin{array}{l}\frac{\partial s}{\partial t} + v\frac{\partial s}{\partial x} = 0 \\ s(t = 0, x) = s_0(x) \quad \forall x \in \mathbb{R}\end{array}\right\} \implies s(t, x) = s_0(x_0(t, x)) \tag{1.19}$$

In Eq. (1.19), $x_0(t, x)$ is the starting (initial) location of the particle whose location at time t is x.

If the initial condition is valid for a part of the real axis, or if some characteristics do not intersect the x-axis, a boundary condition, e.g., $s(t, x = 0) = b_0(t)$, is required. Therefore,

$$\left.\begin{array}{l}\frac{\partial s}{\partial t} + v\frac{\partial s}{\partial x} = 0 \\ s(t=0, x) = s_0(x) \quad \forall x > 0 \\ s(t, x=0) = b_0(t)\end{array}\right\} \implies \begin{array}{l} s(t, x) = s_0[x_0(t, x)], \\ \text{if the particle path starts} \\ \text{from a point on the } x - \text{axes}, x_0, \\ \text{or} \\ s(t, x) = b_0[t_0(t, x)], \\ \text{if the particle path starts} \\ \text{from a point on the } t - \text{axes}, t_0 \end{array}$$

(1.20)

If the velocity v is constant, an explicit expression can be easily derived for the starting point of the particle path

$$\frac{dx}{dt} = v \implies x_0 = x - vt \qquad (1.21)$$

and, thus,

$$s(t, x) = s_0(x - vt) \quad \text{if} \quad \frac{dx}{dt} = v = \text{const.} \qquad (1.22)$$

The entropy remains constant along the curve $\frac{dx}{dt} = v$. This curve is called the characteristic curve because a characteristic curve is a curve along which a particular property is constant. In the case of the entropy equation, $\frac{dx}{dt} = v$ defines the characteristic curves. The slope of the curve is called characteristic speed. In the case of the entropy equation, the velocity v is the characteristic speed. Figure 1.2 depicts an initial entropy distribution that is advected or transported along the curve $\frac{dx}{dt} = v = \text{const.}$. For this reason, equations of the type of the entropy equation (1.17) are called (simple) advection or transport equation.

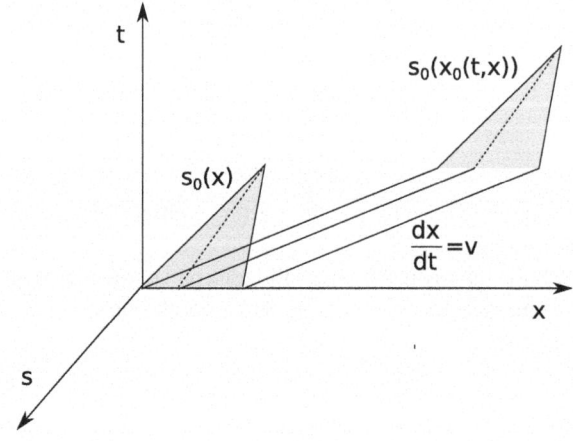

Fig. 1.2 Advection of the initial entropy distribution along the characteristic curve $\frac{dx}{dt} = v = \text{const.}$

1.2 Characteristics of the Compressible Fluid Dynamics

1.2.1.2 Example on Finding the Characteristics of a Transport Equation

An example is helpful. Consider the following equation

$$x\frac{\partial T}{\partial t} + \frac{\partial T}{\partial x} = 0 \tag{1.23}$$

with the initial and boundary conditions

$$T(t = 0, x > 0) = T_0(x) \tag{1.24a}$$
$$T(t, x = 0) = B_0(t) \tag{1.24b}$$

Writing the equation in the advection or transport form

$$\frac{\partial T}{\partial t} + \frac{1}{x}\frac{\partial T}{\partial x} = 0 \tag{1.25}$$

yields the characteristic speed $v = \frac{1}{x}$ and the characteristic curve

$$\frac{dx}{dt} = \frac{1}{x} \tag{1.26}$$

Integrating the characteristic curve equation then yields

$$\frac{1}{2}x^2 = t + C \tag{1.27}$$

where C is a constant of the integration. If $\frac{x^2}{2} \geq t$, then the characteristics curves will originate from axis $t = 0$, and, hence, $C = \frac{x_0^2}{2}$ and $x_0 = \sqrt{x^2 - 2t}$. Otherwise, the curves originate from the axis $x = 0$, and never intersect the x-axis; hence, $C = -t_0$ and $t_0 = t - \frac{1}{2}x^2$. Therefore, the solution is

$$T(t, x) = \begin{cases} T_0(\sqrt{x^2 - 2t}) & t \leq \frac{x^2}{2} \\ B_0(t - \frac{1}{2}x^2) & t > \frac{x^2}{2} \end{cases} \tag{1.28}$$

The two sets of characteristics corresponding to

$$v = \frac{dx}{dt} = \frac{1}{x} \tag{1.29}$$

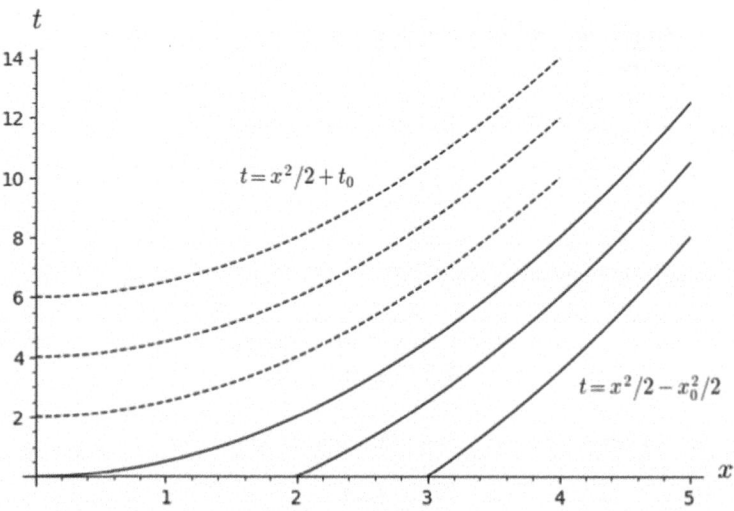

Fig. 1.3 Two sets of characteristics: one originating from the x-axis, and the other from the t-axis for the characteristic curve defined with $dx/dt = 1/x$

are given as

$$t = \frac{x^2}{2} - \frac{x_0^2}{2} \quad \text{originating from } x-\text{axis} \tag{1.30}$$

$$t = \frac{x^2}{2} + t_0 \quad \text{originating from } t-\text{axis} \tag{1.31}$$

and shown in Fig. 1.3. This simple example illustrates how the boundary condition and initial condition affect the solution in the domain's interior. In Chap. 3, we will discuss how proper boundary values can be determined for multi-dimensional compressible fluid dynamics.

1.2.1.3 Euler System

In addition to the entropy characteristics, what other characteristic speeds and curves does the Euler system possess? To answer this question, we need to find the scalar counterpart of the system of equations. Let us begin by inspecting the mass, momentum, and entropy equations.

$$\frac{\partial \rho}{\partial t} + \frac{\partial (\rho v)}{\partial x} = 0 \tag{1.32a}$$

$$\frac{\partial (\rho v)}{\partial t} + \frac{\partial (\rho v^2 + p)}{\partial x} = 0 \tag{1.32b}$$

1.2 Characteristics of the Compressible Fluid Dynamics

$$\frac{\partial s}{\partial t} + v\frac{\partial s}{\partial x} = 0 \quad (1.32c)$$

We can rewrite the mass and momentum equations in quasi-linear form as long as the solution is smooth (solution without shock waves) in the form of

$$\frac{\partial \rho}{\partial t} + \rho\frac{\partial v}{\partial x} + v\frac{\partial \rho}{\partial x} = 0 \quad (1.33a)$$

$$\frac{\partial v}{\partial t} + v\frac{\partial v}{\partial x} + \frac{1}{\rho}\frac{\partial p}{\partial x} = 0 \quad (1.33b)$$

$$\frac{\partial s}{\partial t} + v\frac{\partial s}{\partial x} = 0 \quad (1.33c)$$

There are four unknowns in the above three equations. Knowing that the entropy is a smooth function of density and pressure, $s = s(\rho, p)$ [2], we replace the entropy derivatives with those of density and pressure as

$$\frac{\partial s}{\partial t} = \left(\frac{\partial s}{\partial \rho}\right)_p \frac{\partial \rho}{\partial t} + \left(\frac{\partial s}{\partial p}\right)_\rho \frac{\partial p}{\partial t} \quad (1.34a)$$

$$\frac{\partial s}{\partial x} = \left(\frac{\partial s}{\partial \rho}\right)_p \frac{\partial \rho}{\partial x} + \left(\frac{\partial s}{\partial p}\right)_\rho \frac{\partial p}{\partial x} \quad (1.34b)$$

Substituting the last two equations into the entropy equation and invoking the mass equation yield

$$\frac{\partial p}{\partial t} + v\frac{\partial p}{\partial x} - \rho\left(\frac{\partial s}{\partial \rho}\right)_p \left(\frac{\partial p}{\partial s}\right)_\rho \frac{\partial v}{\partial x} = 0 \quad (1.35)$$

Using the thermodynamic cyclic relation [2], the term $\left(\frac{\partial s}{\partial \rho}\right)_p \left(\frac{\partial p}{\partial s}\right)_\rho$ is simplified as

$$\left(\frac{\partial s}{\partial \rho}\right)_p \left(\frac{\partial p}{\partial s}\right)_\rho = -\left(\frac{\partial p}{\partial \rho}\right)_s = -c^2 \quad (1.36)$$

where c is the local speed of sound. Finally, the Euler system simplifies into the quasi-conservative form

$$\frac{\partial \rho}{\partial t} + \rho\frac{\partial v}{\partial x} + v\frac{\partial \rho}{\partial x} = 0 \quad (1.37a)$$

$$\frac{\partial v}{\partial t} + v\frac{\partial v}{\partial x} + \frac{1}{\rho}\frac{\partial p}{\partial x} = 0 \quad (1.37b)$$

$$\frac{\partial p}{\partial t} + v\frac{\partial p}{\partial x} + \rho c^2 \frac{\partial v}{\partial x} = 0 \quad (1.37c)$$

The system written in a vector form has the form

$$\frac{\partial}{\partial t}\begin{pmatrix}\rho\\v\\p\end{pmatrix}+\begin{pmatrix}v & \rho & 0\\0 & v & \frac{1}{\rho}\\0 & \rho c^2 & v\end{pmatrix}\frac{\partial}{\partial x}\begin{pmatrix}\rho\\v\\p\end{pmatrix}=0 \quad (1.38)$$

Comparing this equation with the advection equation for entropy (1.17), we learn that instead of the scalar entropy, a vector of primitive variables

$$w=\begin{pmatrix}\rho\\v\\p\end{pmatrix} \quad (1.39)$$

appears and instead of a scalar advection velocity, the matrix A

$$A=\begin{pmatrix}v & \rho & 0\\0 & v & \frac{1}{\rho}\\0 & \rho c^2 & v\end{pmatrix} \quad (1.40)$$

is acting on the spatial derivative of the variables. To reduce the effect of a matrix-vector multiplication to a scalar-vector multiplication with the scalar being the characteristic speed, we find the eigenvalues and eigenvectors of the matrix A. The eigenvalues are

$$\lambda_1=v-c,\ \lambda_2=v,\ \lambda_3=v+c \quad (1.41)$$

The matrix A has three linearly independent eigenvectors. The matrix Q whose columns are the eigenvectors of A has the form

$$Q=\begin{pmatrix}1 & 1 & 1\\-\frac{c}{\rho} & 0 & \frac{c}{\rho}\\c^2 & 0 & c^2\end{pmatrix} \quad (1.42)$$

The inverse of matrix Q is

$$Q^{-1}=\begin{pmatrix}0 & \frac{-\rho}{2c} & \frac{1}{2c^2}\\1 & 0 & -\frac{1}{c^2}\\0 & \frac{\rho}{2c} & \frac{1}{2c^2}\end{pmatrix} \quad (1.43)$$

The matrix A decomposes as

$$A=Q\Lambda Q^{-1} \quad (1.44)$$

where Λ is a diagonal matrix with its diagonal values being the characteristic speeds:

$$\Lambda=\begin{pmatrix}v-c & 0 & 0\\0 & v & 0\\0 & 0 & v+c\end{pmatrix} \quad (1.45)$$

1.2 Characteristics of the Compressible Fluid Dynamics

Equation (1.44) is verified by the definition of eigenvectors as

$$AQ = Q\Lambda \tag{1.46}$$
$$AQQ^{-1} = Q\Lambda Q^{-1} \tag{1.47}$$
$$A = Q\Lambda Q^{-1} \tag{1.48}$$

Equation (1.44) is called the eigen-decomposition of the matrix A. Now, replacing the matrix A with its eigendecomposition into Eq. (1.38) yields

$$\frac{\partial \boldsymbol{w}}{\partial t} + Q\Lambda Q^{-1}\frac{\partial \boldsymbol{w}}{\partial x} = 0 \tag{1.49}$$

Let us define the characteristic vector $d\boldsymbol{w}_c$ as

$$d\boldsymbol{w}_c = Q^{-1}d\boldsymbol{w} \tag{1.50}$$

where the differential is taken along the characteristic curves

$$\frac{dx}{dt} = \lambda_i \tag{1.51}$$

Now, multiplying Eq. (1.49) with Q^{-1}, we obtain Euler's equations in the eigenvector space or characteristic space in the form

$$\frac{\partial \boldsymbol{w}_c}{\partial t} + \Lambda \frac{\partial \boldsymbol{w}_c}{\partial x} = 0 \tag{1.52}$$

Equation (1.52) consists of three decoupled equations in the form

$$\frac{\partial w_{c,i}}{\partial t} + \lambda_i \frac{\partial w_{c,i}}{\partial x} = 0 \quad i = 1, \ldots, 3 \tag{1.53}$$

where λ_i are the eigenvalues or the characteristic speeds given in Eq. (1.41). Several points are notable about Eqs. (1.53).

- The only assumption made to obtain the characteristic form of the Euler equations is that solutions are smooth functions of x. For smooth solutions, the conservative form of the Euler system, Eqs. (1.8) or (1.32), is equivalent to characteristic form, Eq. (1.52). In other words, the characteristic form contains all the information of the conservative form of the Euler system as long as the solution is smooth.
- The wave nature of Eq. (1.53) reveals three separate families of characteristics, and along each, a particular characteristic variable remains constant. Specifically,

$$w_{c,1} = \text{const.} \quad \text{along} \quad \frac{dx}{dt} = v - c \tag{1.54a}$$

$$w_{c,2} = \text{const.} \quad \text{along} \quad \frac{dx}{dt} = v \tag{1.54b}$$

$$w_{c,3} = \text{const.} \quad \text{along} \quad \frac{dx}{dt} = v + c \tag{1.54c}$$

While $w_{c,2}$ can be shown to be the entropy, analytical expressions for $w_{c,1}$ and $w_{c,3}$ are not in general possible. Alternatively, the solution (1.54) can be expressed as

$$dw_{c,1} = 0 \quad \text{along} \quad \frac{dx}{dt} = v - c \tag{1.55a}$$

$$dw_{c,2} = 0 \quad \text{along} \quad \frac{dx}{dt} = v \tag{1.55b}$$

$$dw_{c,3} = 0 \quad \text{along} \quad \frac{dx}{dt} = v + c \tag{1.55c}$$

Then, from the definition of the characteristic variables, Eq. (1.50), we have

$$dw_{c,1} = -\frac{\rho}{2c}dv + \frac{1}{2c^2}dp = 0 \quad \text{along} \quad \frac{dx}{dt} = v - c \tag{1.56a}$$

$$dw_{c,2} = d\rho - \frac{1}{c^2}dp = 0 \quad \text{along} \quad \frac{dx}{dt} = v \tag{1.56b}$$

$$dw_{c,3} = \frac{\rho}{2c}dv + \frac{1}{2c^2}dp = 0 \quad \text{along} \quad \frac{dx}{dt} = v + c \tag{1.56c}$$

- Characteristic curves from a single family may be converging, diverging, or parallel.
- Converging characteristic curves result in compression waves. If two converging characteristics from a single family intersect, a discontinuity or a shock wave will develop to resolve the duplicate values of the characteristic variables. In other words, smooth waves travel along the characteristic paths, and non-smooth features develop along the converging characteristic paths. Hence, if we want to examine the smoothness of a solution (or a numerical solution) of a system of conservation laws, it is pertinent to look along the characteristic paths and examine the characteristic variables instead of the primitive or conservative variables.

  In Chap. 2, we will use the insight gained about the characteristics to accurately identify the smoothness of a function from its discrete values and use the computed smoothness indicator to design nonlinear schemes for high-fidelity capturing of discontinuous flow features.
- Diverging characteristic curves result in expansion or rarefaction waves. For example, when a high-pressure gas container empties into a low-pressure gas container, an expansion wave travels into the high-pressure container. On the other hand, a compression wave or shock wave travels inside the low-pressure container. Suppose the initial pressure ratio is above a certain level such that the gas speed exiting the high-pressure container behind the expansion wave is supersonic. As a result,

1.2 Characteristics of the Compressible Fluid Dynamics

the far end of the high-pressure container is deprived of gas and experiences a vacuum. (Similar vacuum state forms in a double-rarefaction wave problems when two rarefaction waves moving in opposite direction with equal supersonic fluid speed behind them collide [3].) Therefore, for the system consisting of the gas inside both containers, there is a significant variation in pressure and density from very high values (behind the shock wave) to very low values (near vacuum).

These large variations pose a challenge in the numerical computations. Near the vacuum state, the approximate numerical solutions can be accurate but still result in negative pressure or density. Negative pressure or density, in turn, results in an imaginary (non-real) speed of sound because for ideal gases, the speed of sound is $c = \sqrt{\gamma p/\rho}$ (with γ being specific heat ratio). Therefore, the positivity preservation of the density and pressure are essential for successful computations because the numerical solution of the compressible flows relies on computations of the speed of sound. The positivity-preserving numerical schemes are discussed for single-fluid compressible flows in Chap. 5 and for two-fluid in Chaps. 6 and 7.

- Parallel characteristic curves arise when velocity and pressure are uniform, but density and other properties are non-uniform. When a density field consists of a single jump, the wave is a simple advection of a contact discontinuity or two-fluid advection of a fluid (material) interface. Hence, the thickness of the contact or interface remains the same over time. In computation of compressible flows, the numerical dissipation, which is necessary for the stability of the scheme, must be kept to a minimum to avoid artificial smearing of the contact or interface. High-order spatial and temporal schemes introduced in the forthcoming chapters address this challenging requirement of the right amount of numerical dissipation. Sharply capturing the fluid interfaces is essential in the computations of critical applications, including those involving Rayleigh-Tayler and Richtmyer-Meshkov instabilities [4].

- This eigenvalue analysis reveals the speed by which various waves propagate in compressible media. The physical time scales by which gradients or changes propagate are linearly proportional to these characteristic speeds. Numerical solution schemes must therefore consist of time steps that are comparable to the physical time scales in size; to resolve a wave within a spatial length h, a time step of $\Delta t = \mathcal{O}(h/c)$, where c is the characteristic speed, is required.

  As discussed in Appendix A, in explicit time integration schemes, the requirement of $\Delta t = \mathcal{O}(h/c)$ is also a stability issue for the numerical scheme.

  Furthermore, the direction and speed of characteristics must also be respected in the approximation of fluxes for stable computations as discussed in Chap. 5.

- The direction of characteristics or waves is also essential for consistent approximations on and near domain boundaries, since for stable computations, those characteristics incoming to the domain must be specified from the boundary data, while those outgoing must be approximated using the data from the interior of the domain. This is the subject of Chap. 3.

1.2.2 Shock Wave in a Nonlinear Conservation Law

An important consequence of nonlinearity in conservation laws, such as Euler's equations, is the spontaneous generation of shock waves. We may demonstrate this using Euler's equations, as often done in a Gas Dynamics course. However, it is easier to show this in the context of a scalar nonlinear equation, namely, Burgers' equation, which can be obtained using the mass and momentum equations with uniform pressure and the assumption of smoothness of the solution. The inviscid Burgers equation is

$$\frac{\partial u}{\partial t} + \frac{\partial (u^2/2)}{\partial x} = 0 \tag{1.57}$$

Given a smooth initial condition $u_0(x)$, the solution of the Burgers equation can be obtained similarly to the transport equation for the entropy. Assuming the solution is smooth for some time, we can write the Burgers equation in a form of a transport or an advection equation, though nonlinear, as

$$\frac{\partial u}{\partial t} + u \frac{\partial u}{\partial x} = 0 \tag{1.58}$$

Since the characteristic speed is the same as the solution and we know that solution is constant along the characteristic curves, the characteristic speed is constant, and thus, the solution is

$$u(t, x) = u_0(x_0(t, x)) = u_0(x - ut) \tag{1.59}$$

Also, the characteristic curves are straight lines since u is constant for each characteristic. If two distinct characteristics are convergent, they will intersect in finite time, leading to a dual-value solution at a single x, thus the generation of a discontinuity or a shock wave. The smooth initial solution and its deformation along the converging and diverging characteristics are shown in Fig. 1.4. The time of shock formation for a smooth initial solution corresponds to the infinite value of the spatial derivative of u, $u_x = \frac{\partial u}{\partial x}$. Using Eq. (1.59), we have

$$u_x = u_0'(1 - u_x t) \implies u_x = \frac{u_0'}{1 + u_0' t} \tag{1.60}$$

where $u_0' = du_0(x)/dx$. Therefore, at the time

$$t_s = -\frac{1}{u_0'} \tag{1.61}$$

u_x goes to infinity, which is the time of shock formation. Note that t_s is positive only if u_0' is negative somewhere in x. Hence, a shock forms only if u_0' is negative somewhere in x. Also, the first shock appears at

1.2 Characteristics of the Compressible Fluid Dynamics

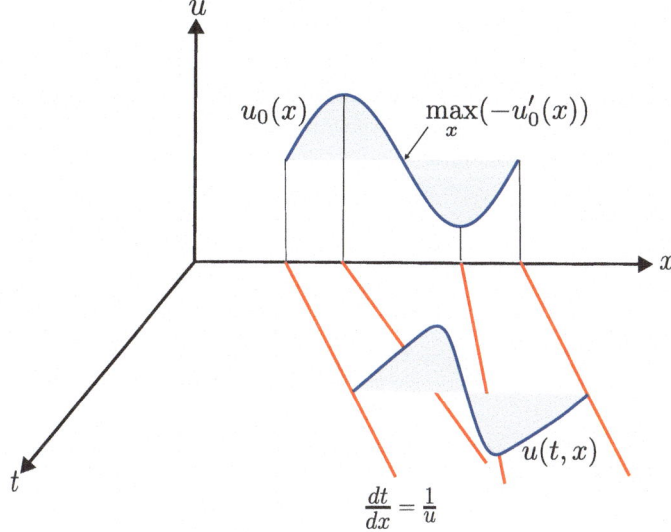

Fig. 1.4 Deformation of a smooth initial profile $u_0(x)$ resulting from converging and diverging characteristics shown in red color

$$t_s = \frac{1}{\max_x(-u_0'(x))} \quad \text{if } u_0' < 0 \tag{1.62}$$

For later times, $t > t_s$, this model predicts at least dual-value solutions. However, this will not occur for the actual scenario where the viscous Burgers' equation is more suitable. This is because as the solution slope increases, the role of the viscous effect becomes larger and prevents the formation of discontinuities and multiple-valued solutions.

A diffusive model of the compressible single-fluid is discussed next.

1.2.3 Shock Wave and Diffusive Model

How does the solution of the system (1.8) support shock waves which are entropy-generating waves? The answer is that for increasingly larger gradients such as shock wave, the model without diffusive effects is no longer applicable as the velocity and temperature gradients become large at the shock. The relations between velocity and temperature gradients with the viscous stress and heat conduction are given using Newton's law of viscosity and Fourier's law of heat conduction; the laws in one dimension have the form

$$\text{1D viscous stress} = \mu \frac{\partial v}{\partial x} \tag{1.63}$$

$$\text{1D heat conduction flux} = -\kappa \frac{\partial T}{\partial x} \quad (1.64)$$

where μ and κ are dynamic viscosity and thermal conductivity. The differential form of the governing equations (1.6) in one-space dimension and with the viscous friction and heat conduction have the form

$$\frac{\partial \rho}{\partial t} + \frac{\partial (\rho v)}{\partial x} = 0 \quad (1.65a)$$

$$\frac{\partial (\rho v)}{\partial t} + \frac{\partial (\rho v^2 + p - \mu \frac{\partial v}{\partial x})}{\partial x} = 0 \quad (1.65b)$$

$$\frac{\partial [\rho(e + \frac{v^2}{2})]}{\partial t} + \frac{\partial [(\rho e + \frac{\rho v^2}{2} + p - \mu \frac{\partial v}{\partial x})v - \kappa \frac{\partial T}{\partial x}]}{\partial x} = 0 \quad (1.65c)$$

Following the procedure for the Euler equations, we can show that the diffusive model (1.65) supports the entropy generation in the form of

$$\frac{ds}{dt} = \frac{\partial s}{\partial t} + v \frac{\partial s}{\partial x} = \frac{\nu}{T} \left(\frac{\partial v}{\partial x} \right)^2 + \frac{\alpha}{T} \frac{\partial^2 T}{\partial x^2} \quad (1.66)$$

where $\nu = \mu/\rho$ is the kinematic viscosity and $\alpha = \kappa/\rho$ is the thermal diffusivity, both having positive magnitudes. Note that the first term on the right-hand side of Eq. (1.17) is entropy generation due to the viscous effects, while the second term is the contribution of heat conduction to the entropy generation.

1.2.4 Characteristics of Diffusion and the Stiffness

In Sect. 1.2.1, we discussed the wave (advective) characteristics of the smooth compressible fluid dynamics in the absence of diffusive effects. Specifically, we introduced the characteristic speeds by which gradients are transported. We now discuss the unique characteristic of the diffusion processes which is different from the wave mechanisms.

The viscous friction and heat conduction are diffusive effects arising from the molecular dissipation of gradients in velocity and temperature fields. The higher the gradient of velocity or temperature, the higher the velocity or temperature gradient dissipation rate. An order of magnitude estimate of the diffusion rate can be derived by matching the unsteady term with the viscous term or the heat conduction term in (1.65) as

$$\frac{\partial v}{\partial t} = \nu \frac{\partial^2 v}{\partial x^2} \quad (1.67)$$

1.2 Characteristics of the Compressible Fluid Dynamics

If a velocity change of V occurs within a spatial length of h, then the time scale of the dissipation, Δt, is estimated as

$$\frac{V}{\Delta t} = \mathcal{O}\left(\frac{\nu V}{h^2}\right) \implies \Delta t = \mathcal{O}\left(\frac{h^2}{\nu}\right) \tag{1.68}$$

Equation (1.68) reveals that the flow properties' high gradients or high wavenumber modes are dissipated much faster than the low gradients or low wavenumber modes. In the numerical solution of the diffusion equation, the spread of spatial scales is estimated using the ratio of the lowest wave number with a length of the order of computational domain to the highest wave number with a length of order of grid spacing. Hence, since the spread of temporal scales is proportional to that of the spatial scales, the high-wave number modes dissipate much faster than the low-wave number modes. (Sect. A.2.2 of Appendix A discusses the eigenvalues of the diffusion operators in more details.)

The implication of this to numerical time integration schemes is significant. In an explicit time integration where solutions in the previous time step are used to approximate the diffusion term in the right-hand side of Eq. (1.67), the time step must be as small as h^2/ν based on the estimate (1.68), where h being the spatial grid spacing. This size for the time step is unnecessarily restrictive as the high-wave number modes' rapid dissipation contributes negligibly to the accuracy of the approximation but dictates a highly restrictive time step size for an explicit time integration algorithm. If a time step larger than h^2/ν is chosen, significant errors are introduced from the approximation of the high-wave number modes leading to erroneous growth (instead of rapid dissipation) and eventual numerical instability and blowup of the computation. This difficulty facing the explicit time integration schemes is called stiffness, which necessitates using implicit time integration of the diffusion terms. An implicit scheme uses the current solution values instead of solely relying on the solution values at previous time steps to approximate derivatives. Appendix A reviews the stiffness concept and the basics of the low-order explicit and implicit schemes, while Chap. 4 discusses high-order temporal schemes.

1.2.5 Conservation and Shock Speed

As discussed in Sect. 1.2.2, the nonlinear conservation law admits discontinuous solutions in the form of shock waves or contact discontinuities. We now derive an expression for shock speed and discuss why the conservative form of the governing equation is the only valid form for determining the shock speed.

We consider a conservation law in the form

$$\frac{\partial u}{\partial t} + \frac{\partial f(u)}{\partial x} = 0 \tag{1.69}$$

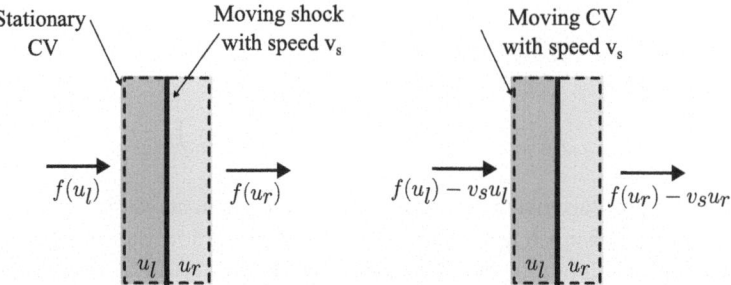

Fig. 1.5 A control volume stationary or moving in a shock reference frame and the corresponding fluxes on the left and right

where $f(u)$ is a nonlinear flux of u; for example $f(u) = u^2/2$ in the Burgers equation. Also, let us consider only a shock wave as the solution to the above equation (Fig. 1.5). With the solution left and right of the shock being u_l and u_r, respectively, the shock speed satisfies a jump condition in the form

$$v_s = \frac{f(u_l) - f(u_r)}{u_l - u_r}. \tag{1.70}$$

(The jump condition in the context of Euler equations of gas dynamics is called the Rankine-Hogoniot condition.) Equation (1.70) is obtained by rewriting the conservation law in the integral form for a control volume around the shock that moves with speed v_s. Specifically, the integral form of Eq. (1.69) is expressed as

$$\frac{\partial}{\partial t} \int u \, dx + \text{(flux exiting from RHS of C.V.)} - \text{(flux entering from LHS of C.V.)} = 0 \tag{1.71}$$

In the shock frame of reference (Fig. 1.5), the unsteady term of the conservation law vanishes, and the flux terms are modified as

$$\underbrace{\frac{\partial}{\partial t} \int u \, dx}_{=0} + \underbrace{[f(u_r) - v_s u_r]}_{\text{flux exiting from RHS of C.V.}} - \underbrace{[f(u_l) - v_s u_l]}_{\text{flux entering from LHS of C.V.}} = 0 \tag{1.72}$$

The shock speed relation Eq. (1.70) follows from this.

We emphasize that the shock speed is only well-defined for the conservative form of the equation in which the fluxes are defined. For the quasi-linear form of the equation

$$\frac{\partial u}{\partial t} + \frac{\partial f(u)}{\partial u} \frac{\partial u}{\partial x} = 0 \tag{1.73}$$

fluxes are not defined, and thus, a solution based on this form does not yield the correct shock speed. To demonstrate this, for $f(u) = u^2/2$, let us find the difference

1.2 Characteristics of the Compressible Fluid Dynamics

between $\partial(u^2/2)/\partial x$ and $u\partial u/\partial x$ at a discontinuity in $u(x)$. The derivative of the conservative form may be interpreted as

$$\frac{\partial(u^2/2)}{\partial x} = \lim_{h \to 0} \left(\frac{u_l^2 - u_r^2}{2h}\right) = \left(\frac{u_l + u_r}{2}\right)\frac{\partial u}{\partial x} \quad (1.74)$$

while the derivative of the quasi-linear from is as

$$u\frac{\partial u}{\partial x} = u_l \frac{\partial u}{\partial x} \quad \text{or} \quad u_r \frac{\partial u}{\partial x} \quad (1.75)$$

where in either case, the difference between the two forms is

$$u\frac{\partial u}{\partial x} - \frac{\partial(u^2/2)}{\partial x} = \frac{|u_l - u_r|}{2}\frac{\partial u}{\partial x} \quad (1.76)$$

If we write the quasi-linear form using (1.76) as the sum of two conservative terms as

$$\tilde{f}(u) = u\frac{\partial u}{\partial x} = \frac{\partial(u^2/2)}{\partial x} + \frac{|u_l - u_r|}{2}\frac{\partial u}{\partial x} \quad (1.77)$$

we can express the shock speed in the quasi-linear form as

$$\frac{\partial}{\partial t}\underbrace{\int u\,dx}_{=0} + \underbrace{[\tilde{f}(u_r) - v_s u_r]}_{\text{flux exiting from RHS of C.V.}} - \underbrace{[\tilde{f}(u_l) - v_s u_l]}_{\text{flux entering from LHS of C.V.}} = 0 \quad (1.78)$$

or

$$\underbrace{[f(u_r) + \frac{|u_l - u_r|}{2}(u_r) - v_s u_r]}_{\text{flux exiting from RHS of C.V.}} - \underbrace{[f(u_l) + \frac{|u_l - u_r|}{2}(u_l) - v_s u_l]}_{\text{flux entering from LHS of C.V.}} = 0 \quad (1.79)$$

Thus, the shock speed in the quasi-linear form is

$$\tilde{v}_s = \underbrace{\frac{f(u_l) - f(u_r)}{u_l - u_r}}_{\text{shock speed from conservative form}} + \frac{|u_l - u_r|}{2} = v_s + \frac{|u_l - u_r|}{2} \quad (1.80)$$

revealing a significant difference in the shock speed compared to the conservative form.

Therefore, we deduce that the conservative form, which is consistent with the conservation laws, must be used in the numerical solutions of PDEs governing compressible transport phenomena. Moreover, the numerical scheme must also possess a conservative form. Chapter 2 addresses the conservative form of discretized governing equations using spatial schemes.

1.3 Exercises

1. Consider a mixture of two fluids. Denoting the mass fraction of the fluids by y_1 and y_2, and the density of the mixture by ρ, derive a statement of mass conservation for each fluid occupying a control volume. Also, write the linear momentum equation and energy conservation for each fluid.
2. Consider the energy equation for a control volume, Eq. (1.4c). (a) Give explicit expressions for the energy rate entering a control volume using conduction and radiation heat transfer. (b) Under the assumption of smooth property, give the differential form of the energy equation with heat conduction and radiation.
3. Consider the mass, momentum, and energy system Eq. (1.4c). (a) Rewrite this system with the gravitational force and work terms. (b) Under the assumption of smooth property, give the differential form of the resultant system.
4. Using the conservation of mass equation (1.4a), rewrite the entropy equation (1.17) in a conservative form.
5. Derive expressions for the solution of the following equation

$$x^3 \frac{\partial c}{\partial t} + 2 \frac{\partial c}{\partial x} = 0 \qquad (1.81)$$

with $c(t = 0, x) = c_0(x)$ and $c(t, x = 0) = b_0(t)$ as its initial and boundary conditions.
6. By direct calculation confirm that $d\mathbf{w} = Q\, d\mathbf{w}_c$, where $d\mathbf{w}$ and $d\mathbf{w}_c$ are vectors of primitive and characteristic variables of the Euler system and Q is the matrix of the eigenvectors of matrix A given in Eq. (1.40).
7. Consider the Burgers equation (1.57) often used for verifying a newly developed numerical scheme or computer code because its exact solution can be determined. (a) Derive an exact solution for the initial condition $u(t = 0, x) = u_0 = ax + b$ (b) For the initial condition $u(t = 0, x) = u_0 = \frac{1+\sin(\pi x)}{2}$, determine the time for a first shock to appear. (c) Explain why for the linear initial condition, no shock wave will ever appear.
8. For the general nonlinear scalar conservation law in the form of $u_t + q_x = 0$ (subscripts "t" and "x" signify partial time and space differentiations) where $q = q(u)$ is a nonlinear function and the initial condition $u(t = 0, x) = u_0$, obtain an expression for the time of shock formation. Also, under what condition will a shock be formed?

1.3 Exercises

9. Find the eigenvalues of the Euler system. (a) Pick density, velocity, and entropy as the unknown primitive variables and consider the Euler equations for these variables. (b) Pick conservative variables as the unknowns. (c) Are the eigenvalues different in both cases? Explain why. (d) Are eigenvectors different? Why?

10. Consider a model of isothermal compressible flow in the form

$$\frac{\partial \rho}{\partial t} + \frac{\partial (\rho v)}{\partial x} = 0 \tag{1.82}$$

$$\frac{\partial (\rho v)}{\partial t} + \frac{\partial (\rho v^2 + p)}{\partial x} = 0 \tag{1.83}$$

$$\frac{\partial T}{\partial t} + v\frac{\partial T}{\partial x} = 0 \tag{1.84}$$

(a) Find all characteristic speeds of this system; specialize to the ideal gas with $p = \rho R T$, where R is the gas-specific constant. (b) Discuss briefly why maintaining positivity of the density and pressure is important in the numerical computations of this system. (c) Show that using the variables $(\rho, v, T)^T$, the vector

$\begin{pmatrix} \frac{1}{TR} \\ 0 \\ -\frac{1}{\rho R} \end{pmatrix}$ is a characteristic direction or an eigenvector of the quasi-linear form

of the above system.

11. Explain why a phenomenon governed by a linear advection equation would never develop a discontinuous solution from a smooth initial solution.

12. Consider the following conservation law

$$\frac{\partial u^3}{\partial t} + \frac{\partial (u^5)}{\partial x} = 0 \tag{1.85}$$

with the initial condition

$$u(t = 0, x) = u_0(x) \tag{1.86}$$

Determine an expression for the shock speed.

13. Consider the Burgers equation

$$\frac{\partial u}{\partial t} + \frac{\partial (u^2/2)}{\partial x} = 0 \tag{1.87}$$

(a) Multiply the equation by $2u$ and write the resultant equation in the conservative form. (b) Find the speed of shock for both equations (the Burgers equation and the equation obtained from the multiplication by $2u$). (c) Discuss what you obtained. Are both shock speeds identical? If not, explain why.

14. Consider the two-dimensional advection of a concentration of a chemical species $C = C(t, x, y)$ governed by

$$\frac{\partial C}{\partial t} + \left(\frac{1}{x-5}\right)\frac{\partial C}{\partial x} + \frac{\partial C}{\partial y} = 0$$

(a) Consider a square spatial domain $[0, 1] \times [0, 1]$. For all boundaries, determine whether the characteristics are incoming or outgoing. (b) Consider a quadrilateral domain with vertices $[0, 0]$, $[1, 1]$, $[1, 2]$, and $[0, 2]$. For all boundaries, determine the components of the characteristics that are incoming or outgoing.

15. Using Eq. (1.49) and the characteristic vector definition given in (1.50) and (1.51), show the advection-type equation for the characteristic vector in Eq. (1.53). Hint: Prove for each element of the characteristic vector by finding the matrix-vector product.

References

1. Menikoff, R., Plohr, B.J.: The Riemann problem for fluid flow of real materials. Rev. Mod. Phys. **61**, 75–130 (1989). Jan
2. Callen, H.B.: Thermodynamics and an Introduction To thermostatistics; 2nd ed. Wiley, New York, NY (1985)
3. Shahbazi, K.: Positivity preservation of a first-order scheme for a quasi-conservative compressible two-material model. SIAM J. Sci. Comput. **43**(4), B1029–B1055 (2021)
4. Zhou, Y., Williams, R.J., Ramaprabhu, P., Groom, M., Thornber, B., Hillier, A., Mostert, W., Rollin, B., Balachandar, S., Powell, P.D., Mahalov, Attal, A.N.: Rayleigh-Taylor and Richtmyer-Meshkov instabilities: A journey through scales. Phys. D: Nonlinear Phenom. **423**, 132838 (2021)

Chapter 2
Interpolation and High-Order Nonlinear Finite Difference Schemes

One of the main ingredients of high-order numerical solutions of PDEs is finding an approximation for the spatial derivative of a function given its point-wise values, to which the linear polynomial interpolation is a simple, efficient approach. We begin this chapter with the linear polynomial interpolation in one-space dimension [1]. We then address the error analysis of the interpolation approximation. For an equispaced grid and an increasingly high number of interpolation points (or increasingly high polynomial degrees), we highlight the *non-uniform* error distribution over the approximation domain. Specifically, we review the Runge phenomenon with the characteristic of high interpolation errors near the boundary of the domain [2] and even higher outside of the domain of approximation. One approach to remedy the Runge phenomenon is to use local interpolations over overlapping grids of smaller sizes (stencils), whose union includes the entire approximation domain. The local interpolation approximation at a point is determined using a stencil that is (nearly) centered at that point; such an algorithm of local approximations on the centers of overlapping stencils is the core of finite difference methods.

Another limitation of linear polynomial interpolation or any linear scheme is the generation of spurious oscillations in approximating non-smooth functions. The erroneous approximations are related to the Gibbs phenomenon, do not improve under increasing polynomial degrees, and may even worsen [3]. We present two subclasses of nonlinear finite difference schemes to overcome this limitation of linear interpolation of non-smooth functions, namely weighted essentially non-oscillatory methods [4–6] and targeted essentially non-oscillatory schemes [7–9]. We describe particular variants of these methods based on solution variable reconstructions rather than flux reconstructions [5, 6, 9]. These variants are general approaches and essential for computations of compressible two-fluid systems since the classical flux-based methods yield non-vanishing spurious oscillations of the order of the discontinuity jump at a fluid interface [6, 10]. We leave further details of this issue to Chap. 6, where we deal with two-fluid models and their numerical approximation.

2.1 Linear Interpolation

In an interpolation problem, given a set of discrete values of a function $u(x)$, $u_i = u(x_i)$ at x_i with $i = 0, \ldots, r$, we seek a polynomial of degree r, $P_r(x)$, that passes through all u_i, i.e., $P_r(x)$ satisfies

$$P_r(x_i) = u_i \quad \forall i = 0, \ldots, r \tag{2.1}$$

An example of an interpolating polynomial for $r = 3$ is shown in Fig. 2.1.

To construct $P_r(x)$, following Ref. [1], we first introduce a polynomial that is zero at all points except at x_i where it is unity. To this end, we introduce a notation that we shall use, namely,

$$\prod_i(x) \equiv \prod_{\substack{0 \leq j \leq r \\ j \neq i}} (x - x_j) = (x - x_0)(x - x_1) \cdots (x - x_{i-1})(x - x_{i+1}) \cdots (x - x_r) \tag{2.2}$$

where since $\prod_i(x)$ is the product of all factors $(x - x_j)$ except the ith factor, $(x - x_i)$, it is zero for all points except at x_i. Now to have a polynomial that returns the value one at x_i, we need to form

$$\frac{\prod_i(x)}{\prod_i(x_i)} \equiv \frac{\prod_{j \neq i}(x - x_i)}{\prod_{j \neq i}(x_i - x_j)} \tag{2.3}$$

Multiplying this polynomial with u_i, i.e., $u_i \frac{\prod_i(x)}{\prod_i(x_i)}$, yields a polynomial of degree r that gives u_i at x_i. Summation for all is then yields the interpolating polynomial in the form

$$P_r(x) = \sum_{i=0}^{r} u_i \frac{\prod_{j \neq i}(x - x_j)}{\prod_{j \neq i}(x_i - x_j)} \tag{2.4}$$

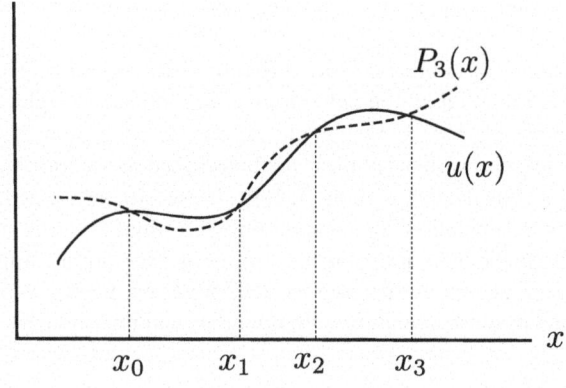

Fig. 2.1 Interpolating polynomial $P_3(x)$ approximating function $f(x)$ using the values of $f(x)$ at four points $x_0, \ldots, x_3$

2.1 Linear Interpolation

Note that for $(r + 1)$ grid points, $P_r(x)$ is a polynomial of degree r. For instance, for the ten-point interpolation, $P_r(x)$ is a polynomial of degree nine. Note also that the interpolation in (2.4) is linearly dependent on the function values; hence, it is a linear approximation method. The polynomials

$$\ell_i(x) = \frac{\prod_{j \neq i}(x - x_j)}{\prod_{j \neq i}(x_i - x_j)} = \prod_{\substack{0 \leq j \leq r \\ j \neq i}} \left(\frac{x - x_j}{x_i - x_j}\right) \quad i = 0, \ldots, r \quad (2.5)$$

are referred to as Lagrange polynomials. The interpolating approximation (2.4) is an expansion of the function $u(x)$ using the Lagrange polynomials expressed as

$$P_r(x) = \sum_{i=0}^{r} u_i \ell_i(x) \quad (2.6)$$

We next discuss the error resulting from the one-dimensional interpolation approximation.

2.1.1 Error in Interpolation

Following Ref. [1], we can derive a theoretical expression for the difference between the original function $u(x)$ and the approximating polynomial $P_r(x)$. Noting that the difference is zero at all the sample points, we may write

$$u(x) - P_r(x) = (x - x_0)(x - x_1) \cdots (x - x_r) K(x) \quad (2.7)$$

where $K(x)$ is suitably chosen. Now, let us consider the function

$$\phi(y) = u(y) - P_r(y) - (y - x_0)(y - x_1) \cdots (y - x_r) K(x) \quad (2.8)$$

If $f(y)$ has all derivatives up to and including $(r + 1)$th derivative, Eq. (2.8) can be differentiated $(r + 1)$ times, and since $P_r(y)$ is a polynomial of degree r and $K(x)$ is a constant, it yields

$$\phi^{(r+1)}(y) = u^{(r+1)}(y) - (r + 1)! K(x) \quad (2.9)$$

However, $\phi(y)$ vanishes at $(r + 2)$ points (at x and $x_0, x_1, \ldots, x_r$). Hence, by the mean value theorem, $\phi'(y)$ vanishes at $(r + 1)$ points and $\phi^{(k)}$ vanishes at $(r + 2 - k)$ points and $\phi^{(r+1)}$ vanishes at least once. Thus, there is an ξ in the interval spanning by the interpolation points and y such that

$$u^{(r+1)}(\xi) = (r+1)!K(x) \implies K(x) = \frac{u^{(r+1)}(\xi)}{(r+1)!} \tag{2.10}$$

We now have a value for $K(x)$, and we can substitute this back in the original expression (2.8) to obtain

$$\phi(y) = u(y) - P_r(y) - (y-x_0)(y-x_1)\cdots(y-x_r)\frac{f^{(r+1)}(\xi)}{(r+1)!} \tag{2.11}$$

Finally, since $\phi(y=x) = 0$, we have

$$u(x) = P_r(x) + (x-x_0)(x-x_1)\cdots(x-x_r)\frac{u^{(r+1)}(\xi)}{(r+1)!} \tag{2.12}$$

The error term is

$$R(x) = (x-x_0)(x-x_1)\cdots(x-x_r)\frac{u^{(r+1)}(\xi)}{(r+1)!} \tag{2.13}$$

which we can rewrite using

$$\prod_{0 \le j \le r}(x-x_j) = (x-x_0)(x-x_1)\cdots(x-x_r) \tag{2.14}$$

as

$$R(x) = \frac{u^{(r+1)}(\xi)}{(r+1)!}\prod_{0 \le j \le r}(x-x_j) \tag{2.15}$$

2.1.2 Interpolation Error for Smooth Function

The error term (2.15) reveals that the magnitude of the interpolation error is proportional to the $(r+1)$th derivative of the function and varies with the grid spacing as $O(h^{r+1})$. We may express (2.15) as

$$R(x) = \frac{u^{(r+1)}(\xi)}{(r+1)!}\prod_{i=0}^{r}(x-x_i) = O(h^{r+1})\frac{u^{(r+1)}(\xi)}{(r+1)!} \tag{2.16}$$

Thus, for functions with large high-order derivatives, increasing the order of polynomial interpolation may not yield higher accuracy and grid refinement at a fixed polynomial degree needed to achieve the desired level of accuracy. In some cases, e.g., in a collision of two shocks, it may even become necessary to reduce the degree of the polynomial. Moreover, from the term $(x-x_0)(x-x_1)\cdots(x-x_r)$, we deduce that for an equispaced set of points, the error is not uniform in the interval; for points near

2.1 Linear Interpolation

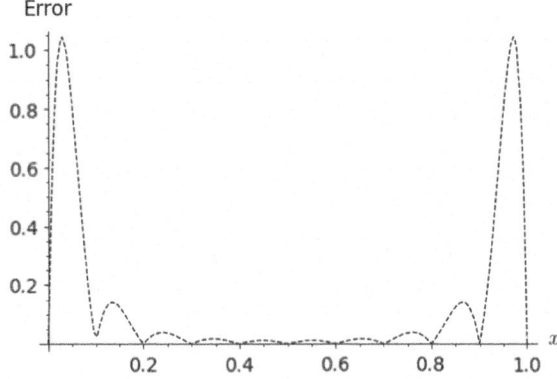

Fig. 2.2 Normalized interpolation error $R(x)/u^{(r+1)}(\bar{x})$ from Eq. (2.13) for approximation in $x \in [0, 1]$ and for $r = 10$

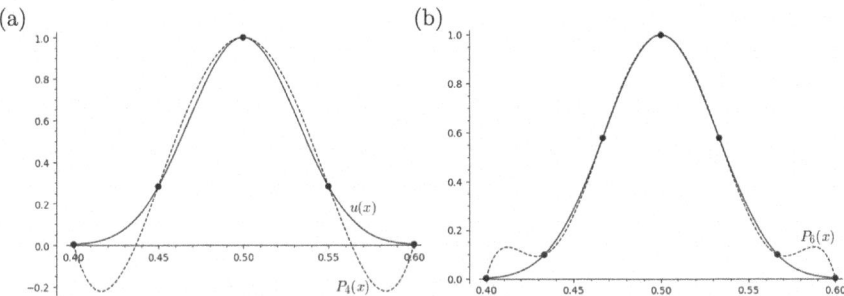

Fig. 2.3 Polynomial interpolation of the smooth function $u(x) = \cos(\pi(\cos(\pi x))^{10}$ with $x \in [0.4, 0.6]$; **a** degree polynomial $r = 4$, and **b** $r = 6$, clearly demonstrating the increased accuracy by increasing the polynomial degrees everywhere except near the boundaries of the interpolation domain

the boundary of the x interval, higher errors result. Figure 2.2 demonstrates the normalized error $R(x)/u^{(r+1)}(\xi)$ from Eq. (2.13) in the interpolation interval of $[0, 1]$ for $r = 10$. It reveals significantly higher errors near the endpoints of the interpolation interval. Also, Fig. 2.3a demonstrates the accuracy of the fourth-degree interpolating polynomials for a smooth function, clearly demonstrating high accuracy near the center of the interpolating domain and lower near the boundaries. Increasing the interpolation degree to $r = 6$ improves the accuracy everywhere, but lower accuracy is still observed near the boundaries as seen in Fig. 2.3b.

In extrapolation, where approximation points fall outside the original x interval, the high-order interpolations are particularly poor. The poor accuracy is not unique to the interpolation, and other schemes also suffer from this.

This phenomenon of high interpolation errors near the boundary of the approximation domain for an equispaced grid is called the Runge phenomenon [2]; in spectral and high-order finite element methods, a non-uniform grid with dramatically reduced spacing near the domain boundary solves the Runge phenomenon; we will not pursue those methods here.

One may alternatively keep the equispaced grid, but use overlapping sub-grids (referred to as local stencils) to determine local interpolations that are targeted for evaluation at or near the center of the stencil. This local interpolation gives rise to the finite difference scheme we focus on and discuss next. However, before doing so, let us illustrate the Runge phenomenon and the spurious oscillations arising from linear interpolation for non-smooth or discontinuous functions using a few examples.

2.1.2.1 Example

Consider the 10th degree interpolating polynomial defined over a grid with a uniform spacing of $h = 0.1$ over $x \in [0, 1]$. We want to compare the interpolation error for two points, one near one of the endpoints, $x = 0.05$, and the other near the center of the interval $x = 0.55$.

Using the equation for the error in the interpolation (2.13), we have

$$R(0.05) = (0.05)(-0.05)(-0.15)(-0.25)\cdots(-0.95)\frac{f^{(r+1)}(\bar{x})}{(r+1)!} = 3.2 \times 10^{-6}\frac{u^{(r+1)}(\bar{x})}{(r+1)!} \tag{2.17}$$

and

$$R(0.55) = (0.55)(0.45)(0.35)\cdots(0.05)(-0.15)(-0.25)\cdots(-0.45)\frac{u^{(r+1)}(\bar{x})}{(r+1)!}$$
$$= -4.8 \times 10^{-8}\frac{f^{(r+1)}(\bar{x})}{(r+1)!} \tag{2.18}$$

The above calculation reveals that the error magnitude closer to the center of the interpolation interval is approximately two orders of magnitude smaller than the error near the endpoints of the interval.

2.1.3 Interpolation for Discontinuous Functions

The error term of the interpolation, Eq. (2.13), clearly reveals that for discontinuous functions, one may not expect convergence using polynomial interpolation. The lack of convergence is clear from Fig. 2.4a, b, where the approximations to a step function using polynomial degrees $r = 7$ and $r = 9$ appear. As long as the interpolation grid crosses a discontinuity, we do not obtain an acceptable convergence rate even away from the discontinuity. In both cases, the maximum error near discontinuity appears to be approximately 10% of the jump. This error will not diminish even with a much higher number of interpolating polynomials. This spurious oscillatory behavior and

2.1 Linear Interpolation

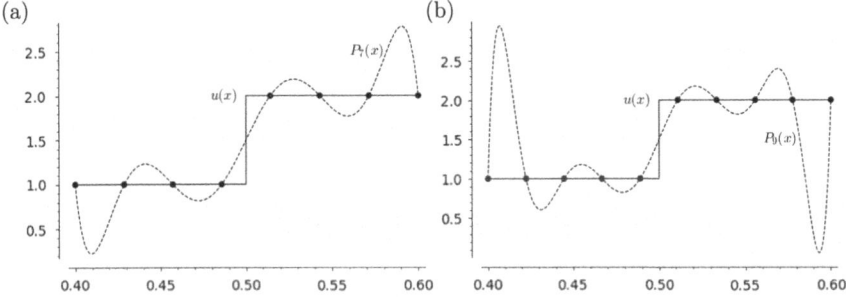

Fig. 2.4 Polynomial interpolation of a step function: **a** degree polynomial $r = 7$, and **b** $r = 9$, clearly demonstrating the persistent oscillatory behavior of the Gibbs phenomenon and lack of convergence despite increasing polynomial degrees

lack of convergence are related to the Gibbs phenomena observed in approximating a discontinuous function using truncated Fourier series. Also, computing the function's derivative using interpolation is even less accurate than the function itself and consists of more dramatic oscillatory behavior. Taking the derivative of Eq. (2.12)

$$\frac{du(x)}{dx} = \frac{dP_r(x)}{dx} + \frac{u^{(r+1)}(\xi)}{(r+1)!} \frac{d}{dx}\left[\prod_{i=0}^{r}(x-x_i)\right] = \frac{dP_r(x)}{dx} + \mathcal{O}(h^r)\frac{u^{(r+1)}(\xi)}{(r+1)!} \tag{2.19}$$

and comparing with that of the function itself, Eq. (2.16), we obtain that the error in the derivative approximation is $1/\mathcal{O}(h)$ larger.

The derivative of the step function

$$u(x) = \begin{cases} 1 & -0.2 \leq x \leq 0 \\ 0 & 0 < x \leq 0.2 \end{cases} \tag{2.20}$$

computed using $r = 7$ and $r = 15$ are shown in Fig. 2.5. The figure clearly shows a lack of convergence for the derivative approximation of a discontinuity using the polynomial interpolation.

Therefore, direct applications of the Lagrange interpolation in approximating conservation laws whose solution can consist of discontinuous features are inappropriate.

To avoid spurious oscillations in the entire domain resulting from isolated discontinuities, we may use local interpolation instead of global interpolation using all spatial grid points.

Striving for interpolations that overcome the Gibbs phenomena in solving the conservation laws motivates us to design nonlinear interpolations discussed later in Sect. 2.3. But, first, we discuss finite difference schemes that help overcome the Runge

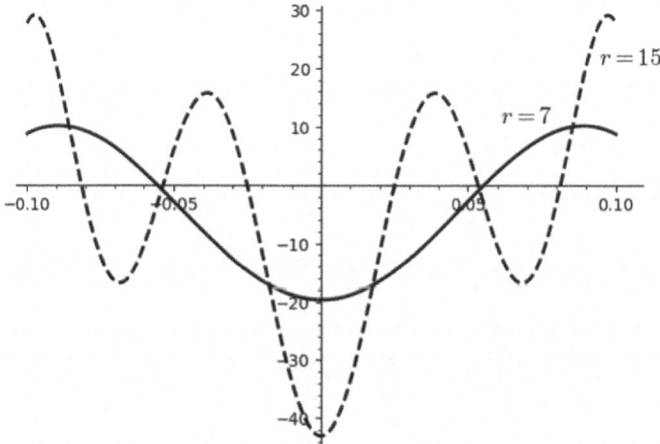

Fig. 2.5 Polynomial interpolation of the first derivative of a step function with the discontinuity at $x = 0$ with a solid line for degree polynomial $r = 7$, and the dashed line for $r = 15$, clearly demonstrating the persistent large oscillatory behavior of the Gibbs phenomenon and lack of convergence despite increasing polynomial degrees

phenomena and pave the way for designing nonlinear methods for discontinuous solutions.

2.2 High-Order Finite Difference Scheme

In finite difference schemes, evaluating the global interpolating polynomials near the endpoints of the interval is avoided by using local interpolation, formed over overlapping sets of grid points called stencils. For high-order finite difference approximations, the polynomial degrees are chosen as high as $r = O(10)$ (not $r = O(100)$ or higher), and interpolating functions are evaluated near the center (not necessarily exactly at the center) of the interval of the $r + 1$ given points.

2.2.1 Why Mid-Point Approximations?

To motivate the design of a high-order finite difference scheme and to demonstrate the need for mid-point or cell-boundary approximations, we need to discuss how local interpolation performs for a discontinuous function approximation. Consider the derivative approximation in advecting a jump discontinuity at x_i using a stencil. For solving advection-type problems, the interpolation must be upwind-biased for stable, optimally accurate approximation [11]. If l grid points to the left and m grid points to

2.2 High-Order Finite Difference Scheme

the right of x_i are used, with the total of $l + m + 1$ points in the stencil, then as shown in Ref. [11], for a positive advection speed, the optimal-order schemes with the order of $r = l + m$ are stable for $m \leq l \leq m + 2$ and unstable otherwise. Consistent with this, let us choose a left-biased stencil $\{x_{i-2}, x_{i-1}, x_i, x_{i+1}\}$ for computing derivative at x_i. Differentiation of the polynomial interpolation yields

$$\left.\frac{d\hat{u}(x)}{dx}\right|_{x_i} = \left.\frac{dP_r(x)}{dx}\right|_{x_i} = \sum_{j=i-2}^{j=i+1} u_j \ell'_j(x_i) \tag{2.21}$$

where $\ell'(x)$ is the derivative of the Lagrange polynomials. We know this approximation is a third-order accurate approximation for a smooth function u. However, if there is a discontinuity, the accuracy may significantly deteriorate. For simplicity, let us say the solution is just a step function located somewhere inside the interval (x_i, x_{i+1}), but not at x_i or x_{i+1}, with values u_1 and u_2 to the left and right of the jump, respectively. Then, the approximation reduces to

$$\left.\frac{d\hat{u}(x)}{dx}\right|_{x_i} = \left.\frac{dP_r(x)}{dx}\right|_{x_i} = u_1 \sum_{j=i-2}^{j=i} \ell'_j(x_i) + u_2 \ell'_{i+1}(x_i) \tag{2.22}$$

Since

$$\sum_{j=i-2}^{j=i+1} \ell'_j(x) = 0 \implies \sum_{j=i-2}^{j=i} \ell'_j(x_i) = -\ell'_{i+1}(x_i) \tag{2.23}$$

then

$$\left.\frac{d\hat{u}(x)}{dx}\right|_{x_i} = \left.\frac{dP_r(x)}{dx}\right|_{x_i} = (u_2 - u_1)\ell'_{i+1}(x_i) = \frac{1}{3h}(u_2 - u_1) \tag{2.24}$$

The derivative approximation consists of significant error because the exact derivative is zero at x_i. However, if the step is sufficiently close to x_i, it is accurate within up to the grid spacing since within $[x_{i-1/2}, x_{i+1/2}]$, the derivative is proportional to the jump discontinuity. On the other hand, if the discontinuity falls in (x_{i-2}, x_{i-1}), then the computed derivative is still proportional to the jump:

$$\left.\frac{d\hat{u}(x)}{dx}\right|_{x_i} = \left.\frac{dP_r(x)}{dx}\right|_{x_i} = u_1 \ell'_{i-2}(x_i) + u_2 \sum_{j=i-1}^{j=i} \ell'_j(x_i) = \frac{1}{6h}(u_2 - u_1) \tag{2.25}$$

This approximation is significantly in error as it yields high derivative values for a jump more than two grid points away. In general, an approximation using $(l + m + 1)$ points can yield erroneously high derivatives for a discontinuity located l points away. Figure 2.6 depicts the spurious solutions near a discontinuity using the third- and ninth-order upwind schemes in the advection of a step function.

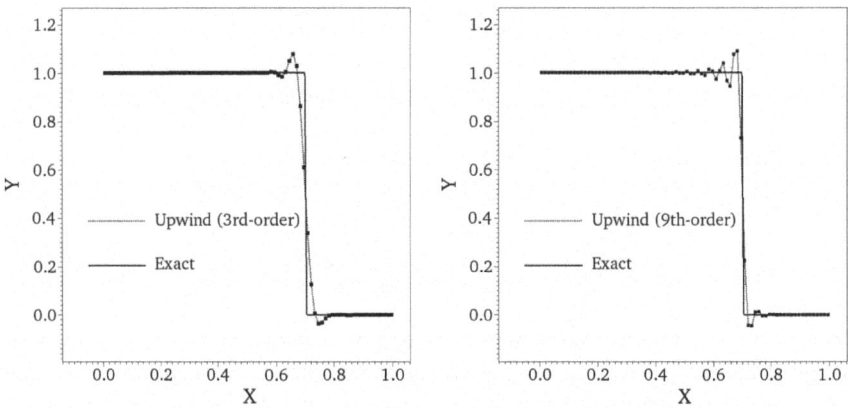

Fig. 2.6 Advection of a step function initially located at $x = 0.5$ using the third-order upwind-biased scheme on the left and the ninth-order on the right, showing the spurious oscillations near the discontinuity

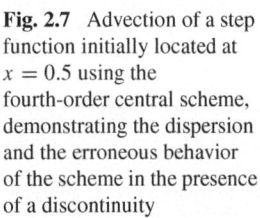

Fig. 2.7 Advection of a step function initially located at $x = 0.5$ using the fourth-order central scheme, demonstrating the dispersion and the erroneous behavior of the scheme in the presence of a discontinuity

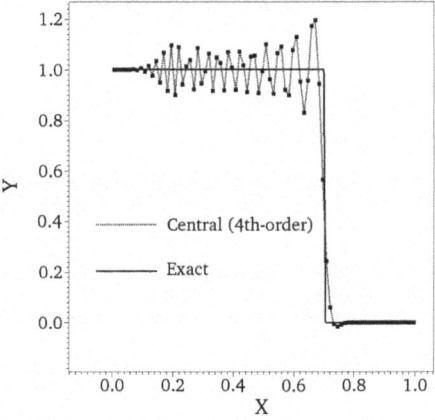

A central scheme in the presence of discontinuity presents a similar error as the upwind scheme, but since there is no numerical dissipation, the spurious oscillations will pollute the entire region left of the discontinuity as depicted in Fig. 2.7.

One remedy is to avoid the discontinuity in the approximation stencil and choose a stencil that is free from the discontinuity. For instance, if instead of stencil $\{x_{i-2}, x_{i-1}, x_i, x_{i+1}\}$, the more compact stencil $\{x_{i-1}, x_{i-1/2}, x_i, x_{i+1/2}, x_{i+1}\}$ that encompasses the mid-points and it is only one grid point away from x_i, the discontinuity is not crossed and exact derivative obtained for the jump in (x_{i-2}, x_{i-1}). This stencil would give a fourth-order accurate scheme for smooth functions. If one uses a sixth-order scheme corresponding to the stencil $\{x_{i-2}, x_{i-1}, x_{i-1/2}, x_i, x_{i+1/2}, x_{i+1}, x_{i+2}\}$, the discontinuity would be crossed. However, the error is significantly smaller than that of using the original grid. For this

2.2 High-Order Finite Difference Scheme

six-order stencil that consists of the two mid-points surrounding x_i, the approximation is

$$\left.\frac{d\hat{u}(x)}{dx}\right|_{x_i} = \frac{1}{h}\left[\frac{64}{45}(u_{i+1/2} - u_{i-1/2}) - \frac{2}{9}(u_{i+1} - u_{i-1}) + \frac{1}{180}(u_{i+2} - u_{i-2})\right] = \frac{1}{180h}(u_2 - u_1) \quad (2.26)$$

yielding a thirty-fold smaller error compared to the approximation (2.25) obtained using the stencil encompassing the original grids.

Therefore, it is sensible to include the left and right mid-points in the approximating stencil. However, since only cell-center (or nodal) values are known, we need to determine the mid-point values using the given nodal values and ensure no discontinuities crossed during the mid-point computations. Analogous to the derivative approximation, the mid-point approximations also inherit a high error if a discontinuity falls within the stencil and is several grid points away from the mid-point. An effective strategy is to adjust the stencil depending on the location of the discontinuity, leading to nonlinear schemes such as essentially non-oscillatory (ENO) schemes and their variants which we will discuss. But first, we need to present the mid-point approximations for various stencils and degree polynomials.

2.2.2 Mid-Point Approximation

To obtain an approximation at each mid-point $x_{i+1/2}$, a stencil surrounding $i + 1/2$ is chosen. For simplicity, the stencil structure relative to the grid point is identical for all grid points. However, the stencil choice may vary from one grid point to another if the solution consists of discontinuities. In this case, it is sensible to locally adjust the stencil such that no point in the stencil crossing the discontinuity. This adaptive adjustment is at the heart of nonlinear, essentially non-oscillatory schemes discussed in Sect. 2.3. Here, we discuss the mid-point approximation using the Lagrange interpolation for various choices of local stencils needed for nonlinear interpolations.

The interpolating polynomials (2.6) evaluated at $x_{i+\frac{1}{2}}$ results in

$$P_{r,k_s,i+1/2} = \sum_{l=0}^{l=r} a_{r,k_s,l}\, u_{i-k_s+l} \quad \text{(for some integer } k_s\text{)} \quad (2.27)$$

where

$$a_{r,k_s,l} = \ell_{i-k_s+l}(x_{i+1/2}) = \prod_{\substack{j=i-k_s \\ j \neq i-k_s+l}}^{j=i-k_s+r} \left(\frac{x_{i+1/2} - x_j}{x_{i-k_s+l} - x_j}\right) \quad (2.28)$$

As mentioned above, there is more than one stencil candidate to reconstruct the function, and k_s signifies the starting point of a stencil relative to x_i. For $r = 1$,

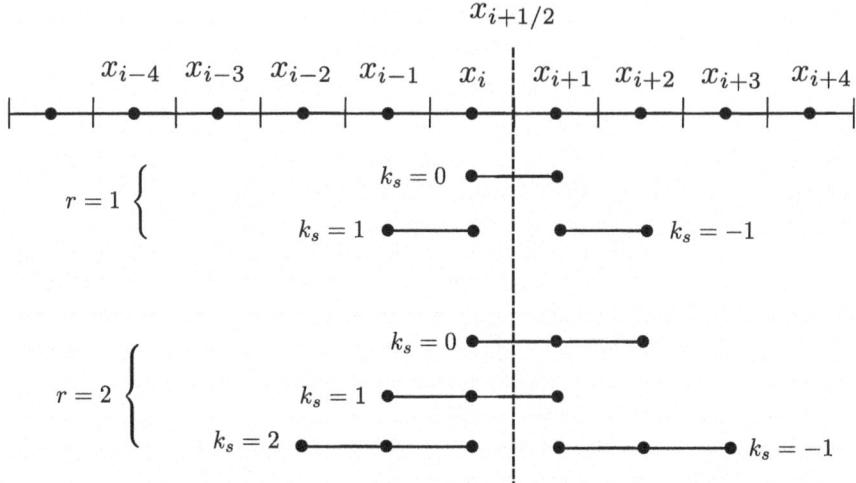

Fig. 2.8 Interpolation stencils for polynomial degrees $r = 1$ and $r = 2$

one may consider three stencils, (x_{i-1}, x_i), (x_i, x_{i+1}) and (x_{i+1}, x_{i+2}) corresponding to $k_s = 1, 0$, and -1, as shown in Fig. 2.8. Other stencils with data points farther away from the evaluation point are also possible; however, they consist of higher interpolation or extrapolation errors not worthy of consideration. The second stencil is symmetric around $x_{i+\frac{1}{2}}$, while the first and the third are pure left and right approximations. For $r = 2$, one may consider four stencils, (x_{i-2}, x_{i-1}, x_i), (x_{i-1}, x_i, x_{i+1}), (x_i, x_{i+1}, x_{i+2}), and $(x_{i+1}, x_{i+2}, x_{i+3})$ corresponding to $k_s = 2, 1, 0$, and -1, as shown in Fig. 2.8.

Owing to symmetry, knowing the coefficients for the pure left scheme, one can easily find the coefficients for the pure right scheme. Therefore, we may only tabulate the coefficients for schemes with at least one point left of $x_{i+1/2}$. Using Eq. (2.28), coefficient $a_{r,k_s,l}$ for $r = 1 - 4$ are given in Table 2.1.

The index i signifies the global numbering of the discrete points, while k_s, the stencil index, assume the values of $k_s = 0, 1, \ldots, r$, with a lower value of k_s signifying a lower level of left-biasedness and a higher k_s a higher level of left-biasedness.

Also, for minimum local truncation errors, it is preferred to have stencils centered around $i + 1/2$. The choice of k_s, however, allows for the stencil to be biased to the left or right of the $i + 1/2$ essential for the approximation of the hyperbolic equations (like advection equation and the equations of compressible flows) where the direction of the characteristics must be taken into account in the construction of the numerical solution to ensure stable schemes.

We also note that for each polynomial degree r, Table 2.1 lists coefficients for $r + 1$ stencils. These stencils are elected such that when they are combined (as discussed in Sect. 2.3) yield a minimally left-biased approximation of order $2r + 1$. The symmetry of the grid points allows the determination of coefficients of another set of stencils

2.2 High-Order Finite Difference Scheme

Table 2.1 Coefficients $a_{r,k_s,l}$ in Eq. (2.28) used for computing left upwind reconstruction with polynomial degree r [order accuracy of $(r+1)$] on stencil k_s

Polynomial degree, r	l	$k_s = 0$	$k_s = 1$	$k_s = 2$	$k_s = 3$	$k_s = 4$
4	0	$\frac{35}{128}$	$-\frac{5}{128}$	$\frac{3}{128}$	$-\frac{5}{128}$	$\frac{35}{128}$
	1	$\frac{35}{32}$	$\frac{15}{32}$	$-\frac{5}{32}$	$\frac{7}{32}$	$-\frac{45}{32}$
	2	$-\frac{35}{64}$	$\frac{45}{64}$	$\frac{45}{64}$	$-\frac{35}{128}$	$\frac{189}{64}$
	3	$\frac{7}{32}$	$-\frac{5}{32}$	$\frac{15}{32}$	$\frac{35}{32}$	$-\frac{105}{32}$
	4	$-\frac{5}{128}$	$\frac{3}{128}$	$-\frac{5}{128}$	$\frac{35}{128}$	$\frac{315}{128}$
3	0	$\frac{5}{16}$	$-\frac{1}{16}$	$\frac{1}{16}$	$-\frac{5}{16}$	
	1	$\frac{15}{16}$	$\frac{9}{16}$	$-\frac{5}{16}$	$\frac{21}{16}$	
	2	$-\frac{5}{16}$	$\frac{9}{16}$	$\frac{15}{16}$	$-\frac{35}{16}$	
	3	$\frac{1}{16}$	$-\frac{1}{16}$	$\frac{5}{16}$	$\frac{35}{16}$	
2	0	$\frac{3}{8}$	$-\frac{1}{8}$	$\frac{3}{8}$		
	1	$\frac{6}{8}$	$\frac{6}{8}$	$-\frac{5}{4}$		
	2	$-\frac{1}{8}$	$\frac{3}{8}$	$\frac{15}{8}$		
1	0	$\frac{1}{2}$	$-\frac{1}{2}$			
	1	$\frac{1}{2}$	$\frac{3}{2}$			

such that their combination can yield a minimally right-biased approximation of order $2r + 1$. Of course, the left- or right-biased approximation is essential for hyperbolic equations by providing proper upwinding needed for stability.

2.2.3 Derivative Approximation

In this section, we describe central schemes for derivative approximations. The central nature of the derivative approximation is sufficient because the biased nature of the overall approximation is handled in the mid-point approximations using various stencils as discussed in Sect. 2.2.2.

The interpolation values at the mid-points are now used in the derivative approximation. For clarity in the following discussion, we introduce the notation

$$u_{i+\frac{1}{2}} \equiv P_{r,k_s,i+\frac{1}{2}} \tag{2.29}$$

For the derivative computation, our first choice is to use the difference of the two consecutive values of $u_{i+\frac{1}{2}}$ as

$$\frac{\partial u}{\partial x}(x_i) \approx \frac{(u_{i+\frac{1}{2}} - u_{i-\frac{1}{2}})}{h} \qquad (2.30)$$

This simple choice, however, yields only a second-order accurate scheme, which can be shown by writing two Taylor series for $u(x)$, both about x_i, but one evaluated at $x_{i+\frac{1}{2}}$ and the other at $x_{i-\frac{1}{2}}$ as

$$u(x_{i+\frac{1}{2}}) = u(x_i) + \frac{h}{2}u^{(1)}(x_i) + u^{(2)}(x_i)\frac{h^2}{4\,2!} + u^{(3)}(x_i)\frac{h^3}{8\,3!} + \cdots + u^{(n)}(x_i)\frac{h^n}{2^n\,n!} + \cdots$$

$$u(x_{i-\frac{1}{2}}) = u(x_i) - \frac{h}{2}u^{(1)}(x_i) + u^{(2)}(x_i)\frac{h^2}{4\,2!} - u^{(3)}(x_i)\frac{h^3}{8\,3!} + \cdots + (-1)^n u^{(n)}(x_i)\frac{h^n}{2^n\,n!} + \cdots$$

where $u^{(i)}$ denotes the ith derivative. Subtracting the second equation from the first and dividing by h then yields

$$u^{(1)}(x_i) = \frac{u(x_{i+\frac{1}{2}}) - u(x_{i-\frac{1}{2}})}{h} + \sum_{n=2}^{\infty}\left[1 - (-1)^n\right]\frac{h^{n-1}}{2^n\,n!}u^{(n)}(x_i) \qquad (2.31)$$

All even terms in the sum are zero. The leading nonzero term is for $n = 3$. The derivative approximation is thus of second-order accuracy, i.e.,

$$u^{(1)}(x_i) = \frac{u(x_{i+\frac{1}{2}}) - u(x_{i-\frac{1}{2}})}{h} + O(h^2) \qquad (2.32)$$

which is only useful for a second-degree interpolating polynomial. How would one then find a higher order approximation? One possible approach is to include, in addition to the immediate mid-point values, other mid-point values as

$$\left(u(x_{i+\frac{1}{2}}) - u(x_{i-\frac{1}{2}}),\ u(x_{i+\frac{3}{2}}) - u(x_{i-\frac{3}{2}}),\ u(x_{i+\frac{5}{2}}) - u(x_{i-\frac{5}{2}}),\ \cdots \right) \qquad (2.33)$$

Using the same Taylor series expansion as above, the differences take the form of

$$u(x_{i+\frac{1}{2}}) - u(x_{i-\frac{1}{2}}) = \sum_{n=0}^{\infty}\left[1 - (-1)^n\right]\frac{h^n}{2^n\,n!}u^{(n)}(x_i)$$

$$u(x_{i+\frac{3}{2}}) - u(x_{i-\frac{3}{2}}) = \sum_{n=0}^{\infty}\left[1 - (-1)^n\right]\frac{3^n h^n}{2^n\,n!}u^{(n)}(x_i)$$

2.2 High-Order Finite Difference Scheme

or for p differences

$$u(x_{i+j-\frac{1}{2}}) - u(x_{i-j+\frac{1}{2}}) = \sum_{n=0}^{\infty} \left[1 - (-1)^n\right] \frac{(2j-1)^n h^n}{2^n n!} u^{(n)}(x_i) \quad \forall j = 1, \ldots, p \tag{2.34}$$

Now, a linear combination of the differences yield a high-order approximation of the derivative which is needed. The linear combination is

$$\sum_{j=1}^{p} d_j \left[u(x_{i+j-\frac{1}{2}}) - u(x_{i-j+\frac{1}{2}})\right] = \sum_{n=0}^{\infty} u^{(n)}(x_i) \left[1 - (-1)^n\right] \frac{h^n}{2^n n!} \sum_{j=1}^{p} d_j (2j-1)^n \tag{2.35}$$

If we choose only one difference, then $d_1 = 1$ recovers the first derivative with the error term of the third derivative, as obtained above. If we pick two differences, we can determine the values of d_1 and d_2 such that we recover the first derivative and the term with the third derivative becoming zero, yielding an algebraic system for d_1 and d_2 in the form

$$d_1 + 3d_2 = 1$$
$$d_1 + 27d_2 = 0$$

with solutions

$$\left(d_1 = \tfrac{9}{8}, \, d_2 = -\tfrac{1}{24}\right) \tag{2.36}$$

With these solutions, we have

$$\sum_{j=1}^{2} d_j \left[u(x_{i+j+\frac{1}{2}}) - u(x_{i-j-\frac{1}{2}})\right] = hu^{(1)}(x_i) + O(h^5) \tag{2.37}$$

A higher number of differences allows determining particular linear combinations such that higher order derivatives are zero, which requires solving larger algebraic systems for d_js. In general, we have

$$\sum_{j=1}^{p} d_j \left[u(x_{i+j+\frac{1}{2}}) - u(x_{i-j-\frac{1}{2}})\right] = hu^{(1)}(x_i) + O(h^{2p+1}) \tag{2.38}$$

or the derivative is $\mathcal{O}(h^{2p})$, i.e.,

$$\frac{1}{h} \sum_{j=1}^{p} d_j \left[u(x_{i+j+\frac{1}{2}}) - u(x_{i-j-\frac{1}{2}})\right] = u^{(1)}(x_i) + O(h^{2p}) \tag{2.39}$$

Table 2.2 Coefficients d_j in Eq. (2.40) used for computing the first-order derivative with approximation order $2p$

Approximation order, $2p$	d_1	d_2	d_3	d_4	d_5
10	$\frac{2047}{1690}$	$-\frac{735}{8192}$	$\frac{567}{40960}$	$-\frac{86}{48707}$	$\frac{35}{294912}$
8	$\frac{1225}{1024}$	$-\frac{245}{3072}$	$\frac{49}{5120}$	$-\frac{5}{7168}$	
6	$\frac{75}{64}$	$-\frac{25}{384}$	$\frac{3}{640}$		
4	$\frac{9}{8}$	$-\frac{1}{24}$			
2	1				

Therefore, if the mid-point interpolation computation is of order q, we require the derivative approximations to be of the same order, $2p = q$ for the even order interpolation, or $2p = q + 1$ for the odd order interpolation, yielding $p = q/2$ for an even order interpolation, and $p = (q + 1)/2$ for an odd order. Note that q order interpolation corresponds to the $(q - 1)$ degree polynomial interpolation.

Given the reconstructed mid-point values, the high-order derivative computations results from

$$\sum_{j=1}^{p} d_j \left[u(x_{i+j-\frac{1}{2}}) - u(x_{i-j+\frac{1}{2}}) \right] \approx h u^{(1)}(x_i) \qquad (2.40)$$

the coefficients, d_j are given in Table 2.2. This approach was developed in [5, 6]. The derivative calculations based on (2.40) always yield an even order accurate approximation. Thus, to be consistent with the accuracy of the reconstruction, one must choose $p = r + 1$ yielding $2r + 2$ accurate derivative calculations, which is sufficient for the $2r + 1$ reconstruction order.

Remark 1 The above analysis of the derivative approximations relies on the exact mid-point values. However, in the actual scheme, the mid-point values are approximations obtained using $r + 1$ interpolation. The use of approximate mid-point values $u_{i+1/2}$ instead of the exact mid-point values in the derivative approximation (2.40) incurs an additional error of $\mathcal{O}(h^{r+1})$.

2.2.3.1 Alternative Scheme for High-Order Derivative Computation

To compute the derivative at the center of cell i with a specific order, the scheme presented in the previous section requires not only the cell-boundary values of the cell i but also the cell-boundary values of the adjacent cells spanning half order away from both sides of the cell i. It implies that the effective stencil for computing the derivative with order $2r$ is $4r$. The enlarged effective stencil can potentially complicate the imposition of boundary conditions at solid walls and non-reflecting boundaries. Moreover, in parallel computations, where the grid is divided into many

2.2 High-Order Finite Difference Scheme

parts and distributed among different computing cores, this numerical scheme for derivative approximations requires either larger messages or twice an many ones, compared to a second-order method, be communicated among adjacent processors.

An alternative approach based on a mix of cell-boundary and cell-center values is possible [12]. The function values used for the derivative approximation at the cell-center i are

$$\left(u(x_{i+\frac{1}{2}}) - u(x_{i-\frac{1}{2}}), u(x_{i+1}) - u(x_{i-1}), u(x_{i+2}) - u(x_{i-2}), \cdots\right) \quad (2.41)$$

The scheme has the following form:

$$\frac{1}{h}\left\{d'_1\left[u(x_{i+1/2}) - u(x_{i-1/2})\right] + \sum_{j=2}^{p} d'_j\left[u(x_{i+j-1}) - u(x_{i-j+1})\right]\right\} = u^{(1)}(x_i) + O(h^{2p}) \quad (2.42)$$

where d'_j constant coefficients are specific to a given approximation order. For $p = 2$ or a fourth-order derivative approximation, the coefficients are

$$\left(d'_1 = \tfrac{8}{6}, d'_2 = -\tfrac{1}{6}\right) \quad (2.43)$$

For $p = 3$ or a six-order approximation, the coefficients are

$$\left(d'_1 = \tfrac{64}{45}, d'_2 = -\tfrac{2}{9}, d'_3 = \tfrac{1}{180}\right) \quad (2.44)$$

Reference [12] provides the coefficients for $p = 4$ or an eight-order approximation. They are

$$\left(d'_1 = \tfrac{256}{175}, d'_2 = -\tfrac{1}{4}, d'_3 = \tfrac{1}{100}, d'_4 = -\tfrac{1}{2100}\right) \quad (2.45)$$

The coefficients d'_j for the derivation computations in Eq. (2.42) for up to orders $2p = 10$ are given in Table 2.3.

We refer to this scheme as Scheme 2 and the former scheme introduced in Sect. 2.2.3 as Scheme 1.

Table 2.3 Coefficients d'_j in Eq. (2.42) used for computing the first-order derivative with approximation order $2p$

Approximation order, $2p$	d'_1	d'_2	d'_3	d'_4	d'_5
10	$\tfrac{16384}{11025}$	$-\tfrac{4}{15}$	$\tfrac{1}{75}$	$-\tfrac{4}{3675}$	$\tfrac{1}{17640}$
8	$\tfrac{256}{175}$	$-\tfrac{1}{4}$	$\tfrac{1}{100}$	$-\tfrac{1}{2100}$	
6	$\tfrac{64}{45}$	$-\tfrac{2}{9}$	$\tfrac{1}{180}$		
4	$\tfrac{8}{6}$	$-\tfrac{1}{6}$			
2	1				

Remark 2 In the second approach, the stencil of the derivative computation coincides with that of the mid-point approximation, while in the first approach, the effective stencil for the derivative approximation is larger than the interpolation stencil. Therefore, as pointed out at the beginning of this section, the second approach is more compact and may be more efficient for computations on parallel computers and more straightforward in implementing various boundary conditions.

2.2.4 Conservative Form of High-Order Derivative Approximation

The derivative approximation schemes presented above, Eqs. (2.40) and (2.42), can always be cast into a conservative form as the following theorem establishes.

Theorem 1 *Consider a spatial derivative approximation in the form*

$$\left.\frac{\partial f}{\partial x}\right|_{x_i} \approx \frac{1}{h} \sum_{j=q}^{j=q+r} d_j f_{i+j} \tag{2.46}$$

where r being a positive integer, and f_{i+j} a representation of $f(x_{i+j+\alpha})$, $0 \leq \alpha < 1$, and $d_j s$ are constants chosen to yield a specific order of accuracy and satisfy the consistency

$$\sum_{j=q}^{j=q+r} d_j = 0 \tag{2.47}$$

The derivative approximation (2.46) can always be cast into the conservation form

$$\left.\frac{\partial f}{\partial x}\right|_{x_i} \approx \frac{1}{h} \sum_{j=q}^{j=q+r} d_j f_{i+j} = \frac{1}{h}(\hat{F}_{i+\frac{1}{2}} - \hat{F}_{i-\frac{1}{2}}) \tag{2.48}$$

Proof Denoting the right flux as

$$\hat{F}_{i+\frac{1}{2}} = \sum_{j=q}^{j=q+r} c_j f_{i+j} \tag{2.49}$$

with c_j being constant, we require the left flux to be in the form

$$\hat{F}_{i-\frac{1}{2}} = \sum_{j=q}^{j=q+r} c_j f_{i+j-1}. \tag{2.50}$$

to satisfy the flux continuity or the conservation. Equations (2.48), (2.49), and (2.50) then yields

$$\sum_{j=q}^{j=q+r} d_j f_{i+j} = \sum_{j=q}^{j=q+r} c_j f_{i+j} - \sum_{j=q}^{j=q+r} c_j f_{i+j-1} \qquad (2.51)$$

which, in turn, yields the coefficients c_js as

$$c_q = 0 \qquad (2.52a)$$
$$c_{q+1} = c_q - d_q = -d_q \qquad (2.52b)$$
$$c_{q+2} = c_{q+1} - d_{q+1} = -(d_q + d_{q+1}) \qquad (2.52c)$$
$$\dots$$
$$c_{q+r} = c_{q+r-1} - d_{q+r-1} = \underbrace{-(d_q + d_{q+1} + \dots + d_{q+r-1})}_{\text{from Eq.(2.47)}} = d_{q+r} \qquad (2.52d)$$

where the last equality results from the consistency of the approximation, Eq. (2.47). □

The explicit expressions for the conservative fluxes using high-order derivative approximations in Scheme 1 and 2, Eqs. (2.40) and (2.42) are left to an end-of-chapter exercise. Section 2.3.5 further discusses the conservative forms of Scheme 1 and 2.

2.3 High-Order Finite Difference Method for Discontinuous Solutions

Several methods are available for overcoming the difficulty of approximating discontinuous functions. A straightforward strategy is to choose, among all possible stencils, the stencil that does not contain any discontinuity. To this end, we need to answer two questions: (1) how to determine whether a stencil contains a discontinuity, and (2) how to switch to the stencil with the smooth solution. For the former, some measure of the derivatives of the reconstructed interpolating polynomial $P_{r,k_s,i+\frac{1}{2}}(x)$, referred to as smoothness indicator, would be a reasonable choice. For the latter, there are two possibilities. The first approach uses the stencil with the smallest smoothness measure to compute the mid-point values, while the second approach forms a weighted combination of the approximations based on all stencils. The weight or coefficient of each stencil controls the contribution of each stencil to the approximation. The weights need to be inversely proportional to the size of the smoothness indicator while yielding the maximum order of accuracy possible from the combination of all stencils for a smooth solution. These strategies lead to the popular scheme of weighted essentially non-oscillatory (WENO) methods [4, 5] and their recent variants, targeted weighted essentially non-oscillatory (TENO) schemes [7–9], that we will discuss. Ref. [13] provides an algorithm for generating arbitrary order WENO method. But first, let us describe the design of the smoothness indicator's main ingredients in both schemes.

2.3.1 Smoothness Indicator

For a function $u(x)$ only known on a set of discrete points, u_is, the derivative of the interpolating polynomial $P_r(x)$ is a sensible measure of the smoothness of the function because according to (2.12)

$$P_r(x) = u(x) - \underbrace{(x - x_0)(x - x_1) \cdots (x - x_r) \frac{u^{(r+1)}(\bar{x})}{(r+1)!}}_{R(x)}, \quad (2.53)$$

and, hence,

$$\frac{dP_r(x)}{dx} = \frac{du(x)}{dx} - \frac{dR(x)}{dx} \quad (2.54)$$

This analysis shows that non-smoothness in the unknown function $u(x)$ appears in the interpolating polynomial's derivatives. However, for every grid point i, we do not know the exact location of the non-smoothness. To remedy this difficulty, we may consider the integral of the derivative of the interpolating polynomial for a cell centered at x_i to be the measure of the non-smoothness (the smoothness indicator) as

$$\int_{x_{i-\frac{1}{2}}}^{x_{i+\frac{1}{2}}} \frac{d}{dx} P_{r,k_s,i}(x) dx \quad (2.55)$$

However, in this integral measure, a large positive derivative can cancel out large negative derivatives giving rise to an erroneously small smoothness indicator. We may consider the absolute value of the derivative or an even power of the derivative. However, absolute value introduces a discontinuity in the derivative of the smoothness indicator (recall, for instance, the function $|x|$ has a discontinuous derivative at the origin). Alternatively, an even power of the derivative seems a sensible choice. In particular, the square of the derivative yields a computationally inexpensive smooth indicator in the form

$$\int_{x_{i-\frac{1}{2}}}^{x_{i+\frac{1}{2}}} \left[\frac{d}{dx} P_{r,k_s,i}(x) \right]^2 dx \quad (2.56)$$

Moreover, our smoothness indicator needs to capture the non-smoothness in the higher derivatives. Hence, we need to include higher order derivatives and sum over all derivatives up to the order r as

2.3 High-Order Finite Difference Method for Discontinuous Solutions

$$\sum_{l=1}^{r} \int_{x_{i-\frac{1}{2}}}^{x_{i+\frac{1}{2}}} \left[\frac{d^l}{dx^l} P_{r,k_s,i}(x) \right]^2 dx \tag{2.57}$$

However, different orders of derivatives have different dimensions, i.e.,

$$\text{Dimension of } \left[\frac{d^l}{dx^l} P_{r,k_s,i}(x) \right] = \frac{\text{Dimension of } u}{h^l} \tag{2.58}$$

while

$$\text{Dimension of } \left[\frac{d^{l+1}}{dx^{l+1}} P_{r,k_s,i}(x) \right] = \frac{\text{Dimension of } u}{h^{l+1}} \tag{2.59}$$

Therefore, we need to scale the derivatives before summation as

$$\sum_{l=1}^{r} h^{2l-1} \int_{x_{i-\frac{1}{2}}}^{x_{i+\frac{1}{2}}} \left[\frac{d^l}{dx^l} P_{r,k_s,i}(x) \right]^2 dx \tag{2.60}$$

The above smoothness indicator was derived in Ref. [4]. It is often shown by $\beta_{r,k_s,i}$. The smoothness indicators can thus be determined as

$$\beta_{r,k_s,i} = \sum_{l=1}^{r} h^{2l-1} \int_{x_{i-\frac{1}{2}}}^{x_{i+\frac{1}{2}}} \left[\frac{d^l}{dx^l} P_{r,k_s,i}(x) \right]^2 dx. \tag{2.61}$$

Recall the polynomial $P_{r,k_s,i}(x)$ are determined using Eq. (2.6) and Eq. (2.4) as

$$P_{r,k_s,i}(x) = \sum_{l=0}^{l=r} u_{i-k_s+l}\, \ell_{i-k_s+l}(x) \tag{2.62}$$

where $\ell_j(x)$ are the Lagrange interpolating polynomials defined in (2.5).

Several points are notable about (2.61):

- The smoothness indicators depend on the known discrete values of the generally unknown function $u(x)$.
- The smoothness indicators do not depend on grid spacing.
- The smoothness indicators depend on the degree of interpolating polynomial r and the stencil k_s.
- The determination of the smoothness indicator is necessary for every grid i.

The smoothness indicators $\beta_{r,k_s,i+\frac{1}{2}}$ up to 17th order of accuracy are given in Ref. [14]. For instance, for $r = 1$,

$$\beta_{1,0,i} = (u_{i+1} - u_i)^2 \tag{2.63a}$$
$$\beta_{1,1,i} = (u_i - u_{i-1})^2 \tag{2.63b}$$

For $r = 2$,

$$\beta_{2,0,i} = \frac{13}{12}(u_i - 2u_{i+1} + u_{i+2})^2 + \frac{1}{4}(3u_i - 4u_{i+1} + u_{i+2})^2 \qquad (2.64a)$$

$$\beta_{2,1,i} = \frac{13}{12}(u_{i-1} - 2u_i + u_{i+1})^2 + \frac{1}{4}(u_{i-1} - u_{i+1})^2 \qquad (2.64b)$$

$$\beta_{2,2,i} = \frac{13}{12}(u_{i-2} - 2u_{i-1} + u_i)^2 + \frac{1}{4}(u_{i-2} - 4u_{i-1} + 3u_i)^2 \qquad (2.64c)$$

2.3.2 Essentially Non-oscillatory Scheme

The idea of an essentially non-oscillatory (ENO) scheme is to choose the smoothest interpolating polynomial among all the candidate stencils [15]. This means the cell-boundary (or mid-point) values are approximated as

$$u_{i+1/2} = P_{r,k_s^*,i+1/2}(x), \qquad \beta_{r,k_s^*,i} = \min_{k_s=0}^{r}(\beta_{r,k_s,i}) \qquad (2.65)$$

The ENO scheme uses only one sub-stencil to compute an approximation of the function at each cell boundary while utilizing a larger stencil to determine the smoothness, i.e., the smoothness indicator is determined for all sub-stencils. Alternatively, a properly weighted sum of the approximation from each sub-stencil may be used, giving rise to weighted essentially non-oscillatory (WENO) schemes.

2.3.3 Weighted Essentially Non-oscillatory Scheme

To tame each sub-stencil approximation which may contain non-smoothness, we need to scale each sub-stencil approximation $P_{r,k_s,i+1/2}(x)$ by the inverse of its smoothness indicator $\beta_{r,k_s,i}$ as

$$\sum_{k_s=0}^{r} \frac{1}{\beta_{r,k_s,i}} P_{r,k_s,i+1/2}(x) \qquad (2.66)$$

Also, to avoid division by zero for the case where $\beta_{r,k_s,i} = 0$, we may add a small parameter ϵ (of the order of machine zero) to the denominator as

$$\sum_{k_s=0}^{r} \frac{1}{\beta_{r,k_s,i} + \epsilon} P_{r,k_s,i+1/2}(x) \qquad (2.67)$$

2.3 High-Order Finite Difference Method for Discontinuous Solutions

We note that we can use higher powers of $\beta_{r,k_s,i}$ in the denominators leading to the higher separation of the contribution of each stencil. For instance $\beta^2_{r,k_s,i}$ and $\beta^r_{r,k_s,i}$ are often used in the literature giving rise to

$$\sum_{k_s=0}^{r} \frac{1}{\beta^2_{r,k_s,i} + \epsilon} P_{r,k_s,i+1/2}(x) \tag{2.68}$$

and

$$\sum_{k_s=0}^{r} \frac{1}{\beta^r_{r,k_s,i} + \epsilon} P_{r,k_s,i+1/2}(x) \tag{2.69}$$

Examining these two expressions, we notice that they do not yield an approximation comparable to the one from the full stencil. The lack of agreement is because when the function $u(x)$ is smooth, the differences in the smoothness indicators $\beta_{r,k_s,i}$ are small, and, hence, the expressions yield an approximation with almost equal contribution from each sub-stencil which is not necessarily of higher order accuracy. We thus need to find the proper coefficients b_{r,k_s} such that

$$\sum_{k_s=0}^{r} b_{r,k_s} P_{r,k_s,i+\frac{1}{2}} = \sum_{l=0}^{2r} a_{2r,r,l}\, u_{i-r+l} \tag{2.70}$$

where coefficients $a_{2r,r,l}$ are defined in Eq. (2.28). The expression (2.69) is then modified to

$$\sum_{k_s=0}^{r} \frac{b_{r,k_s}}{\beta^r_{r,k_s,i} + \epsilon} P_{r,k_s,i+1/2}(x) \tag{2.71}$$

The weights,

$$\alpha_{r,k_s,i+1/2} = \frac{b_{r,k_s}}{\beta^r_{r,k_s,i} + \epsilon}, \quad \text{or} \quad \alpha_{r,k_s,i+1/2} = \frac{b_{r,k_s}}{\beta^2_{r,k_s,i} + \epsilon} \tag{2.72}$$

do not add up to 1; hence, a normalization in the form

$$w_{r,k_s,i+\frac{1}{2}} = \frac{\alpha_{r,k_s,i+\frac{1}{2}}}{\sum_{j=0}^{r} \alpha_{r,j,i+\frac{1}{2}}} \tag{2.73}$$

is required. Using the weights $w_{r,k_s,i+\frac{1}{2}}$, the mid-point approximation becomes

$$\sum_{k_s=0}^{r} w_{r,k_s,i+\frac{1}{2}}\, P_{r,k_s,i+\frac{1}{2}} \tag{2.74}$$

Table 2.4 Optimal weights, b_{r,k_s}, in Eq. (2.72) obtained from the system (2.70) and used for computing the WENO weights

Polynomial degree, r	$k_s = 0$	$k_s = 1$	$k_s = 2$	$k_s = 3$	$k_s = 4$
4	$\frac{9}{256}$	$\frac{21}{64}$	$\frac{63}{128}$	$\frac{9}{64}$	$\frac{1}{256}$
3	$\frac{7}{64}$	$\frac{35}{64}$	$\frac{21}{64}$	$\frac{1}{64}$	
2	$\frac{5}{16}$	$\frac{10}{16}$	$\frac{1}{16}$		
1	$\frac{3}{4}$	$\frac{1}{4}$			

which is called the WENO approximation. Therefore, the WENO approximation has the form

$$u_{r,\text{WENO},i+\frac{1}{2}} = \sum_{k_s=0}^{r} w_{r,k_s,i+\frac{1}{2}} \, P_{r,k_s,i+\frac{1}{2}} \tag{2.75}$$

Coefficients b_{r,k_s} are defined in Ref. [6] and given in Table 2.4 for the polynomial degree $r = 1$ to $r = 4$.

Note that rth-degree polynomial approximation on sub-stencils yields an overall WENO scheme of a formal accuracy of $2r + 1$ order for smooth functions in the L_1 and L_2 norms. In the L_∞ norm, the order of accuracy drops at the critical points of the solution. See the next section for the analysis.

Stencils for WENO schemes of orders 3, 5, and 7 corresponding to polynomial degrees $r = 1, 2$, and 3 on each sub-stencil for the $i + 1/2$ cell-boundary interpolation appear in Fig. 2.9.

Remark 3 Unlike the linear interpolation approximation in Eq. (2.27), the WENO scheme is a nonlinear scheme because the weights in Eq. (2.75) do depend on the solution values via the smoothness indicator. Also, when all smoothness indicators are large, except for one, the WENO scheme approaches the ENO scheme. This observation reveals that the ENO scheme is also a nonlinear scheme.

2.3.4 WENO Algorithm

To summarize, the WENO approximation of a formal accuracy of $2r + 1$ is determined using the following five-step algorithm

1. Determine the $(r + 1)$th order linear approximation of the mid-point values using $r + 1$ stencils for each mid-point value:

$$P_{r,k_s,i+\frac{1}{2}} = \sum_{l=0}^{l=r} a_{r,k_s,l} \, u_{i-k_s+l} \tag{2.76}$$

2.3 High-Order Finite Difference Method for Discontinuous Solutions

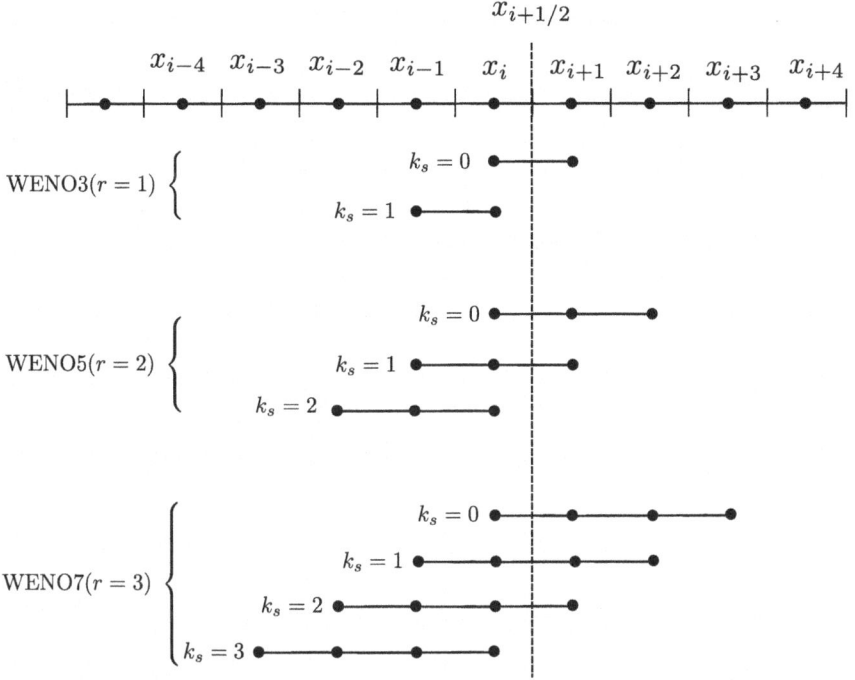

Fig. 2.9 Stencils used for left-biased WENO schemes of orders 3, 5 and 7 corresponding to $r = 1$, 2, and 3 for $i + 1/2$ cell-boundary interpolation

2. Determine the smoothness indicator $\beta_{r,k_s,i}$.
3. Determine the unnormalized nonlinear weights using the optimal weights b_{r,k_s} and the smoothness indicators $\beta_{r,k_s,i}$ as

$$\alpha_{r,k_s,i+\frac{1}{2}} = \frac{b_{r,k_s}}{\beta^2_{r,k_s,i} + \epsilon} \quad \text{or} \quad \alpha_{r,k_s,i+\frac{1}{2}} = \frac{b_{r,k_s}}{\beta^r_{r,k_s,i} + \epsilon} \quad (2.77)$$

4. Compute the normalized nonlinear weights or the WENO weights using

$$w_{r,k_s,i+\frac{1}{2}} = \frac{\alpha_{r,k_s,i+\frac{1}{2}}}{\sum_{j=0}^{r} \alpha_{r,j,i+\frac{1}{2}}} \quad (2.78)$$

5. Finally, compute the WENO approximation of the mid-point values as

$$u_{r,\text{WENO},i+\frac{1}{2}} = \sum_{k_s=0}^{r} w_{r,k_s,i+\frac{1}{2}} \, P_{r,k_s,i+\frac{1}{2}} \quad (2.79)$$

The interpolation coefficients $a_{r,k_s,l}$ and the optimal weights b_{r,k_s} are given in Tables 2.1 and 2.4, respectively, and smoothness indicators $\beta_{r,k_s,i+\frac{1}{2}}$ are defined in Eq. (2.61) and explicit expressions are given for $r = 1$ and $r = 2$ in Eqs. (2.63) and (2.64), respectively. Smoothness indicators for orders $r > 2$ are given in Ref. [14]. The parameter ϵ is a small number near machine zero. For double-precision computations, the machine zero is 10^{-153}, and thus $\epsilon > 10^{-153}$ is acceptable. In the double-precision computations, $\epsilon > 10^{-16}$ may be a more robust choice.

Note that using the interpolation coefficients $a_{r,k_s,l}$ given in Tables 2.1 yields a left-biased WENO approximation. For a right-biased WENO scheme, the same coefficients are valid on the right-biased stencil points symmetric to the left-biased stencil points.

Lemma 1 *The WENO reconstruction operator denoted by $\mathcal{R}$ has a particular linear property in the sense that for a small value of parameter ϵ, e.g., $\epsilon = \mathcal{O}(10^{-16})$, the following is valid up to a small term of the order of ϵ*

$$\mathcal{R}\{\alpha u(x)\} = \alpha \mathcal{R}\{u(x)\} \tag{2.80}$$

Remark 4 This property enables the proof of the consistency of a compressible two-fluid model in Chap. 6.

Proof The proof is the subject of an exercise at the end of the chapter. □

Remark 5 Equations (2.76), (2.77), (2.78), and (2.79) can be combined to express the left-biased WENO interpolation in one single equation as

$$u^+_{r,\text{WENO},i+\frac{1}{2}} = \sum_{l=1}^{2r-1} B^+_{i-r+l}(u)\, u_{i-r+l} \tag{2.81}$$

and the right-biased WENO scheme as

$$u^-_{r,\text{WENO},i+\frac{1}{2}} = \sum_{l=2}^{2r} B^-_{i-r+l}(u)\, u_{i-r+l} \tag{2.82}$$

The average of the left- and right-biased gives the central WENO interpolation:

$$u^c_{r,\text{WENO},i+\frac{1}{2}} = \frac{1}{2}\left(u^+_{r,\text{WENO},i+\frac{1}{2}} + u^-_{r,\text{WENO},i+\frac{1}{2}}\right) \tag{2.83}$$

Coefficients $B^+_{i-r+l}(u)$ and $B^-_{i-r+l}(u)$ are functions of u_ls. Their derivations are the subject of an end-of-chapter exercise.

2.3.4.1 WENO Scheme Applied to Characteristic Variables

In the case of a nonlinear system of conservation laws, the non-smoothness develops in characteristic variables along the characteristic paths. Hence, if we want to know

2.3 High-Order Finite Difference Method for Discontinuous Solutions

if the solution vector is smooth over a set of grid points at a specific time, we need to look at these grid points on the same family of characteristic paths. Across the same family of characteristic curves spanning the spatial grid points, it makes sense to consider the smoothness of the corresponding characteristic variables instead of the primitive or conservative variables. The process is repeated for all families of characteristics spanning the grid points.

To further appreciate the difference between the ENO/WENO of the characteristic and non-characteristic variables, we point out that in a spatial neighborhood of a stencil of grid points, examining the smoothness of characteristic variables is more consistent with the physics of the underlying conservation laws than the smoothness of the primitive or conservative variables. While the first spatial derivative of the characteristic variables is directly related to the first spatial derivatives of the primitive or conservative variables, their nth derivatives of the characteristics involve all nth derivatives and lower derivatives of the primitive or conservative variables. Therefore, the higher the derivative of the characteristic variables, the greater the separation from those of the non-characteristic (conservative or primitive) variables. As a result, while low-order characteristic and non-characteristic ENO/WENO schemes are comparable, the high-order versions differ significantly. In a nutshell, the following points are worth emphasizing:

- Characteristic variables are more consistent with the physics than non-characteristic variables because the signals are carried out and evolved along the characteristic paths are the characteristic variables.
- While the derivatives of the characteristic variables are related to those of the non-characteristic variables, the higher the derivatives, the greater the difference between the characteristics and non-characteristics. Hence, a smoothness indicator that involves computations of all derivatives up to the order of the scheme is more appropriate when applied to the characteristic variables.
- The difference between the characteristic and non-characteristic WENO reconstructions is only significant when the solution is non-smooth.

Therefore, in a system of conservation laws, it is important to apply the ENO or WENO schemes to the characteristic variables instead of primitive or conservative variables. Let us see how. Using Eq. (1.50), for all points in the overall stencil surrounding point $x_{i+1/2}$, we first determine the characteristic vector $\boldsymbol{u}_{c,i}$ from the primitive vector $\boldsymbol{u}_i$

$$d\bar{u}_{c,l} = [Q^{-1}]d\bar{u}_l, \quad l = i - r + j, \quad j = 1, \ldots, 2r \quad (2.84)$$

Here $\bar{u}_l$ and $\bar{u}_{c,l}$ are vectors of the dimension D, the number of equations in the system of conservation laws. Furthermore, $[Q^{-1}]$ is a matrix whose rows are the left eigenvectors of the matrix A associated with the quasi-linear form of the system of PDEs under consideration, and $[Q]$ is a matrix whose columns are the right eigenvectors of the matrix A. These matrices appear in the case of the Euler system in Chap. 1. In practice, all points in the stencil are mapped using a single matrix $[Q^{-1}]$ determined using $\bar{u}_{i+1/2} = (\bar{u}_i + \bar{u}_{i+1})/2$. Also, if we choose $d\bar{u}_l$ as

$$d\bar{u}_l = \bar{u}_l - \bar{u}_{i+1/2} \tag{2.85}$$

since $\bar{u}_{i+1/2}$ is a fixed vector, the WENO interpolation of $d\bar{u}_{c,l}$ and that of $\bar{u}_{c,l}$ are only different by $\bar{u}_{i+1/2}$ term. Hence, for simplicity we may use the transformation of $\bar{u}_l$ instead of $d\bar{u}_l$ as

$$\bar{u}_{c,l} = [Q^{-1}_{i+1/2}]\bar{u}_l \quad l = i - r + j, \quad j = 1, \ldots, 2r \tag{2.86}$$

and apply the WENO procedure to every component of the vector $\bar{u}_{c,l}$ to determine the vector $\bar{u}_{c,\text{WENO},i+1/2}$. The WENO approximation of the primitive variable vector is finally obtained by an inverse transformation as

$$\bar{u}_{\text{WENO},i+1/2} = [Q_{i+1/2}]\bar{u}_{c,\text{WENO},i+1/2} \tag{2.87}$$

The left- and right-biased WENO approximations at $x_{i+1/2}$ are both constructed as they are needed for the solution of the system of conservation laws as will be discussed in Chap. 5.

Using Eqs. (2.81), (2.86), and (2.87), the characteristic WENO interpolation are written in a matrix form as

$$u_{\text{WENO},i+1/2} = Q \, \text{diag}([Q^{-1}][u][B^t_c]) \tag{2.88}$$

and the non-characteristic WENO interpolation as

$$\bar{u}_{\text{WENO},i+1/2} = \text{diag}([u][B^t]) \tag{2.89}$$

In the last two expressions, B^t and B^t_c are matrices of size $(2r - 1) \times D$ obtained using Eq. (2.81). Matrices B^t and B^t_c are respectively functions of matrices $[u]$ and $[u_c]$ of dimensions $D \times (2r - 1)$. The columns of matrices $[u]$ and $[u_c]$ are formed by the non-characteristic and characteristic variables at points on the approximation stencil. The notation "diag" of a matrix is a vector formed by the diagonal entries of the matrix.

Equations (2.88) and (2.89) reveal the characteristic interpolations differ from the non-characteristic ones when a nonlinear interpolation like a WENO scheme is used, and the solution is non-smooth. For the case of a smooth solution, as discussed in the next section, the WENO scheme approaches the linear scheme, and, hence, the interpolation from the characteristic ENO/WENO scheme does not differ significantly from the non-characteristic method.

Figure 2.10 depicts the computed pressure for a Mach 3 shock moving from the left to right in the water with a ninth-order WENO scheme and with and without characteristic decomposition. The zoomed-in view right behind the shock in Fig. 2.10b shows that the scheme with characteristics has significantly smaller oscillations.

2.3 High-Order Finite Difference Method for Discontinuous Solutions

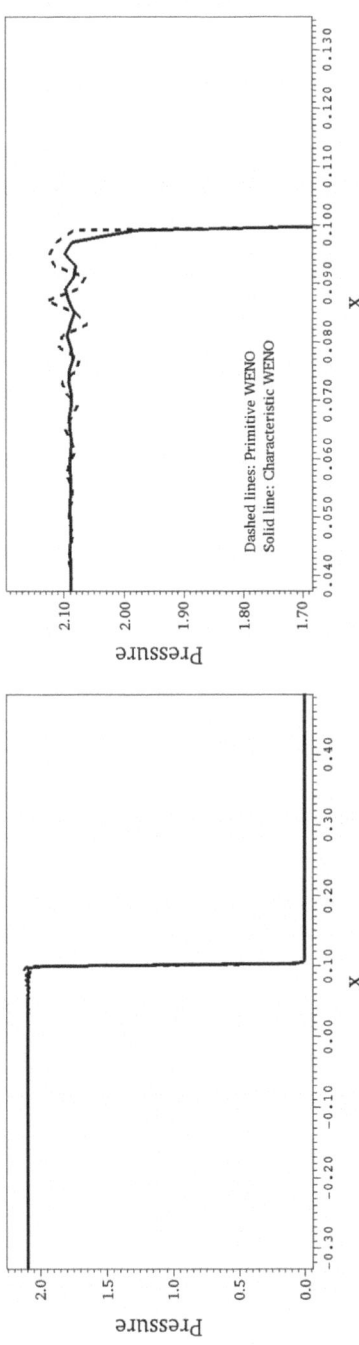

Fig. 2.10 Left: Pressure in a Mach 3 shock wave moving through liquid water computed using the ninth-order WENO scheme on characteristic variables (characteristic WENO) and primitive variables (primitive WENO); right: The zoomed-in view right behind the shock shows reduced oscillations with the characteristic WENO scheme

2.3.5 Conservative Derivative Approximation and WENO Interpolation

As the WENO scheme enables capturing discontinuities such as shock waves, and since the shock speed is defined only in the conservation form, it is sensible to derive the conservative form of the two high-order derivative approximations Eqs. (2.40) and (2.42). Specifically, we require to determine $\hat{F}_{i+\frac{1}{2}}$ and $\hat{F}_{i-\frac{1}{2}}$ such that

$$\left.\frac{\partial f}{\partial x}\right|_{x_i} = \frac{1}{h}(\hat{F}_{i+\frac{1}{2}} - \hat{F}_{i-\frac{1}{2}}) + \mathcal{O}(h^{2r+1}) \tag{2.90}$$

using the first method

$$(\hat{F}_{i+\frac{1}{2}} - \hat{F}_{i-\frac{1}{2}}) = \sum_{j=1}^{j=r+1} d_j(\hat{f}_{i+j-\frac{1}{2}} - \hat{f}_{i-j+\frac{1}{2}}) \tag{2.91}$$

or the second method

$$(\hat{F}_{i+\frac{1}{2}} - \hat{F}_{i-\frac{1}{2}}) = d'_1(\hat{f}_{i+\frac{1}{2}} - \hat{f}_{i-\frac{1}{2}}) + \sum_{j=2}^{j=r+1} d'_j(\hat{f}_{i+j} - \hat{f}_{i-j}) \tag{2.92}$$

Here, for brevity, we denote the $2r+1$ order cell-boundary fluxes, $f(\hat{u}_{i\pm\frac{1}{2}})$, by $\hat{f}_{i\pm\frac{1}{2}}$. In Eqs. (2.91) and (2.92), if we denote

$$f_q = \hat{f}_{j+\alpha} \quad 0 < \alpha < 1 \tag{2.93}$$

and the derivative formulas (2.91) and (2.92) as

$$\left.\frac{\partial f}{\partial x}\right|_{x_i} = \sum_{j=q}^{j=q+r} \delta_j f_{i+j} \tag{2.94}$$

then we can use Theorem 1 to find an expression for $\hat{F}_{i+\frac{1}{2}}$. Specifically, using Theorem 1, we have

$$\hat{F}_{i+\frac{1}{2}} = \sum_{j=q}^{j=q+r} c_j f_{i+j} \tag{2.95}$$

where

$$c_q = 0 \tag{2.96a}$$

$$c_{q+1} = -\delta_q \tag{2.96b}$$

2.4 Targeted Essentially Non-oscillatory Schemes

$$c_{q+2} = -(\delta_q + \delta_{q+1}) \tag{2.96c}$$

$$\cdots$$

$$c_{q+r} = \underbrace{-(\delta_q + \delta_{q+1} + \cdots + \delta_{q+r-1}) = \delta_{q+r}}_{\text{consistency of the approximation}} \tag{2.96d}$$

In the second method (2.92), specializing to the fourth-order derivative approximation, we have

$$\sum_{i=q}^{i=q+r} \delta_i f_i = \frac{1}{6}\hat{f}_{i-1} - \frac{4}{3}\hat{f}_{i-1/2} + \frac{4}{3}\hat{f}_{i+1/2} - \frac{1}{6}\hat{f}_{i+1} \tag{2.97}$$

and

$$\hat{F}_{i+\frac{1}{2}} = \sum_{j=q}^{j=q+r} c_i f_i = \frac{1}{6}\hat{f}_{i-1/2} + \frac{7}{8}\hat{f}_{i+1/2} - \frac{1}{6}\hat{f}_{i+1} \tag{2.98}$$

$$= (d_1' + d_2')\hat{f}_{i+1/2} + d_2'(\hat{f}_{i+1} - \hat{f}_{i-1/2}) \tag{2.99}$$

Or for a $2(r+1)$-th order derivative approximation using the second method, we have

$$\hat{F}_{i+\frac{1}{2}} = \sum_{j=1}^{r+1} d_j' \hat{f}_{i+1/2} + \sum_{j=2}^{r+1} d_j'(\hat{f}_{i+1} - \hat{f}_{i-1/2})$$

$$+ \sum_{j=3}^{r+1} d_j'(\hat{f}_{i+2} - \hat{f}_{i-1}) + \cdots + d_{r+1}'(\hat{f}_{i+r} - \hat{f}_{i-r+1}) \tag{2.100}$$

Similarly, for the first method, (2.92), we have

$$\hat{F}_{i+\frac{1}{2}} = \sum_{j=1}^{r+1} d_j \hat{f}_{i+1/2} + \sum_{j=2}^{r+1} d_j(\hat{f}_{i+3/2} - \hat{f}_{i-1/2})$$

$$+ \sum_{j=3}^{r+1} d_j(\hat{f}_{i+5/2} - \hat{f}_{i-3/2}) + \cdots + d_{r+1}(\hat{f}_{i+2(r+1)/2} - \hat{f}_{i-2r/2}) \tag{2.101}$$

2.4 Targeted Essentially Non-oscillatory Schemes

In WENO schemes, all sub-stencils consist of the same number of points. Hence, when two discontinuities are close, WENO schemes may yield no smooth sub-stencil, as seen in Fig. 2.11. On the other hand, if the substencil's interpolations are formed

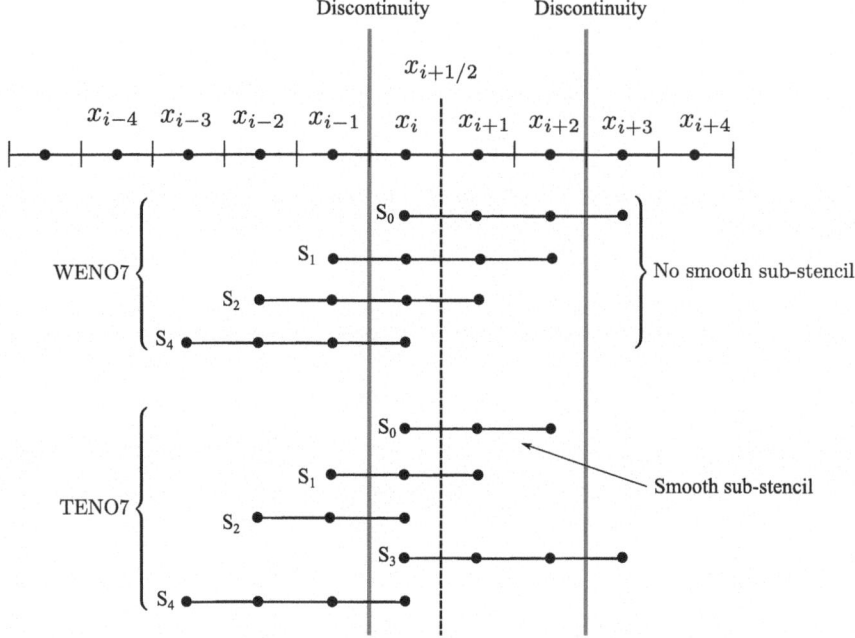

Fig. 2.11 Comparison of WENO7 to TENO7 in the case of two close discontinuity; WENO7 consists of no smooth stencil, while TENO7 with its incremental sub-stencils consists on one smooth sub-stencil

incrementally with an increasing number of grid points (or approximation orders), there is a higher possibility for a smooth sub-stencil. This strategy is at the heart of the targeted essentially non-oscillatory (TENO) scheme [7–9].

Also, in the traditional WENO scheme, the genuine discontinuity can be mistaken for high-wave number smooth physical features. TENO schemes overcome this difficulty by utilizing a more effective scale separation strategy. Unlike WENO schemes that use all interpolations from the candidate stencils by assigning a weight inversely proportional to the smoothness indicator to each interpolation, TENO schemes eliminate the stencils with genuine discontinuities analogous to ENO and assign the optimal weights to the interpolations from the remaining candidate stencils with smooth solutions. Hence, TENO schemes enable robust shock-capturing capability while eliminating the dissipation up to the limit of the underlying linear scheme.

2.4.1 Incremental Stencil Interpolation

Different from the classical WENO scheme in which candidate stencils have the same width, a TENO scheme creates a high-order interpolation by assembling a set of low-order (higher than third) upwind-biased candidate stencils having incremental

2.4 Targeted Essentially Non-oscillatory Schemes

widths. The scheme can gradually degenerate to third order, according to local flow features, thus avoiding the problem of multiple discontinuities and recovering the robustness of the classical fifth-order WENO scheme.

A TENO scheme of order K ($K \geq 5$) consists of the incremental stencils with widths (r_k) defined as

$$\{r_k\} = \begin{cases} \underbrace{\{3, 3, 3, 4, 4, 5, \ldots, \frac{K+2}{2}\}}_{k=0,\ldots,K-3} & \text{if} \quad \text{mod}\,(K, 2) = 0 \\ \underbrace{\{3, 3, 3, 4, 4, 5, \ldots, \frac{K+1}{2}\}}_{k=0,\ldots,K-3} & \text{if} \quad \text{mod}\,(K, 2) = 1 \end{cases}$$

Figure 2.12 depicts the incremental stencils of the TENO scheme up to the eighth order of accuracy.

The interpolated values from a candidate stencils for a TENO scheme with the order of accuracy of $K = 7$ appear in Ref. [9]

$$S_0: \quad P'_{0,i+1/2} = P_{2,0,i+1/2} = \frac{3}{8}u_i + \frac{3}{4}u_{i+1} - \frac{1}{8}u_{i+2} \tag{2.102}$$

$$S_1: \quad P'_{1,i+1/2} = P_{2,1,i+1/2} = -\frac{1}{8}u_{i-1} + \frac{3}{4}u_i + \frac{3}{8}u_{i+1} \tag{2.103}$$

$$S_2: \quad P'_{2,i+1/2} = P_{2,2,i+1/2} = \frac{3}{8}u_{i-2} - \frac{5}{4}u_{i-1} + \frac{15}{8}u_i \tag{2.104}$$

$$S_3: \quad P'_{3,i+1/2} = P_{3,1,i+1/2} = \frac{5}{16}u_i + \frac{15}{16}u_{i+1} - \frac{5}{16}u_{i+2} + \frac{1}{16}u_{i+3} \tag{2.105}$$

$$S_4: \quad P'_{4,i+1/2} = P_{3,3,i+1/2} = -\frac{5}{16}u_{i-3} + \frac{21}{16}u_{i-2} - \frac{35}{16}u_{i+1} + \frac{35}{16}u_i \tag{2.106}$$

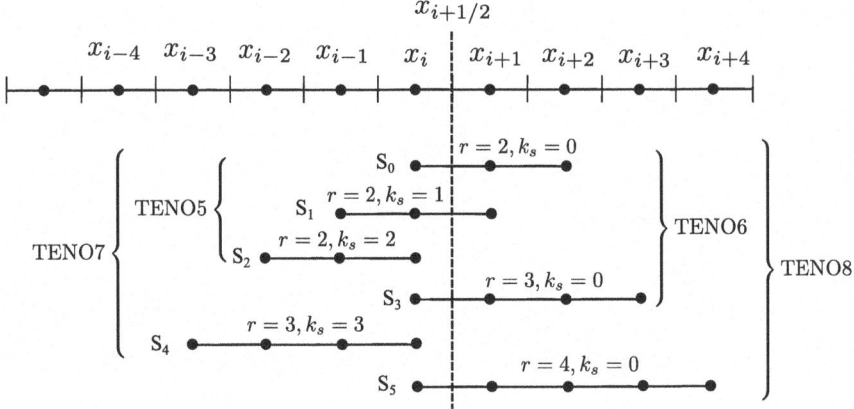

Fig. 2.12 Incremental stencil of TENO schemes

The linear combination of the interpolation from each candidate stencil then yields the Kth-order interpolation over the full stencil using

$$g_{i+1/2}^K = \sum_{k=0}^{K-3} b'_{K,k} P_{k,i+1/2} \qquad (2.107)$$

where for interpolation on the full stencil with order $K = 5$, the coefficients b'_k are given as

$$(b'_{5,0} = \frac{5}{16}, b'_{5,1} = \frac{5}{8}, b'_{5,2} = \frac{1}{16}) \qquad (2.108)$$

and for interpolation on the full stencil with order $K = 7$, the coefficients b'_k are given as

$$(b'_{7,0} = \frac{9}{35}, b'_{7,1} = \frac{18}{35}, b'_{7,2} = \frac{3}{35}, b'_{7,3} = \frac{4}{35}, b'_{7,4} = \frac{1}{35}) \qquad (2.109)$$

2.4.2 Strong Scale Separation

In a TENO scheme, the diagnosis of the presence of a discontinuity is based on the smoothness indicators computed in the WENO scheme, but a different relation is adopted to determine the degree of smoothness or smoothness of a solution in a candidate stencil. Inspired by Ref. [16], Fu et al. proposed the following measure for separating the scales [7]:

$$\gamma_k = \left(C + \frac{\tau_K}{\beta_{k,r}}\right)^q \quad k = 0, \ldots, K - 3 \qquad (2.110)$$

where $C = 1, q = 6, \beta_{k,r}$ are smoothness indicators used for the WENO scheme and τ_K is a full stencil smoothness indicator which is determined using Eq. (2.61) with polynomial approximation over the full stencil [7]. For a fifth-order TENO scheme $\tau_K = |\beta_2 - \beta_1|$.

2.4.3 Stencil Selection

With a strong scale separation in a TENO scheme, an interpolation from a candidate stencil is discarded if the corresponding smoothness measure (2.110) reveals a certain degree of discontinuity in the candidate stencil. To this end, the smoothness measures are first normalized as

$$\mathcal{X}_k = \frac{\gamma_k}{\sum_{m=0}^{K-3} \gamma_m} \qquad (2.111)$$

Using the normalized smoothness measure and a threshold C_T, we define a sharp cutoff function is as

$$\delta_k = \begin{cases} 0, & \text{if } \mathcal{X}_k < C_T \\ 1, & \text{otherwise} \end{cases} \quad (2.112)$$

We compute the threshold C_T by a quasi-linear spectral analysis and numerical experiments such that the computed solution has a desirable spectral property while having a sufficient dissipation for non-oscillatory capturing of discontinuities [7].

2.4.4 Assembled High-Order Scheme

From the optimal weights $b'_{K,k}$, and the sharp cutoff function δ_k, the weights for the assemble of all interpolation from the candidate stencils results as

$$w'_{K,k} = \frac{b'_{K,k}\delta_k}{\sum_{m=0}^{K-3} b'_{K,m}\delta_m} \quad (2.113)$$

and the high-order cell-boundary solution is determined as

$$u_{K,\text{TENO},i+\frac{1}{2}} = \sum_{k=0}^{K-3} w'_{K,k} P'_{k,i+\frac{1}{2}} \quad (2.114)$$

2.5 Exercises

1. Discuss why spatial schemes with orders higher than one are essential.
2. Values of a function $f(x)$ at a uniformly spaced grid $\{x_i\} = \{0, 0.1, 0.2, 0.3, 0.4, 0.5, 0.6\}$ is given as $\{f(x_i)\} = \{0, 0.0998, 0.1987, 0.2955, 0.3894, 0.4794, 0.5646\}$. Determine a fifth-order approximation of $f(0.225)$.
3. Using interpolating polynomial, an approximation to $f_{i+\frac{1}{2}}$ of $r+1$ order of accuracy can be written as

$$f_{i+\frac{1}{2}} = \sum_{j=0}^{r} \alpha_j f_j \quad (2.115)$$

(a) Determine α_js for schemes of orders two. How many different schemes can you obtain? What is the difference between them. (b) Determine α_js for schemes of orders three and four. How many different schemes can you obtain for each order? What is the difference between different schemes of the same order of accuracy?

4. Determine all possible second-order approximations of the first-order spatial derivative. Hint: determine all a_ls in Eq. (2.27) and all d_js in Eq. (2.35).
5. Determine all possible third-order approximations of the first-order spatial derivative.
6. Determine all possible fourth-order approximations of the first-order spatial derivative.
7. Determine the six-order central approximations of the first-order spatial derivative. Write the final expression as

$$f'_i = \frac{1}{h} \sum_{l=0}^{6} \beta_l f_{i-3+l} \qquad (2.116)$$

where i is the global numbering index.
8. Verify Remark 2 for fourth-order derivative approximations.
9. In this exercise, we want to confirm that only a minimally biased scheme is required for stability. To this end, consider an example of spatial derivative approximation consisting of a minimally biased third-order mid-point approximation combined with a fourth-order (central) derivative approximation using Scheme 2. The approximation of the spatial derivative at grid x_i has the form

$$\frac{\partial u}{\partial x} \approx \frac{1}{6} u_{i-2} - u_{i-1} + \frac{1}{2} u_i + \frac{1}{3} u_{i+1} \qquad (2.117)$$

(a) Derive the above expression by choosing an appropriate third-order stencil for the mid-points, and an appropriate Scheme 2 for the derivative. (b) By calculating the eigenvalues of the corresponding circulant matrix, show that this scheme yields stable discretization.
10. Consider a set of data defined on a uniform grid. (a) Write down a fourth-order central scheme for a mid-point value. (b) Write down a fourth-order scheme for a mid-point value with the most amount of upwinding. The stencil boundary can be only half-grid spacing away from the mid-point.
11. Derive the optimal coefficients $b_{r,ks}$ for $r = 1, 2$.
12. Derive the smoothness indicators $\beta_{r,k_s,i+\frac{1}{2}}$ for $r = 1$ and 2.
13. Consider pure advection equation with either a smooth or discontinuous solution. The discontinuous solution can be a constant-speed advection of the interface between two different fluids. (a) Program a linear first and linear third-order accurate scheme for the spatial derivative. The temporal scheme must have comparable order of accuracy. (b) Verify and compare the accuracy of both the first- and third-order schemes using the exact solution that is smooth. (c) Use your linear schemes (the first and third order schemes) to capture the exact solution with a discontinuity. Compare the results and discuss what you obtain. (d) Program the nonlinear third-order scheme, third-order WENO scheme. (e) Verify the accuracy of the WENO scheme with the smooth solution. (f) Verify the accuracy of the WENO scheme with the discontinuous solution.

14. Consider a discrete values of a jump discontinuity function $f(x)$ where the jump is located at $x_s = x_i + 3h/4$, and h is the grid spacing. (**a**) Using the stencil $\{x_{i-1}, x_i, x_{i+1}\}$, provide a third-order linear approximation of $f(x_i + h/2)$. (**b**) Determine the corresponding third-order WENO approximation of $f(x_i + h/2)$. (**c**) Compare the linear and the WENO approximation error as $h \to 0$. Discuss.
15. Prove Lemma 1.
16. Derive expressions for the coefficients B^+_{i-r+l} and B^-_{i-r+l} in the overall equations for the left- and right-biased WENO interpolations, Eqs. (2.81) and (2.82).
17. Using Theorem 1, derive explicit expressions for the conservative fluxes in the high-order derivative approximations using Scheme 1 and Scheme 2, Eqs. (2.40) and (2.42), for derivative approximation orders of 4, 6, 8 and 10.

References

1. Hamming, R.W.: Numerical Methods for Scientists and Engineers, 2nd edn. Dover, Mineola, NY (1986)
2. Epperson, J.F.: On the runge example. Am. Math. Mon. **94**(4), 329–341 (1987)
3. Hewitt, E., Hewitt, R.E.: The Gibbs-Wilbraham phenomenon: an episode in fourier analysis. Arch. Hist. Exact Sci. **21**(2), 129–160 (1979). Jun
4. Jiang, G.-S., Shu, C.-W.: Efficient implementation of weighted eno schemes. J. Comput. Phys. **126**, 202–228 (1996)
5. Deng, X., Zhang, H.: Developing high-order weighted compact nonlinear schemes. J. Comput. Phys. **165**(1), 22–44 (2000)
6. Shahbazi, K.: High-order finite difference scheme for compressible multi-component flow computations. Comput. Fluids **190**, 425–439 (2019)
7. Lin, F., Hu, X.Y., Adams, N.A.: A family of high-order targeted eno schemes for compressible-fluid simulations. J. Comput. Phys. **305**, 333–359 (2016)
8. Lin, F., Hu, X.Y., Adams, N.A.: Targeted eno schemes with tailored resolution property for hyperbolic conservation laws. J. Comput. Phys. **349**, 97–121 (2017)
9. Ye, C.-C., Zhang, P.-J.-Y., Wan, Z.-H., Sun, D.-J.: An alternative formulation of targeted eno scheme for hyperbolic conservation laws. Comput. Fluids **238**, 105368 (2022)
10. Johnsen, E., Colonius, T.: Implementation of WENO schemes in compressible multicomponent flow problems. J. Comput. Phys. **219**(2), 715–732 (2006)
11. Iserles, A., Strang, G.: The optimal accuracy of difference schemes. Trans. Am. Math. Soc. **277**(2), 779–803 (1983)
12. Wong, M.L., Angel, J.B., Barad, M.F., Kiris, C.C.: A positivity-preserving high-order weighted compact nonlinear scheme for compressible gas-liquid flows. J. Comput. Phys. **444**, 110569 (2021)
13. Hesthaven, J.S.: Numerical Methods for Conservation Laws. Society for Industrial and Applied Mathematics. Philadelphia, PA (2018)
14. Gerolymos, G.A., Sénéchal, D., Vallet, I.: Very-high-order WENO schemes. J. Comput. Phys. **228**(23), 8481–8524 (2009)
15. Harten, A., Engquist, B., Osher, S., Chakravarthy, S.R.: Uniformly high order accurate essentially non-oscillatory schemes, iii. J. Comput. Phys. **71**(2), 231–303 (1987)
16. Borges, R., Carmona, M., Costa, B., Don, W.S.: An improved weighted essentially non-oscillatory scheme for hyperbolic conservation laws. J. Comput. Phys. **227**(6), 3191–3211 (2008)

Chapter 3
Approximation On and Near Boundary

In Chap. 2, to approximate a function value at a mid-point or cell boundary with a specific order of accuracy, we assume that the function values at nodes or cell centers surrounding the mid-point (the function values on the stencil surrounding the mid-point) are all known. This assumption, while valid in the interior of a domain, is not necessarily valid for approximations near the domain's boundaries. Here, we discuss approximations on and near boundaries in the context of numerical solutions of PDEs governing compressible fluid dynamics. To explain how a proper approximation near domain boundaries can be accomplished in a finite difference scheme, we first need to discuss the type of grid points we intend to use. Given a spatial domain, we may choose two types of grid points: (1) a set of grid points that overlap the domain's boundary surfaces or (2) a set of grid points interior to the domain's boundary surfaces. Let us choose the second type of grid where the grid points are interior to the domain's boundary surfaces. This type of grid points facilitates the conservative approximation of discontinuous features such as shock waves near boundaries. In this chapter, we describe boundary treatments for Cartesian domains (or those easily transformed to Cartesian ones) and general non-Cartesian domains. We first discuss a general approach for imposing boundary conditions for equations such as Euler and Navier-Stokes equations [1, 2]. The approach is consistent with the dimensionality and physics of the underlying system, unlike the common one-dimensional, non-diffusive boundary condition enforcement [3]. This general characteristic approach will be used in boundary treatments for both Cartesian and non-Cartesian domains.

For Cartesian domains, we discuss the discretization schemes for the (near) boundary values in the case of four different boundary condition types, namely periodic, reflecting, Dirichlet where function values are given at the boundaries, and Neumann where the derivative of the function is specified at the boundaries. For the Cartesian domains, our proposed discretization is inspired by the work of Deng and Zhang [4], where we use inward-biased spatial schemes at and near the boundaries.

For general non-Cartesian domains, we present an approach that uses flux values at a set of ghost points (points outside of the domain); as a result, the same flux derivative schemes of the interior points can be used for points near the boundary. However, the extrapolation of the data from the interior and boundary of the domain to the ghost points must be accomplished carefully to avoid numerical instability [5]. Tan and Shu demonstrated that stable and accurate boundary treatments in non-Cartesian domains are possible if the suitably computed boundary-normal derivatives are used in the extrapolation of data to the ghost points. The process of computing boundary-normal derivatives is called the inverse Lax-Wendroff method and uses the governing PDEs and the boundary data to determine the boundary-normal derivatives [6]. Our discussion of the boundary treatment on the non-Cartesian domains is based on the inverse Lax-Wendroff (ILW) method introduced in [5, 7] and recently further enhanced in Ref. [8, 9].

Section 3.1 introduces the general boundary treatments based on the multi-dimensional viscous characteristic method. Section 3.2 introduces discretization schemes for various boundary condition types on Cartesian domains, while Sect. 3.3 addresses the ILW scheme for boundary treatments on non-Cartesian domains.

3.1 General Characteristic Boundary Treatments for System of Equations

Consider a system of PDEs such as the three-dimensional Euler or Navier-Stokes system. As discussed in Chap. 1, owing to their nature, these systems support incoming and outgoing characteristics (or waves) at the domain boundaries. The characteristic speeds of the Euler and Navier-Stokes system at a boundary with a local outward unit normal vector $\boldsymbol{n}$ are $v_n - c$, v_n, v_n, v_n, and $v_n + c$ with c being the sound speed and v_n the fluid velocity in the normal direction. At a supersonic inlet, all characteristics are incoming, while at a supersonic outlet, all characteristics are outgoing (Fig. 3.1). At a subsonic inlet, only one characteristic corresponding to the speed $v_n + c$ is outgoing (Fig. 3.2), and the rest are incoming. For a subsonic outlet, only one characteristic corresponding to the speed $v_n - c$ is incoming, and the rest are outgoing (Fig. 3.2). Figure 3.3 depicts various inflow and outflow conditions scenarios and their corresponding local characteristics on non-cartesian domains.

For consistency, the incoming characteristics at the domain boundaries must be imposed using the boundary data; this means that in the numerical schemes, the solution from the domain's interior cannot be used for the boundary values for those incoming characteristics. If it is used, it will lead to instability.

The question is how to impose the incoming characteristics. Here, we present a general approach that is multi-dimensional and applies to both viscous (diffusive) and inviscid (non-diffusive) flows [1, 2]. This approach improves the classical characteristic boundary condition method, which is locally one-dimensional and inviscid [3].

3.1 General Characteristic Boundary Treatments for System of Equations 65

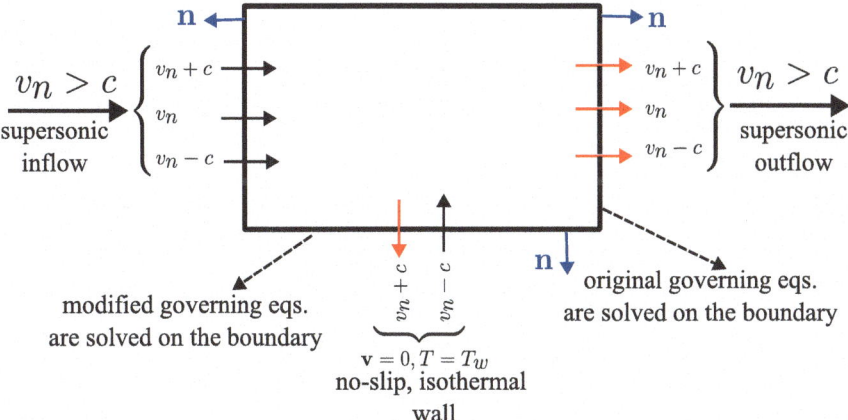

Fig. 3.1 Characteristic boundary conditions for supersonic inflow and outflow, and no-slip isothermal wall. Black and Red arrows signify the incoming and outgoing characteristics, respectively. Repeated characteristics (with speed (v_n)) and those with zero speed at no-slip wall ($v_n = 0$) are not shown. Modified boundary conditions are solved on the non-slip wall, while the original governing equations are solved on the outlet boundary. There is no need to solve the governing equation on the supersonic inflow boundary, as all variables are specified using the inflow condition

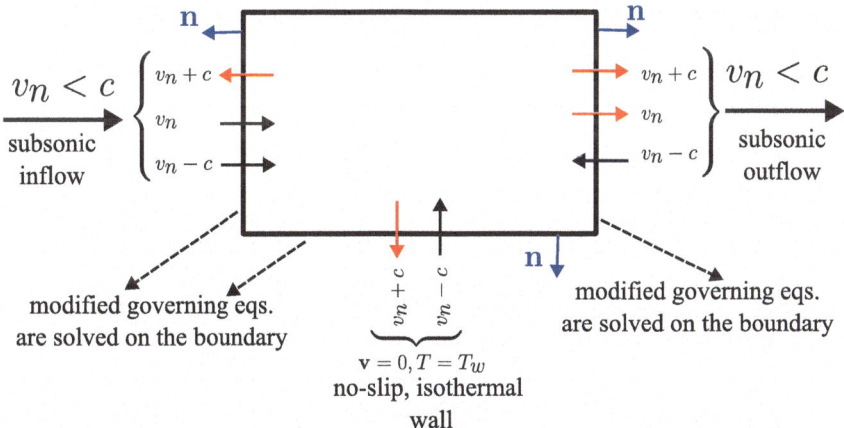

Fig. 3.2 Characteristic boundary conditions for subsonic inflow and outflow, and no-slip isothermal wall. Black and Red arrows signify the incoming and outgoing characteristics, respectively. Repeated characteristics (with speed (v_n)) and those with zero speed at no-slip wall ($v_n = 0$) are not shown. Modified governing equations are solved on the boundaries

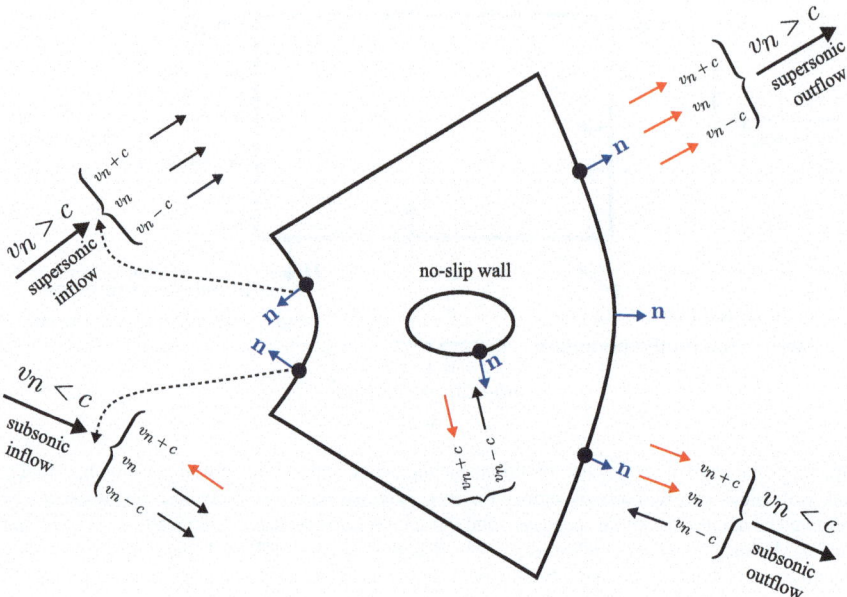

Fig. 3.3 Characteristic boundary conditions for on non-Cartesian domains. Black and Red arrows signify the incoming and outgoing characteristics, respectively. Repeated characteristics (with speed (v_n) and those with zero speed at no-slip wall $(v_n = 0)$ are not shown. The inverse Lax-Wendroff method uses the governing equations modified by the incoming characteristics to compute boundary-normal derivatives, which are used to determine data on the ghost points

The outline of this general approach is as follows:

- We impose boundary data corresponding to incoming characteristics; if all characteristics are incoming, we impose boundary data for all variables; if all characteristics are outgoing, we impose no boundary data. For partial imposition of boundary data, we derive modified governing equations.
- For Cartesian domains, we solve modified governing equations using inward-biased spatial discretization. If no boundary data are imposed, we solve the original governing equation using the inward-biased discretization on the boundary. If all variables are imposed at the boundary, no governing equations are solved on the boundary.
- For non-Cartesian domains, the modified or original governing equations are used to compute the boundary-normal derivatives via the ILW approach to determine the ghost point data. The ILW method avoids the "small-cell issue".
- The ILW approach uses the governing equations or its modified version (which is multi-dimensional and viscous) and a multi-dimensional least-squares approach to determine the data on the boundary and the boundary-normal direction, which are, in turn, used to compute the ghost point values.

Next, we derive the modified governing equations on the boundary and give explicit modification terms for various common boundary conditions in the compressible Euler and Navier-Stokes systems. Finally, we describe the boundary treatment for general non-Cartesian domains using the ILW approach.

In the remainder of this section, we denote a vector quantity by an overline and enclose a matrix quantity inside a pair of brackets. For example, $\overline{u}$ signifies a vector quantity, while $[A]$ signifies a two-dimensional matrix quantity.

3.1.1 Modified Governing Equations on the Boundary

Considering the boundary surface normal to the n-direction, let us write the governing equations in the form

$$\frac{\partial \overline{u}}{\partial t} + \frac{\partial \overline{F^a}}{\partial n} + \overline{C} = 0 \tag{3.1}$$

where $\overline{u}$ is the conservative vector, $\overline{F^a}$ the advective flux in the n-direction and $\overline{C}$ is the sum of advective fluxes in other directions and all diffusive fluxes. Rewriting the advective flux term as a quasi-linear form using the characteristic decomposition yields

$$\frac{\partial \overline{u}}{\partial t} + [S][\Lambda][S]^{-1} \frac{\partial \overline{u}}{\partial n} + \overline{C} = 0 \tag{3.2}$$

$[S]$ is a matrix which its columns are the eigenvectors of the matrix associated with the quasi-linear form of (3.1) with respect to the conservative variables. The diagonal matrix $[\Lambda]$ holds the system's eigenvalues (or the characteristic speeds) on its diagonal entries. Multiplying by $[S]^{-1}$ then leads to

$$[S]^{-1} \frac{\partial \overline{u}}{\partial t} + [\Lambda]\overline{S}^{-1} \frac{\partial \overline{u}}{\partial n} + [S]^{-1}\overline{C} = 0 \tag{3.3}$$

The characteristic vector $d\overline{L} = S^{-1}d\overline{u}$ is responsible for transporting data to and away from the domain according to the advection equations

$$\frac{\partial \overline{L}}{\partial t} + [\Lambda]\frac{\partial \overline{L}}{\partial n} = 0 \tag{3.4}$$

The corresponding characteristic components must be determined from the boundary data for the incoming characteristics. In contrast, the outgoing ones must remain unchanged as they originate from the domain's interior. Specifically, using the boundary data $\overline{b}$,

$$L_i^* = (1 - m_i)L_i + m_i b_i, \quad \begin{cases} m_i = 1 & \text{for incoming waves} \\ m_i = 0 & \text{for outgoing waves} \end{cases} \tag{3.5}$$

Let us define the diagonal matrix $[M]$ whose diagonal entries are zero except for the entries corresponding to the incoming characteristics, which are one. Now the use of $[M]$ yields the modified the characteristic vector in the form

$$\overline{L}^* = \overline{L} - [M](\overline{L} - \overline{b}) \tag{3.6}$$

which is the original characteristic variable minus the modification $[M](\overline{L} - \overline{b})$. Transferring the modified characteristic $\overline{L}^*$ back to the conservative vector space by multiplying with $[S]$, and substituting the result into the original governing equation (3.1), we obtain

$$\underbrace{\frac{\partial \overline{u}}{\partial t} - \overline{R}}_{\text{original}} - \underbrace{[S]\overline{X}}_{\text{modification}} = 0 \tag{3.7}$$

where

$$\overline{R} = -\frac{\partial \overline{F^a}}{\partial n} - \overline{C} \tag{3.8}$$

Moreover, $\overline{X}$ is the modification resulting from the imposition of incoming characteristics using the boundary data. Equation (3.7) is the original governing equation with a modification equaling $[S]\overline{X}$. Vector $\overline{X}$ has nonzero entries only for the incoming characteristics and needs to be determined for each boundary condition type. Equation (3.7) is solved on the domain's boundary with one-sided spatial discretizations for Cartesian domains, or used to determined boundary normal derivative using the inverse Lax-Wendroff method for non-Cartesian domains. We next describe the exact form of entries of $\overline{X}$ for the most common types of physical boundary conditions.

3.1.2 Explicit Modification Terms for Various Boundary Conditions

The physical boundary conditions are often easily expressed using primitive variables, while the governing equations or their boundary modifications are expressed for conservative variables. We hence need the mapping from the conservative variables to the primitive variables. For the Euler or Navier-Stokes system with an ideal gas law, we can show that the matrix $[H]$, which maps the conservative variables

$$d\overline{u} = [d\rho, d(\rho u), d(\rho v), d(\rho w), dE]^T \tag{3.9}$$

to the primitive variables

$$d\overline{w} = [d\rho, \rho du, \rho dv, \rho dw, dp/(\gamma - 1)]^T \tag{3.10}$$

3.1 General Characteristic Boundary Treatments for System of Equations

as
$$d\overline{w} = [H]d\overline{u} \tag{3.11}$$

has the form
$$[H] = \begin{pmatrix} 1 & 0 & 0 & 0 & 0 \\ -u & 1 & 0 & 0 & 0 \\ -v & 0 & 1 & 0 & 0 \\ -w & 0 & 0 & 1 & 0 \\ e_k & -u & -v & -w & 1 \end{pmatrix} \tag{3.12}$$

where e_k is the kinetic energy per unit mass defined as
$$e_k = \frac{1}{2}(u^2 + v^2 + w^2) \tag{3.13}$$

and u, v and w are velocity components in the x, y, and z-direction, respectively. Details of the derivation of $[H]$ are left to Exercise 1.

To apply particular physical boundary conditions and determine $\overline{X}$ in Eq. (3.7), we need to map the modified governing equation to the primitive space by multiplying with the matrix $[H]$ as
$$[H]\frac{\partial \overline{u}}{\partial t} = [H]\overline{R} + [H][S]\overline{X} \tag{3.14}$$

We can show that the matrix $[H][S]$ is identical to the matrix $[Q]$ whose columns are the eigenvectors of the matrix associated with the quasi-linear form of the governing equations based on the primitive variables (see Exercise 1), as discussed in Subsection 1.2.1, i.e.,
$$[Q] = [H][S] \tag{3.15}$$

The vector $[H]\overline{R}$ is the mapping of the residual to the primitive space, and we denote $\overline{R}_p = [H]\overline{R}$ with components
$$[H]\overline{R} = \overline{R}_p = [R_{p,1}\ R_{p,2}\ R_{p,3}\ R_{p,4}\ R_{p,5}]^T \tag{3.16}$$

Using the mapping (3.11), Eq. (3.14) is rewritten as
$$\frac{\partial \overline{w}}{\partial t} = [H]\frac{\partial \overline{u}}{\partial t} = [H]\overline{R} + [H][S]\overline{X} \tag{3.17}$$

Or, with the above notation
$$\frac{\partial \overline{w}}{\partial t} = [H]\frac{\partial \overline{u}}{\partial t} = \overline{R}_p + [Q]\overline{X} \tag{3.18}$$

Using (3.18), we next determine the vector $\overline{X}$ for each type of boundary condition. However, before doing so, we give explicit expressions for matrices $[S]$ and $[Q]$ when the governing equations are Euler or Navier-Stokes.

Let us denote the characteristic speed vector as

$$\overline{\lambda} = \left(v_n - c \ v_n \ v_n + c \ v_n \ v_n\right)^T \tag{3.19}$$

where

$$v_n = un_x + vn_y + wn_z \tag{3.20}$$

with $\overline{n} = [n_x \ n_y \ n_z]^T$ being the unit outward normal vector on the boundary surface. For instance, if the boundary surface is normal to the x-direction on $\xi = \xi_{max}$, $\overline{n} = [1 \ 0 \ 0]^T$. Then, the matrix of right eigenvectors for the quasi-linear form of the governing equation in the conservative variables has the form [10]

$$[S] = \left(\overline{S}_1 \ \overline{S}_2 \ \overline{S}_3 \ \overline{S}_4 \ \overline{S}_5\right) = \begin{pmatrix} 1 & 1 & 1 & 0 & 0 \\ u - cn_x & u & u + cn_x & n_y & -n_z \\ v - cn_y & v & v + cn_y & -n_x & 0 \\ w - cn_z & w & w + cn_z & 0 & n_x \\ h_0 - cv_n & e_k & h_0 + cv_n & un_y - vn_x & wn_x - un_z \end{pmatrix} \tag{3.21}$$

where the last two eigenvectors correspond to the repeated eigenvalues $\lambda_4 = \lambda_5 = v_n$. In Eq. (3.21), h_0 is the stagnation enthalpy defined as

$$h_0 = h + e_k \tag{3.22}$$

In ideal gases, the static enthalpy h is related to the speed of sound as

$$h = \frac{c^2}{\gamma - 1} \tag{3.23}$$

The matrix $[Q] = [H][S]$ is given as

$$[Q] = [H][S] = \begin{pmatrix} 1 & 1 & 1 & 0 & 0 \\ -cn_x & 0 & cn_x & n_y & -n_z \\ -cn_y & 0 & cn_y & -n_x & 0 \\ -cn_z & 0 & cn_z & 1 & n_x \\ h & 0 & h & 0 & 0 \end{pmatrix} \tag{3.24}$$

Supersonic Inflow

For supersonic inflow, all characteristics are incoming. Hence, five conditions are imposed using the physical boundary conditions, and there is no need to solve (3.7) as all variables are specified on the boundary (Fig. 3.1).

3.1 General Characteristic Boundary Treatments for System of Equations

Supersonic Outflow

For supersonic outflow, all characteristics are outgoing. Hence, no physical conditions are imposed, and $\overline{X} = 0$ and the original governing equations (3.1) are solved without any modification on the boundary (Fig. 3.1).

Subsonic Outflow

For subsonic outflow, the characteristics corresponding to $v_n - c$ are incoming on the general curved boundaries (Fig. 3.3). On the Cartesian boundaries (with positive direction pointing to the right), the characteristics corresponding to $v + c$ and $v - c$ are respectively incoming on the left and right boundaries (Fig. 3.2). Hence, we need to set physical boundary conditions corresponding to these characteristics. A constant pressure is often assumed at the outflow, leading to

$$\frac{\partial p}{\partial t} = 0 \quad t > 0 \quad \text{for subsonic outflow} \tag{3.25}$$

We need to impose Eq. (3.25) on Eq. (3.18) in order to determine the corresponding component of vector $\bar{X}$. On the left boundary, only the third component of $\overline{X}$ is nonzero, while on the right boundary, only the first component is nonzero. Hence, we have

$$\left. \begin{array}{l} \frac{\partial \overline{w}}{\partial t} = \overline{R}_p + [Q]\overline{X} \\ \frac{\partial \overline{w}_5}{\partial t} = \frac{\partial p}{\partial t} = 0 \end{array} \right\} \implies \begin{cases} X_3 = -\frac{R_{p,5}}{Q_{5,3}} & \text{on the left boundary, } \xi = \xi_{\min} \\ X_1 = -\frac{R_{p,5}}{Q_{5,1}} & \text{on the right boundary, } \xi = \xi_{\max} \end{cases} \tag{3.26}$$

With these, the vector $\overline{X}$ for the subsonic outflow boundary condition on the left and right boundaries become

$$\overline{X} = \left(0\ 0\ -\frac{R_{p,5}}{h}\ 0\ 0 \right) \quad \text{on the left boundary, } \xi = \xi_{\min} \tag{3.27a}$$

$$\overline{X} = \left(-\frac{R_{p,5}}{h}\ 0\ 0\ 0\ 0 \right) \quad \text{on the right boundary, } \xi = \xi_{\max} \tag{3.27b}$$

Note that here h is the static enthalpy and related to the speed of sound as in Eq. (3.23).

For implementation, we need the vector $[S]\bar{X}$ in the modified governing equations (3.7). Using (3.27) and (3.21), the vector $[S]\overline{X}$ is determined as

$$[S]\overline{X} = -\frac{R_{p,5}}{Q_{5,3}}\overline{S}_3 = -\frac{R_{p,5}}{h}\overline{S}_3 \quad \text{on the left boundary, } \xi = \xi_{\min} \tag{3.28a}$$

$$[S]\overline{X} = -\frac{R_{p,5}}{Q_{5,1}}\overline{S}_3 = -\frac{R_{p,5}}{h}\overline{S}_1 \quad \text{on the right boundary, } \xi = \xi_{\max} \tag{3.28b}$$

For the Navier-Stokes equations, the ξ-derivatives of the heat flux and the viscous shear stresses are also set to zero [3]. Specifically,

$$\frac{\partial q_\xi}{\partial \xi} = 0, \quad \frac{\partial \tau_{\xi y}}{\partial \xi} = 0, \quad \frac{\partial \tau_{\xi z}}{\partial \xi} = 0 \tag{3.29}$$

Subsonic Inflow

For subsonic inflow, only the characteristics corresponding to $v_n + c$ is outgoing on the general curved boundaries (Fig. 3.3). On the Cartesian boundaries (with the positive direction pointing to the right), the outgoing characteristic is the wave with the speed $v - c$ at the left boundary (Fig. 3.2), $\xi = \xi_{\min}$ and the wave with the speed $v + c$ at the right boundary, $\xi = \xi_{\max}$. As a result, the zero component of $\overline{X}$ is X_1 or X_3. Other entries of $\overline{X}$ are nonzero and must be imposed using the physical boundary conditions. The physical boundary conditions can be the stagnation temperature and pressure along with the direction of the velocity [2]. Alternatively, velocity components and temperature or density are specified [3].

Let us choose the physical boundary condition consisting of the density and velocity components and impose them on Eq. (3.18) in order to determine the corresponding component of vector $\overline{X}$. We have

$$\left.\begin{array}{l} \frac{\partial \overline{w}}{\partial t} = \overline{R}_p + [Q]\overline{X} \\ \frac{\partial w_1}{\partial t} = \frac{\partial \rho}{\partial t} = 0 \\ \frac{\partial w_2}{\partial t} = \frac{\partial u}{\partial t} = 0 \\ \frac{\partial w_3}{\partial t} = \frac{\partial v}{\partial t} = 0 \\ \frac{\partial w_4}{\partial t} = \frac{\partial w}{\partial t} = 0 \end{array}\right\} \Longrightarrow \begin{cases} \overline{X}_l = -[Q_l^{-1}]\overline{R}_{p,l} & \text{on the left boundary, } \xi = \xi_{\min} \\ \overline{X}_r = -[Q_r^{-1}]\overline{R}_{p,r} & \text{on the right boundary, } \xi = \xi_{\max} \end{cases} \tag{3.30}$$

where

$$\overline{X}_l = [X_2 \ X_3 \ X_4 \ X_5]^T \quad \overline{R}_{p,l} = [R_{p,2} \ R_{p,3} \ R_{p,4} \ R_{p,5}]^T \tag{3.31}$$
$$\overline{X}_r = [X_1 \ X_2 \ X_4 \ X_5]^T \quad \overline{R}_{p,r} = [R_{p,1} \ R_{p,2} \ R_{p,4} \ R_{p,5}]^T \tag{3.32}$$

and

$$[Q_l] = \begin{pmatrix} 1 & 1 & 0 & 0 \\ 0 & cn_x & n_y & -n_z \\ 0 & cn_y & -n_x & 0 \\ 0 & cn_z & 1 & n_x \end{pmatrix} \tag{3.33}$$

and

$$[Q_r] = \begin{pmatrix} 1 & 1 & 0 & 0 \\ -cn_x & 0 & n_y & -n_z \\ -cn_y & 0 & -n_x & 0 \\ -cn_z & 0 & 1 & n_x \end{pmatrix} \tag{3.34}$$

3.1 General Characteristic Boundary Treatments for System of Equations

The matrix $[Q_l]$ is obtained from the elimination of the first column and the last row of matrix $[Q]$, while the matrix $[Q_l]$ is obtained from the elimination of the third column and the last row of matrix $[Q]$.

For implementation, we need the vector $[S]\overline{X}$ in the modified governing equations (3.7). The matrix $[S]$ is given in (3.21) and the vector $\overline{X}$ is determined using $\overline{X}_l$ or $\overline{X}_r$ as

$$\overline{X} = [0 \; X_2 \; X_3 \; X_4 \; X_5]^T \quad \text{on the left boundary}, \xi = \xi_{\min} \quad (3.35a)$$

$$\overline{X} = [X_1 \; X_2 \; 0 \; X_4 \; X_5]^T \quad \text{on the right boundary}, \xi = \xi_{\max} \quad (3.35b)$$

Determination of vector $\overline{X}$ when the stagnation pressure and temperature and the direction of velocity are specified at the boundary is left to Exercise 5.

In the Navier-Stokes equations, the ξ-derivatives of the viscous tensile stress is also set to zero [3] on the subsonic inflow boundary. Specifically,

$$\frac{\partial \tau_{\xi\xi}}{\partial \xi} = 0 \quad (3.36)$$

Inviscid wall

At an inviscid wall, the normal component of the velocity is zero. Specifically,

$$v_n = 0 \quad t = 0 \quad (3.37a)$$

$$\frac{\partial v_n}{\partial t} = 0 \quad t > 0 \quad (3.37b)$$

Based on the characteristic vector speed (3.19), at the left boundary, $\xi = \xi_{\min}$, the incoming wave is $v_n + c$, while at the right boundary, $\xi = \xi_{\max}$, the incoming wave is $v_n - c$. The corresponding nonzero components of the vector $\overline{X}$ are X_3 and X_1 on the left and right boundaries, respectively. Hence, imposing the zero normal velocity on Eq. (3.18) yields

$$X_3 = -\frac{1}{c}\left(n_x R_{p,2} + n_y R_{p,3} + n_z R_{p,4}\right) \quad \text{on } \xi = \xi_{\min}, \overline{n} = [-1\;0\;0] \quad (3.38)$$

$$X_1 = \frac{1}{c}\left(n_x R_{p,2} + n_y R_{p,3} + n_z R_{p,4}\right) \quad \text{on } \xi = \xi_{\max}, \overline{n} = [1\;0\;0] \quad (3.39)$$

Isothermal No-Slip Wall

For isothermal no-slip wall, the characteristics for a general non-Cartesian boundaries are shown in Fig. 3.3. For Cartesian domains, a schematic of characteristics at an isothermal no-slip wall are shown in Fig. 3.2. For an isothermal no-slip wall, the temperature is constant, and all components of the velocity is zero:

$$\frac{\partial T}{\partial t} = 0, \quad \frac{\partial u}{\partial t} = 0, \quad \frac{\partial v}{\partial t} = 0, \quad \frac{\partial w}{\partial t} = 0 \quad t > 0 \quad (3.40)$$

At the left boundary, $\xi = \xi_{\min}$, the only outgoing wave is $\lambda_1 = v_n - c$; hence, $X_1 = 0$ and other components of $\overline{X}$ must be determined using the physical conditions given in (3.40). Similarly, at the right boundary, $\xi = \xi_{\max}$, the only outgoing wave is $v_n + c$; hence, $X_3 = 0$ and other components of $\overline{X}$ must be determined using the physical conditions.

The ideal gas law yields

$$\frac{\partial T}{\partial t} = 0 \implies \frac{1}{\gamma - 1}\frac{\partial p}{\partial t} - \frac{RT}{\gamma - 1}\frac{\partial \rho}{\partial t} = 0 \tag{3.41}$$

Imposing this condition and the no-slip condition on Eq. (3.18) then yields

$$\overline{X}_l = -[Q_l^{-1}]\bar{R}_{p,l} \quad \text{on the left boundary, } \xi = \xi_{\min} \tag{3.42}$$
$$\overline{X}_r = -[Q_r^{-1}]\bar{R}_{p,r} \quad \text{on the right boundary, } \xi = \xi_{\max} \tag{3.43}$$

where

$$\overline{X}_l = [X_2 \ X_3 \ X_4 \ X_5]^T \tag{3.44}$$
$$\overline{X}_r = [X_1 \ X_2 \ X_4 \ X_5]^T \tag{3.45}$$

and

$$\bar{R}_{p,l} = \bar{R}_{p,r} = [R_T \ R_{p,2} \ R_{p,3} \ R_{p,4}]^T \tag{3.46}$$

with

$$R_T = R_{p,5} - \frac{RT}{\gamma - 1}R_{p,1} \tag{3.47}$$

The matrix $[Q_l]$ or $[Q_r]$ is obtained by eliminating the first or third column of the matrix $[Q]$ and an appropriate combination of the first and the last row of the matrix $[Q]$ and the elimination of the last row. Specifically, denoting

$$Q_{T,i} = Q_{5,i} - \frac{RT}{\gamma - 1}Q_{1,i} \quad i = 2, \ldots, 5 \tag{3.48}$$

we have

$$[Q_l] = \begin{pmatrix} Q_{T,2} & Q_{T,3} & Q_{T,4} & Q_{T,5} \\ Q_{2,2} & Q_{2,3} & Q_{2,4} & Q_{2,5} \\ Q_{3,2} & Q_{3,3} & Q_{3,4} & Q_{3,5} \\ Q_{4,2} & Q_{4,3} & Q_{4,4} & Q_{4,5} \end{pmatrix} \tag{3.49}$$

and similarly denoting

$$Q_{T,i} = Q_{5,i} - \frac{RT}{\gamma - 1}Q_{1,i} \quad i = 1, 2, 4, 5 \tag{3.50}$$

we have

$$[Q_r] = \begin{pmatrix} Q_{T,1} & Q_{T,2} & Q_{T,4} & Q_{T,5} \\ Q_{2,1} & Q_{2,2} & Q_{2,4} & Q_{2,5} \\ Q_{3,1} & Q_{3,2} & Q_{3,4} & Q_{3,5} \\ Q_{4,1} & Q_{4,2} & Q_{4,4} & Q_{4,5} \end{pmatrix} \quad (3.51)$$

Adiabatic No-Slip Wall

In an adiabatic no-slip wall, the normal temperature derivative is zero, along with all velocity components. The normal derivative at the wall is computed using a one-sided scheme as

$$\frac{\partial T}{\partial \xi} = \sum_{i=0}^{N} a_i T_i = 0 \quad (3.52)$$

Therefore,

$$\frac{\partial T_0}{\partial t} = -\frac{1}{a_0} \sum_{i=1}^{N} a_i \frac{\partial T_i}{\partial t} \quad (3.53)$$

The imposition of the adiabatic no-slip wall is hence the same at the isothermal no-slip wall except for R_T is replaced with

$$+\frac{1}{a_0} \sum_{i=1}^{N} a_i \frac{\partial T_i}{\partial t} + R_T \quad (3.54)$$

and

$$\bar{R}_{p,l} = \bar{R}_{p,r} = \left[\frac{1}{a_0} \sum_{i=1}^{N} a_i \frac{\partial T_i}{\partial t} + R_T \quad R_{p,2} \quad R_{p,3} \quad R_{p,4} \right]^T \quad (3.55)$$

The order of approximation in Eq. (3.52) depends on the smoothness of the solution near the boundary. Section 3.2.6 addresses the smoothness of the solution near the boundary relevant to imposing a spatial derivative at the boundary.

3.2 Distretization of Various Boundary Conditions on Cartesian Domains

In this section, we present various boundary conditions and the corresponding discretization for Cartesian domains. We assume the boundary surface under discussion is aligned parallel to the y-direction with the left boundary surface denoted by x_L, and the right boundary surface denoted by x_R. Moreover, the first interior point, node, or cell-center is denoted by $x_0 = x_L + h/2$ and the last interior nodes with $x_N = x_R - h/2$. Correspondingly, the solutions at the first and last interior points are denoted as $u_0 = u(x_0) = u(x_L + h/2)$ and $u_N = u(x_N) = u(x_R - h/2)$, respectively. Using high-order interpolation to estimate the boundary values may not have sufficient accuracy due to the Runge phenomena (large oscillations in interpolating polynomials near the boundaries of evenly spaced grids), which could lead to numerical instability [11]. In the following sections, the boundary treatments schemes are presented for an accuracy as high as five, except for the periodic and reflective boundary conditions where stable and accurate arbitrary order approximations are possible.

3.2.1 Periodic Boundary Condition

The statement of the periodic boundary condition for a solution u in the continuous form is

$$u(x_L - x') = u(x_R - x') \qquad (3.56a)$$
$$u(x_R + x') = u(x_L + x') \qquad (3.56b)$$

where x' is the normal distance from the boundary surface. In the discrete case, x' is replaced with the grid spacing h. For a higher order approximation, the periodic condition is imposed as

$$u(x_L - h) = u(x_R - h), \quad u(x_L - 2h) = u(x_R - 2h), \quad \ldots \qquad (3.57a)$$
$$u(x_R + h) = u(x_L + h), \quad u(x_R + 2h) = u(x_L + 2h), \quad \ldots \qquad (3.57b)$$

For instance, a fourth-order central approximation at the cell boundary that coincides with the domain boundary requires solutions at two points outside the domain.

With the solution values over the approximating stencil near the boundary known, the cell-boundary approximation and the cell-center derivative approximation near domain boundaries can be computed using the same schemes as those for the interior points. Obviously, there is no need for solving a modified governing equation, Eq. (3.7), on the domain boundaries with periodic conditions.

3.2.2 Reflecting Boundary Condition

A reflecting boundary condition arises from a symmetry of the solution. For instance, in a laminar flow in a pipe, the flow properties are symmetric around the pipe's center line; hence, the reflecting condition is suitable along the pipe's center line. In a reflecting boundary condition, the required solutions outside the domain boundary are

3.2 Distretization of Various Boundary Conditions on Cartesian Domains

obtained by mirroring the solutions from the interior side of the boundary. Considering the left boundary as a reflecting boundary for the thermodynamic fluid properties (temperature, pressure, density, concentration, volume fraction), this means

$$T(x_L - x') = T(x_L + x'), \quad p(x_L - x') = p(x_L + x'), \quad \ldots \tag{3.58}$$

Moreover, the velocity components (u, v, w) with u and w being the tangential component and v the normal component are enforced as

$$u(x_L - x') = u(x_L + x'), \quad w(x_L - x') = w(x_L + x'), \quad v(x_L - x') = -v(x_L + x') \tag{3.59}$$

Or, in the discrete case, for the temperature and velocity, the treatment of the reflecting boundary conditions are

$$T(x_L - h) = T(x_L + h), \quad T(x_L - 2h) = T(x_L + 2h), \quad \ldots \tag{3.60a}$$
$$u(x_L - h) = u(x_L + h), \quad u(x_L - 2h) = u(x_L + 2h), \quad \ldots \tag{3.60b}$$
$$w(x_L - h) = w(x_L + h), \quad w(x_L - 2h) = w(x_L + 2h), \quad \ldots \tag{3.60c}$$
$$v(x_L - h) = -v(x_L + h), \quad v(x_L - 2h) = -v(x_L + 2h), \quad \ldots \tag{3.60d}$$

Note that the statement of the reflecting boundary conditions for the tangential velocity components and the thermodynamic properties yield the odd derivatives to be zero on the boundary surface. To show this, we can take the derivative of the reflecting condition as

$$T(x_L - x') = T(x_L + x') \implies \left.\frac{dT(x_L - x')}{dx'}\right|_{x'=0} = \left.\frac{dT(x_L + x')}{dx'}\right|_{x'=0} \tag{3.61}$$

which, in turn, implies

$$-\left.\frac{dT(x)}{dx}\right|_{x=x_L} = \left.\frac{dT(x)}{dx}\right|_{x=x_L} \implies \left.\frac{dT(x)}{dx}\right|_{x=x_L} = 0 \tag{3.62}$$

Generally, all odd derivatives are zero:

$$\left.\frac{d^{(2n+1)}T(x)}{dx^{(2n+1)}}\right|_{x=x_L} = 0 \quad \forall n \geq 0 \tag{3.63}$$

Similarly, one can show that all even derivatives of the normal component of the velocity at the reflecting surface are zero, i.e.,

$$\left.\frac{d^{(2n)}v(x)}{dx^{(2n)}}\right|_{x=x_L} = 0 \quad \forall n \geq 0 \tag{3.64}$$

3.2.3 Dirichlet Boundary Condition

Let us consider the Dirichlet boundary condition where the value of the solution at the left boundary $u_{-1/2} = u(x_L) = u_B$ is given. We need to compute the solution at the cell boundary and the solution derivative at the cell centers near the boundary.

Unlike the periodic and reflecting boundary conditions, the solutions at nodes outside the domain boundary (the so-called ghost nodes) are unknown for Dirichlet boundary conditions. One approach that several researchers have used [12] is to extend the solution to the ghost points using reflection as described above. Specifically,

$$[u(x_L - x') - u_B] = -[u(x_L + x') - u_B] \tag{3.65}$$

which clearly yields $u(x_L) = u_B$ for $x' \to 0$. However, as discussed above, this reflecting boundary condition yields zero even derivatives at the boundary, which are not necessarily consistent with the true even derivatives at the boundary as the solution in the domain's interior dictates them. In other words, imposing the Dirichlet condition using reflection Eq. (3.65), yields only a second-order accurate scheme. Therefore, it is not suitable for higher order approximations.

Realizing the limitation of applying the reflecting condition to the Dirichlet condition, we use the Dirichlet condition and the interior nodes near the boundary to approximate the cell-boundary solutions and solution derivative near the boundary. This approach is used by Direnzo et al. [13] and Modesti and Pirozzoli [14]. They noted that this approach is vital for properly imposing boundary fluxes for high-fidelity computations of wall-bounded turbulent flows.

Gustaffson [15] showed that an overall pth-order accurate scheme requires only $(p - 1)$th-order accurate scheme at the boundary. Thus, a fourth-order boundary treatment is sufficient for a fifth-order interior scheme.

We use a stencil of the boundary point and nodes near the boundary to determine the approximations. Specifically, as shown in Fig. 3.4, we use the stencil

$$\{x_L = x_{-1/2}, x_0, x_1, x_2, x_3, \ldots\} \tag{3.66}$$

with solution values

$$\{u_{-1/2} = u_B, u_0, u_1, u_2, u_3 \cdots\} \tag{3.67}$$

3.2 Distretization of Various Boundary Conditions on Cartesian Domains

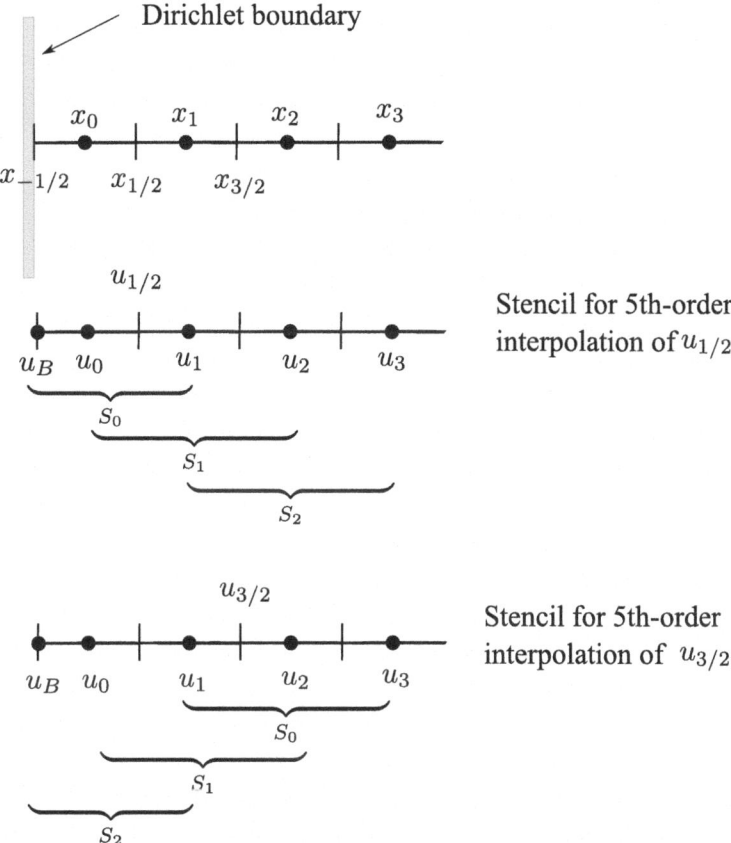

Fig. 3.4 Dirichlet boundary condition and the stencil used for compute the $u_{1/2}$ and $u_{3/2}$ using the specified boundary value u_B as the left-most data

For instance, for a fifth-order approximation, to approximate $u_{1/2}$, we choose the following four-point stencil

$$\{x_{-1/2} = x_L, x_0, x_1, x_2, x_3\} \tag{3.68}$$

as shown in Fig. 3.4.

We can use the TENO5 or WENO5 approach for computing the mid-point values $u_{1/2}$ and $u_{3/2}$ at $x_{1/2}$ and $x_{3/2}$, respectively. However, due to the non-uniform spacing between the grid points in the 5-point stencil, shown in Fig. 3.4, the standard TENO5 or WENO5 scheme is not applicable. We need to determine the new linear interpolation coefficients and smoothness indicator for one of the sub-stencils comprising the boundary node. Also, we need to determine new optimal linear coefficients.

For $u_{1/2}$, the interpolation coefficients using the sub-stencils S_0, S_1, and S_2, shown in Fig. 3.4, are given as

$$S_0: \quad u_{0,1/2} = -\frac{1}{3}u_{-1/2} + u_0 + \frac{1}{3}u_1 \tag{3.69}$$

$$S_1: \quad u_{1,1/2} = \frac{3}{8}u_0 + \frac{6}{8}u_1 - \frac{1}{8}u_2 \tag{3.70}$$

$$S_2: \quad u_{2,1/2} = \frac{15}{8}u_1 - \frac{5}{4}u_2 + \frac{3}{8}u_3 \tag{3.71}$$

The optimal linear coefficients are given as

$$b_{0,1/2} = \frac{3}{7}, \quad b_{1,1/2} = \frac{11}{21}, \quad b_{2,1/2} = \frac{1}{21} \tag{3.72}$$

and the smoothness indicators are expressed according to Eq. (2.61) as

$$\beta_{k,1/2} = \sum_{j=1}^{2} h^{2j-1} \int_{x_{1/2}}^{x_{3/2}} \left[\frac{d^j u_{k,1/2}(x)}{dx^j} \right]^2 dx \tag{3.73}$$

leading to

$$\beta_{0,1/2} = \frac{79}{3}u_0^2 - \frac{194}{9}u_0 u_1 + \frac{127}{27}u_1^2 - \frac{8}{27}(105\,u_0 - 41\,u_1)u_{-1/2} + \frac{256}{27}u_{-1/2}^2 \tag{3.74a}$$

$$\beta_{1,1/2} = \frac{13}{12}(u_0 - 2u_1 + u_2)^2 + \frac{1}{4}(u_2 - u_0)^2 \tag{3.74b}$$

$$\beta_{2,1/2} = \frac{13}{12}(u_3 - 2u_2 + u_1)^2 + \frac{1}{4}(u_3 - 4u_2 + 3u_1)^2 \tag{3.74c}$$

For $u_{3/2}$, the interpolation coefficients using the sub-stencils S_0, S_1, and S_2 shown in Fig. 3.4 are given as

$$S_0: \quad u_{0,3/2} = \frac{3}{8}u_1 + \frac{6}{8}u_2 - \frac{1}{8}u_3 \tag{3.75}$$

$$S_1: \quad u_{1,3/2} = -\frac{1}{8}u_0 + \frac{6}{8}u_1 + \frac{3}{8}u_2 \tag{3.76}$$

$$S_2: \quad u_{2,3/2} = u_{-1/2} - 2u_0 + 2u_1 \tag{3.77}$$

The optimal linear coefficients are given as

$$b_{0,3/2} = \frac{2}{7}, \quad b_{1,3/2} = \frac{22}{35}, \quad b_{2,3/2} = \frac{3}{35}. \tag{3.78}$$

3.2 Distretization of Various Boundary Conditions on Cartesian Domains

The smoothness indicators are expressed according to Eq. (2.61) as

$$\beta_{k,3/2} = \sum_{j=1}^{2} h^{2j-1} \int_{x_{1/2}}^{x_{3/2}} \left[\frac{d^j u_{k,3/2}(x)}{dx^j} \right]^2 dx \quad (3.79)$$

leading to

$$\beta_{0,3/2} = \frac{13}{12}(u_1 - 2u_2 + u_3)^2 + \frac{1}{4}(3u_1 - 4u_2 + u_3)^2 \quad (3.80a)$$

$$\beta_{1,3/2} = \frac{13}{12}(u_0 - 2u_1 + u_2)^2 + \frac{1}{4}(u_0 - u_2)^2 \quad (3.80b)$$

$$\beta_{2,3/2} = \frac{79}{3}u_0^2 - \frac{194}{9}u_0 u_1 + \frac{127}{27}u_1^2 - \frac{8}{27}(105 u_0 - 41 u_1)u_{-1/2} + \frac{256}{27}u_{-1/2}^2 \quad (3.80c)$$

Based on Ref. [16], τ_5 in the TENO scaling Eq. (2.110) is defined as

$$\tau_5 = \beta_5 - \frac{1}{6}(\beta_0 + \beta_1 + \beta_2) \quad (3.81)$$

with β_5, the global smoothness indicator for the 5-point full stencil is defined as

$$\beta_5 = \frac{4953797}{60480} u_0^2 - \frac{921703}{6048} u_0 u_1 + \frac{4625189}{60480} u_1^2 -$$
$$\frac{2}{33075}(1260683 u_0 - 1129433 u_1)u_{-1/2} + \frac{21085184}{1157625} u_{-1/2}^2 +$$
$$\frac{1}{1764000}(144487595 u_0 - 155211245 u_1 - 63242176 u_{-1/2})u_2 + \frac{4584701}{168000} u_2^2$$
$$- \frac{1}{7408800}(126656005 u_0 - 141816115 u_1 - 54526784 u_{-1/2} + 93714159 u_2)u_3 +$$
$$\frac{4805453}{2963520} u_3^2 \quad (3.82)$$

3.2.4 Derivative Approximation Near the Boundary

Using the cell-interface values near the domain boundary, we determine the derivatives at the cell centers near the boundary using the appropriate high-order scheme. For instance, using a fourth-order scheme, the derivative at the cell-center x_0 is determined as

$$\tilde{u}_0^{(1)} = \frac{1}{h}\left(-\frac{22}{24}u_{-1/2} + \frac{17}{24}u_{1/2} + \frac{9}{24}u_{3/2} - \frac{5}{24}u_{5/2} + \frac{1}{24}u_{7/2}\right) \quad (3.83)$$

The boundary value $u_{-1/2} = u_B$ is either given as a physical boundary condition or is determined as the solution of the (modified) governing PDEs, Eq. (3.7), on the boundary as described in Sect. 3.1. The cell-boundary values near the boundary, $u_{1/2}$ and $u_{3/2}$, are determined using the boundary schemes described above, and the remaining cell-boundary values are determined using the interior scheme described in Chap. 2.

The alternative approximation can be used to combine the mid-point values and the node values. However, they result in an effective grid spacing that is half of that of the scheme (3.83). Hence, the scheme (3.83) is preferable.

The derivative at the node x_1 can be determined with a fourth-order accuracy using the interior schemes described in Subsection 2.2.3.1 as

$$\tilde{u}_1^{(1)} = \frac{1}{h}\left[\frac{8}{6}(u_{3/2} - u_{1/2}) - \frac{1}{6}(u_2 - u_0)\right] \quad (3.84)$$

3.2.5 Derivative Approximation on the Boundary

To solve (modified) governing PDEs, Eq. (3.7), on the boundaries of the physical domain as discussed in Sect. 3.1, we need to compute solution derivatives on the boundaries. For the first x-derivatives, we may use either of the following two expressions for a fourth-order accuracy:

$$\tilde{u}_{-1/2}^{(1)} = \frac{1}{h}\left(-\frac{611}{420}u_{-1/2} + \frac{9}{5}u_{1/2} - \frac{6}{35}u_{3/2} - \frac{5}{21}u_{5/2} + \frac{9}{40}u_{7/2}\right) \quad (3.85a)$$

$$\tilde{u}_{-1/2}^{(1)} = \frac{1}{h}\left(-\frac{352}{105}u_{-1/2} + \frac{35}{8}u_0 - \frac{35}{24}u_1 + \frac{21}{40}u_2 - \frac{5}{56}u_3\right) \quad (3.85b)$$

Since the effective grid spacing is halved in the second approach (3.85b), the second approach results in a time steps size that is half of the first approach (3.85a). For this reason, the first approach is preferable.

Since we have the first x−derivatives at the boundary and all cell centers using the schemes for near the boundary or interior domain, we can use a similar expression as Eq. (3.85b) to compute the second x−derivative in the normal direction on the boundary as

$$\tilde{u}_{-1/2}^{(2)} = \frac{1}{h}\left(-\frac{352}{105}\tilde{u}_{-1/2}^{(1)} + \frac{35}{8}\tilde{u}_0^{(1)} - \frac{35}{24}\tilde{u}_1^{(1)} + \frac{21}{40}\tilde{u}_2^{(1)} - \frac{5}{56}\tilde{u}_3^{(1)}\right) \quad (3.86)$$

3.2.6 Neumann Boundary Condition

A Neumann boundary condition is a boundary condition that specifies a solution variable's derivative. A typical example is an adiabatic wall where the heat flux is set to zero or when the heat flux at the wall is set to a nonzero level. The first derivative at the domain boundary is determined for various approximation orders $r = 1, \ldots, 4$ by taking the derivative of (2.4) and evaluating at $x_{-1/2}$ for the five-point stencil (3.68) to obtain

$$h\tilde{u}^{(1)}_{1,-1/2} = 2u_0 - 2u_{-1/2} \tag{3.87}$$

$$h\tilde{u}^{(1)}_{2,-1/2} = 3u_0 - \frac{1}{3}u_1 - \frac{8}{3}u_{-1/2} \tag{3.88}$$

$$h\tilde{u}^{(1)}_{3,-1/2} = \frac{15}{4}u_0 - \frac{5}{6}u_1 - \frac{46}{15}u_{-1/2} + \frac{3}{20}u_2 \tag{3.89}$$

$$h\tilde{u}^{(1)}_{4,-1/2} = \frac{35}{8}u_0 - \frac{35}{24}u_1 - \frac{352}{105}u_{-1/2} + \frac{21}{40}u_2 - \frac{5}{56}u_3 \tag{3.90}$$

Now, knowing the value of the first derivative $\tilde{u}^{(1)}_{-1/2}$, the boundary value with different order of accuracy is determined as

$$u_{1,-1/2} = \frac{1}{2}\left(u_0 - h\tilde{u}^{(1)}_{-1/2}\right) \tag{3.91a}$$

$$u_{2,-1/2} = \frac{3}{8}\left(3u_0 - \frac{1}{3}u_1 - h\tilde{u}^{(1)}_{-1/2}\right) \tag{3.91b}$$

$$u_{3,-1/2} = \frac{15}{46}\left(\frac{15}{4}u_0 - \frac{5}{6}u_1 + \frac{3}{20}u_2 - h\tilde{u}^{(1)}_{-1/2}\right) \tag{3.91c}$$

$$u_{4,-1/2} = \frac{105}{352}\left(\frac{35}{8}u_0 - \frac{35}{24}u_1 + \frac{21}{40}u_2 - \frac{5}{56}u_3 - h\tilde{u}^{(1)}_{-1/2}\right) \tag{3.91d}$$

Before the derivative calculation, the smoothness of the solution near the boundary must be established. Based on that, the width of the stencil is adjusted to avoid discontinuities within the interpolating stencil. Smoothness is established using TENO schemes as described in Sects. 2.4 and 3.2.3.

For example, expressions in (3.91) are used for imposing the adiabatic condition in Eq. (3.53) which is, in turn, used to determine the vector $\overline{X}$ in the modified governing equation (3.7).

3.3 Boundary Condition Approach for General Non-Cartesian Domains

In non-Cartesian domains, we aim to use a Cartesian grid with irregular (curved) boundaries embedded inside. Unlike the boundary-fitted approach, which requires sophisticated grid generations, generating a Cartesian grid is trivial. However, in the boundary-embedded approach, applying the boundary conditions in a stable and accurate fashion is non-trivial. Unlike our approach in Cartesian domains, directly solving the governing equations (or the modified ones) on the boundaries is not possible because, due to the non-Cartesian nature of the domain, the spatial spacing between the boundary point and the first interior point can be arbitrarily small requiring arbitrarily small time steps to ensure stability, which is undesirable. Here, we describe an approach that avoids directly solving the equations on the boundary and uses solutions at points outside of the domain (called ghost points) to carry out the flux and derivatives for interior points near the boundaries. Figure 3.5 illustrates a general non-Cartesian domain, interior computational points, and ghost points. Ghost points allow the same discretization schemes for all near-boundary interior points as the rest of the interior points. The challenge is determining the solution variables at the ghost points so that stability and accuracy are maintained. To this end, instead of directly solving governing equations on the boundary, we use the governing equations to determine the spatial derivatives on the boundaries using an approach called

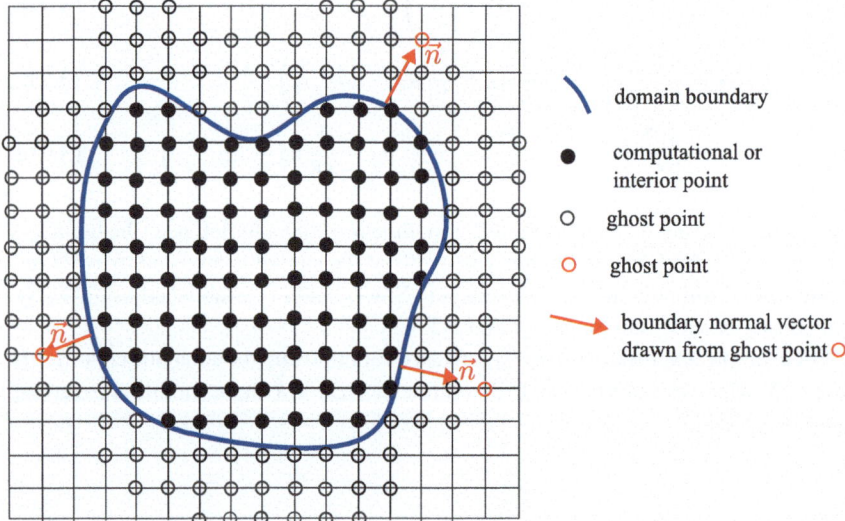

Fig. 3.5 A sketch of a non-Cartesian, irregular domain discretized with a Cartesian grid. Three ghost point layers correspond to a fifth-order WENO reconstruction at the domain's interior. The boundary's normal direction viewed from a ghost point signifies the direction in which a Hermit polynomial extrapolates the boundary and interior data to the ghost point

the inverse Lax-Wendroff (ILW) method and use the given solution values and their spatial derivatives to determine flux or primitive variables on the ghost points. This approach and its enhancements have been introduced in several articles [5, 7, 8, 17]. Our description of the approach is based on the latest enhancement of the scheme, which promises accuracy and stability with a high level of efficiency as outlined in Ref. [17].

Before describing the details of the algorithm, we outline various steps to determine the ghost point data.

- The ghost point fluxes or primitive variables are determined using the boundary and interior data. To ensure stability, the boundary data must include fluxes and their normal derivatives, and the interior data must be chosen from suitable locations. The location of the interior points in the normal direction is determined based on linear stability analysis. These boundary and interior data are then used to reconstruct a Hermit polynomial, which, in turn, is used to determine the ghost point data.
- The location of the boundary where the data is needed for the Hermit polynomial is the intersection of the normal line drawn to the boundary curve from the ghost point. This intersection is called a boundary-intercept point. Hence, the boundary-intercept point and the normal direction are required for each ghost point.
- The fluxes at the boundary-intercept point are determined using the given boundary data for the incoming characteristics. For the remaining data at the boundary, extrapolation using the internal points are used.
- The normal derivatives at the boundary-intercept point are determined using the ILW scheme and the modified governing equations obtained from the general characteristic boundary treatment explained in Section 3.1.
- The tangential derivative values (used to compute the normal derivatives via the ILW method) and the flux values on the boundary-normal line (used to construct the Hermit polynomial) are determined using the least-squares method over a sufficiently large multi-dimensional stencil. The ILW is multi-dimensional approach in the sense that the Hermit polynomial reconstruction in the normal direction involves multi-dimensional least-squares approach.

3.3.1 Computing the Boundary-Intercept Point and the Normal Direction

For every ghost point (labeled $q_{i,j}$ in Fig. 3.6), the approach requires the determination of the normal direction drawn from the ghost point on the boundary and the intercept of the normal line with the boundary (boundary-intercept point, labeled by $q^*_{i,j}$ in Fig. 3.6). The boundary tangential direction at the boundary-intercept point is also required. In two-space dimensions, given the boundary curve as the set

$$\partial \Omega = \{(x, y) | \Psi(x, y) = 0\} \tag{3.92}$$

following [18], the algorithm for determining the boundary-intercept point is as follows:

Algorithm 1 Algorithm for computing the boundary-intercept point $q^*_{i,j}$ for the ghost point $q_{i,j}$ using the Newton method

---
1: Given: $q_{i,j}$
2: Choose $q_{n+1} = q_{i,j}$ and $q_n = q_{n+1}(x_i + \Delta x, y_j + \Delta y)$
3: **while** $|q_{n+1} - q_n| \leq 10^{-15}$ **do**
4:    $n \leftarrow n + 1$
5:    $q_{n+1} \leftarrow q_n - \frac{\Psi(x_i, y_j)}{\|\nabla\Psi(x_i, y_j)\|^2} \nabla\Psi(x_i, y_j)$
6: **end while**
7: $q*_{i,j} \leftarrow q_{n+1}$
8: **return** point $q*_{i,j}$

The normal and tangential directions are then computed using the boundary-intercept point $q^*_{i,j} = (x^*_i, x^*_j)$ as

$$n = \frac{\nabla\Psi(x^*_i, y^*_j)}{\|\nabla\Psi(x^*_i, y*_j)\|} \tag{3.93a}$$

$$\tau = \begin{pmatrix} 0 & 1 \\ -1 & 0 \end{pmatrix} n \tag{3.93b}$$

3.3.2 Ghost Data Approximation: Fluxes or Primitive Variables?

To reconstruct primitive variables at the ghost point, we need a spatial derivative of these variables using the inverse Lax-Wendroff method. This process requires the inversion of the Jacobian matrix, $\frac{\partial F(\bar{u})}{\partial \bar{u}}$. The Jacobian matrix may, however, become singular when the eigenvalues of the matrix are zero or close to zero. This singularity occurs when gas velocity is zero, or the gas goes through a sonic point, i.e., $v_n + c = 0$ or $v_n - c = 0$. To avoid a singular or nearly singular Jacobian matrix, we can reconstruct the flux values instead. This approach will avoid the inversion of the Jacobian matrix and can use the values to directly compute the mid-point flux values, which are, in turn, used for the upwind flux splitting and the flux derivative calculations to complete the spatial discretization.

3.3.3 Hermit Polynomial Reconstruction

To determine the flux data on the ghost points, we first need to construct the Hermit polynomial using the flux and its first normal derivative at the boundary-intercept point and flux values at some suitably chosen points in the normal direction (points shown in red squares in Fig. 3.6). We call these points the Hermit points or nodes. They are chosen so that the stability of the extrapolation to the ghost points is ensured by using only the first normal derivative of the fluxes obtained via the ILW method. Using eigenvalues analysis of a linear advection equation, Ref. [17] proposed the Hermits nodes for approximation order up to seven that appear to perform stably even for a nonlinear system of equations such as the Euler equations of gas dynamics with only the first normal derivative computed using the ILW approach. The location of the Hermit nodes for the third-, fifth-, or seventh-order polynomial reconstruction adopted from Ref. [17] are given as

$$q_k = q^*_{i,j} - k\alpha\delta \boldsymbol{n} \quad k = 1, 2, \ldots \tag{3.94}$$

where $\delta = \sqrt{(\Delta x)^2 + (\Delta y)^2}$ and the suitable range of α are given in Table 3.1.

Linear analysis has shown that for orders smaller and equal to seven, without compromising the stability, only the first spatial derivative at the boundary-intercept point needs to be computed using the ILW method [17].

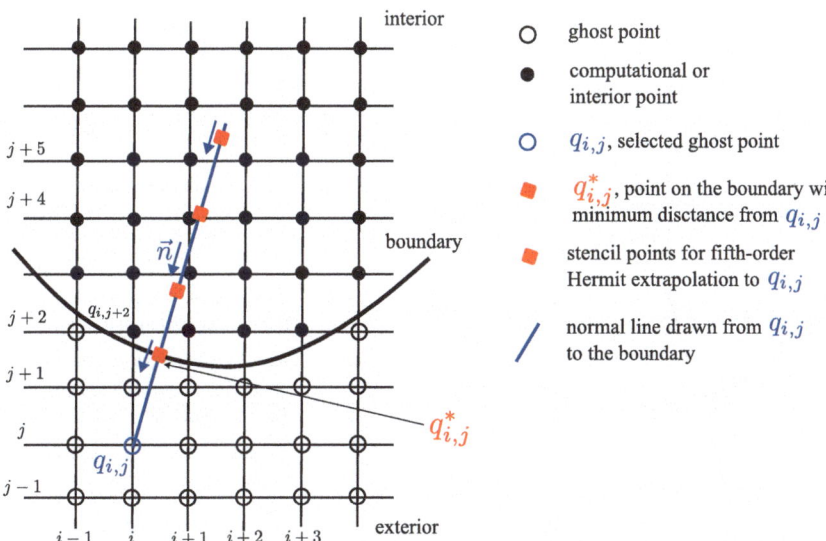

Fig. 3.6 A non-Cartesian domain discretized with a Cartesian grid, and the normal direction and stencil points used for Hermit extrapolation to the ghost point $q_{i,j}$ are shown

Table 3.1 Hermit node location parameter α given in Eq. (3.94) for polynomial degree r

$r+1=3$	$r+1=5$	$r+1=7$
$[0.61\frac{\max(\Delta x,\Delta y)}{\delta}, 10\frac{\min(\Delta x,\Delta y)}{\delta}]$	$[0.92\frac{\max(\Delta x,\Delta y)}{\delta}, 5.11\frac{\min(\Delta x,\Delta y)}{\delta}]$	$[1.34\frac{\max(\Delta x,\Delta y)}{\delta}, 1.99\frac{\min(\Delta x,\Delta y)}{\delta}]$

After obtaining the Hermit points, the Hermit polynomial reconstruction consists of two remaining steps: determining the flux values at the Hermit nodes and finding the flux and its normal derivative at the boundary-intercept point. We describe the algorithm addressing each step next.

Flux Value at Hermit Nodes

The ghost point's normal direction may pass through the interior domain at any angle; hence, we need to use a two-dimensional stencil of the interior nodes to construct a two-dimensional polynomial of total degree r. We use a stencil with its points having a maximum distance $R = r \max(\Delta x, \Delta y)$ from the boundary-intercept point given as

$$S = \left\{ (x_m, y_n) \Big| \sqrt{(x_m - x_i^*)^2 + (y_n - y_j^*)^2} \leq R \right\} \quad (3.95)$$

Figure 3.7 depicts the Stencil S for the boundary point $q_{i,j}^*$. We use the least-squares method to construct two-dimensional polynomials of degree r, $P_r(x, y)$ using the Stencil S. The Stencil S is determined and stored for each ghost point in a preprocessing stage. One trivial algorithm to determine the Stencil S is that for each ghost

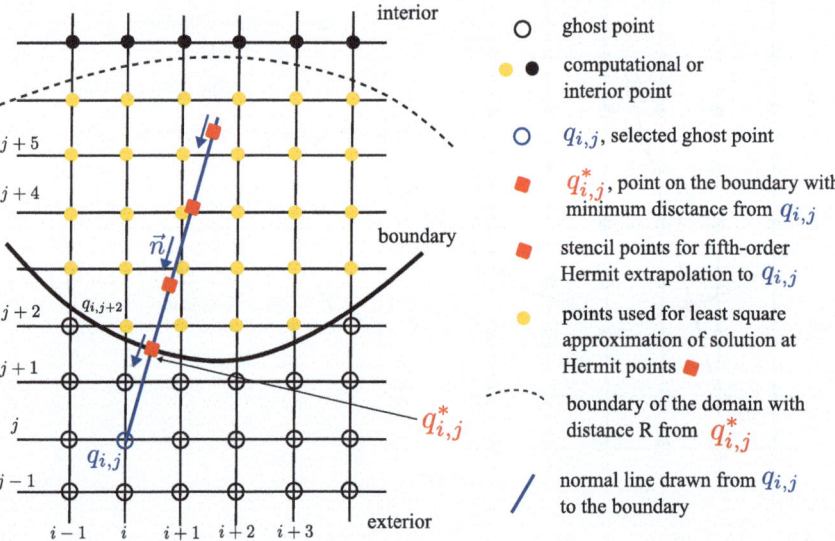

Fig. 3.7 Stencil used to compute flux at Hermit nodes, in turn, used for reconstruction of the Hermit polynomial

3.3 Boundary Condition Approach for General Non-Cartesian Domains

point, all interior points are examined to classify if they are within the radius R of the boundary-intercept point. Alternatively, a significantly more efficient algorithm is to examine a fraction of interior points that are $(3/2)R$ away from the ghost points to determine if they fall within a circle with center $q_{i,j}^*$ and radius of R.

Flux Value at the Boundary-Intercept Point

To determine flux values at the boundary-intercept point, the same approach of General Boundary Treatment described in Sect. 3.1 is used except for the x, and y coordinates are replaced by the local coordinates of (n, τ) in the normal and tangential directions at the boundary-intercept point. For $\boldsymbol{n} = (n_1, n_2)$, the normal and tangential derivatives in the governing equation are defined as

$$\frac{\partial}{\partial n} = n_1 \frac{\partial}{\partial x} + n_2 \frac{\partial}{\partial y} \tag{3.96a}$$

$$\frac{\partial}{\partial \tau} = -n_2 \frac{\partial}{\partial x} + n_1 \frac{\partial}{\partial y} \tag{3.96b}$$

The fluxes corresponding to the incoming characteristics are imposed based on the boundary data. In contrast, the fluxes corresponding to the outgoing characteristics are determined using the least squares polynomial $P_r(x, y)$ formed using the Stencil S.

Derivative of Flux at the Boundary-Intercept Point

The normal derivatives of fluxes at the boundary-intercept point are derived using the ILW method. We begin with Eq. (3.1) in the local (n, τ) as

$$\frac{\partial \overline{u}}{\partial t} + \frac{\partial \overline{F^a}}{\partial n} + \overline{C} = 0 \tag{3.97}$$

where C is all tangential derivatives and viscous terms. The temporal derivative and C are all determined using the boundary data and the least square extrapolation outlined above. The normal derivative is then computed as

$$\frac{\partial \overline{F^a}}{\partial n} = -\frac{\partial \overline{u}}{\partial t} - \overline{C} \tag{3.98}$$

Hermit Polynomial Reconstruction

The Hermit polynomial of degree r, $H_r(n)$, in the normal direction, is determined using the r flux values and one flux derivative value. Specifically, $H_r(n)$ is reconstructed using $r + 1$ data as

$$\begin{cases} \left. \frac{\partial F^a}{\partial n} \right|_{q_{i,j}^*} \\ \left. F^a \right|_{q_{i,j}^*} \\ \left. F^a \right|_{q_k} \quad k = 1, 2, \ldots, r - 1 \end{cases} \tag{3.99}$$

Finally, the reconstructed Hermit polynomial is used to determine the flux values at the ghost points. The Hermit polynomials display a less severe Runge phenomenon and since derivatives contains information about the direction of the characteristics, the Hermit extrapolations are expected to display better stability properties than the scheme based on the Lagrange interpolations. An end-of-chapter exercise reinforces this property.

3.4 Exercises

1. For the Euler or Navier-Stokes system with the ideal gas law, denote the conservative and primitive variable vectors by $\overline{u}$ and $\overline{w}$. The mapping from the conservative and primitive variables can be expressed as in Eq. (3.11). (a) Derive the matrix $[H]$. (b) Denoting the characteristic variable by $d\overline{w}_c = [Q]^{-1}d\overline{w} = [S]^{-1}d\overline{u}$. Determine the relationship between $[Q]$ and $[S]$ matrices.
2. For the Robin boundary condition in the form $\alpha_1 T + \alpha_2 \partial T/\partial x = \alpha_3$, determine the modified Navier-Stokes equation, i.e., determine the vector $\overline{X}$ in Eq. (3.7).
3. Derive cell-boundary solution approximations and cell-center derivative approximations of at least seventh order of accuracy at and near boundary.
4. For a simple advection equation, write a computer code to verify the stability of the fifth and seventh-order boundary schemes.
5. For subsonic inflow condition, assume the stagnation pressure and temperature and velocity direction are specified. Determine the vector $\overline{X}$ required in the modified governing Eq. (3.7).
6. This exercise compares the extrapolation properties of Hermit and Lagrange polynomials. Consider the function $u(x) = \sin(2\pi x)$ defined over all real numbers. Pick $r + 1$ points as

$$x_i = ih \quad i = 0, \ldots, 4 \tag{3.100}$$

where h is the grid spacing. (a) Construct a fifth-order accurate Lagrange polynomial using the grid points given above and $u(x_i)$. Verify its order of accuracy with grid refinement. (b) Construct a fifth-order accurate Hermit polynomial using the grid points $x_0, \ldots x_3$ and the function values at these points, $u(x_i)$, and the derivative of the function $u(x_0)$. Verify its order of accuracy with grid refinement (c) Approximate $u(x = -h)$, $u(x = -2h)$ and $u(x = -3h)$ using the Lagrange and Hermit polynomials for exceedingly refined grids. Which scheme is more accurate? Which location displays the largest error? Explain why?

References

1. Du, Y.: Implicit boundary equations for conservative Navier-Stokes equations. J. Comput. Phys. **375**, 641–658 (2018)
2. Du, Y.: Generalized boundary equations for conservative Navier-Stokes equations. Aerosp. Sci. Technol. **86**, 836–849 (2019)
3. Poinsot, T.J., Lelef, S.K.: Boundary conditions for direct simulations of compressible viscous flows. J. Comput. Phys. **101**(1), 104–129 (1992)
4. Deng, X., Zhang, H.: Developing high-order weighted compact nonlinear schemes. J. Comput. Phys. **165**(1), 22–44 (2000)
5. Tan, S., Shu, C.-W.: Inverse lax-wendroff procedure for numerical boundary conditions of conservation laws. J. Comput. Phys. **229**(21), 8144–8166 (2010)
6. Gustafsson, B.: High Order Difference Methods for Time Dependent PDE, Springer Series in Computational Mathematics, vol. 38. Springer (2008)
7. Tan, S., Wang, C., Shu, C.-W., Ning, J.: Efficient implementation of high order inverse Lax-Wendroff boundary treatment for conservation laws. J. Comput. Phys. **231**(6), 2510–2527 (2012)
8. Jianfang, L., Shu, C.-W., Tan, S., Zhang, M.: An inverse Lax-Wendroff procedure for hyperbolic conservation laws with changing wind direction on the boundary. J. Comput. Phys. **426**, 109940 (2021)
9. Liu, S., Li, T., Cheng, Z., Jiang, Y., Shu, C-W., Zhang, M.: A new type of simplified inverse Lax-Wendroff boundary treatment i: hyperbolic conservation laws (2024)
10. Rohde, A.: Eigenvalues and eigenvectors of the Euler equations in general geometries. In: 15th AIAA Computational Fluid Dynamics Conference (2001)
11. Hagstrom, T.: Experiments with stable, high-order difference approximations to hyperbolic initial-boundary value problems. In: Proceedings of Fifth International Conference on Mathematics Numerical Asp. Wave Propag. Phenom., Santiago de Compostela, Spain, July 10-14 2000. SIAM (2010)
12. Lin, F.: Very-high-order teno schemes with adaptive accuracy order and adaptive dissipation control. Comput. Methods Appl. Mech. Eng. **387**, 114193 (2021)
13. Di Renzo, M., Lin, F., Urzay, J.: Htr solver: An open-source exascale-oriented task-based multi-gpu high-order code for hypersonic aerothermodynamics. Comput. Phys. Commun. **255**, 107262 (2020)
14. Modesti, D., Pirozzoli, S.: Reynolds and mach number effects in compressible turbulent channel flow. Int. J. Heat Fluid Flow **59**, 33–49 (2016)
15. Gustafsson, B.: The convergence rate for difference approximations to mixed initial boundary value problems. Math. Comput. **29**(130), 396–406 (1975)
16. Lin, F., Hu, X.Y., Adams, N.A.: A new class of adaptive high-order targeted eno schemes for hyperbolic conservation laws. J. Comput. Phys. **374**, 724–751 (2018)
17. Liu, S., Li, T., Cheng, Z., Jiang, Y., Shu, C-W., Zhang, M.: A new type of simplified inverse Lax-Wendroff boundary treatment i: hyperbolic conservation laws (2024)
18. Fernández-Fidalgo, J., Clain, S., Ramírez, L., Colominas, I., Nogueira, X.: Very high-order method on immersed curved domains for finite difference schemes with regular cartesian grids. Comput. Methods Appl. Mech. Eng. **360**, 112782 (2020)

Chapter 4
High-Order Time Integration Methods

A numerical solution of a partial differential equation (PDE) with expected large solution gradients demands a high level of resolution in approximating temporal and spatial derivatives. Low-order methods, first- and second-order ones (reviewed in Appendix A), provide limited resolution for a given grid spacing since their errors scale as $\mathcal{O}(\tau)$ or $\mathcal{O}(\tau^2)$ for a temporal scheme and $\mathcal{O}(h)$ or $\mathcal{O}(h^2)$ for a spatial one. In other words, the rate of convergence of the first- and second-order methods is low; hence, high-fidelity computations require an exceedingly high level of grid refinements. The high number of grid points in space and time can lead to prohibitively expensive computations.

The total number of grid points N_x directly influences the total CPU time of the computation. Considering the explicit temporal discretization, the stability condition also demands that the time step size scales with $\mathcal{O}(h)$ in the case of an advection-dominant problem or $\mathcal{O}(h^2)$ in the case of a diffusion-dominant problem. Hence, the total number of time steps required for a specific final time scales as $\mathcal{O}(N_x)$ or $\mathcal{O}(N_x^2)$ for an advection- or diffusion-dominant problem, respectively. Since the total CPU time of a PDE computation using an explicit scheme in an asymptotic regime is proportional to the number of grid points in space multiplied by the number of grid points in time, we have for a d-dimensional computations

$$\text{CPU time} \propto N_x^{1+d} \quad \text{for advection PDE computation} \quad (4.1a)$$

$$\text{CPU time} \propto N_x^{2+d} \quad \text{for diffusion PDE computation} \quad (4.1b)$$

While, for one-dimensional calculations, the CPU time or computational cost scales as N_x^2 or N_x^3, for three-dimensional calculations, the cost scales as N_x^4 or N_x^5, which can be prohibitively high; for instance, a ten-fold increase in the resolution can yield an increase of 10^4 or 10^5 in the CPU time.

We substantiate the above back-of-the-envelop estimate by a numerical experiments, where we compare the first-order solution with a fifth-order solution of a

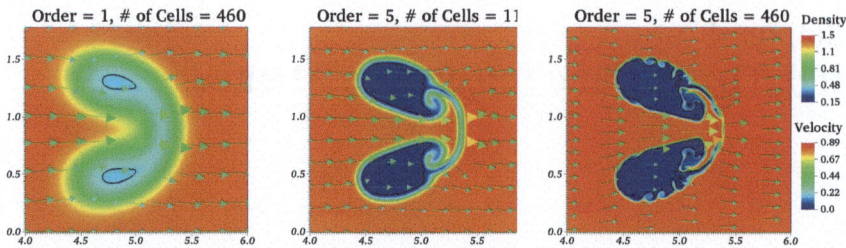

Fig. 4.1 Helium bubble shock interaction at time $t/(R_0/c_{air}) = 3.84$ (left) first-order method with 460 cells per unit square, (middle) fifth-order WENO method with 115 cells per unit square, and (right) high-resolution solution with first-order method with 460 cells per unit square. "Order=5" in the legend of the middle and right panel refer to as the fifth-order WENO reconstruction. These results were first reported in the author's earlier work [1]

shock-bubble interaction problems as performed in Ref. [1]. Both high-order and low-order computations are carried out with a CFL condition of 0.5, which is the CFL value ensuring the positivity-preservation of low and high-order schemes. (Positivity preservation is discussed in Chaps. 5 and 6 for single and two-fluid systems.) Specifically, for the problem of interaction of a Mach 1.22 shock with a helium bubble placed in a body of air for the dimensionless time of $t/(R_0/C_{air}) = 3.84$, Fig. 4.1 depicts the result of three computations: (left) density, velocity and interface location computed using the first-order method with 460 cells per unit square (fine resolution), (middle) those of the high-order method with only 115 cells (coarse resolution, four-fold lower resolution), and (right) those of the high resolution high-order method (460 cells). As clearly seen from the figures, the results of coarse high-order computation (middle figure) is much closer to the results of the fine high-order than the solution of the fine first-order method (with four-fold higher number of cells). To quantify the superior performance of the high-order method over the first order, Table 4.1 lists the errors in the minimum density, the maximum velocity, and the total CPU times. It is clear that the high-order calculation not only is significantly more accurate, but also it is at least hundred-fold faster. In a nutshell, although it is true that the first-order method is sufficient for isolated shock wave or isolated interfaces due to their discontinuous nature, for the complex scenarios of interaction and the ensuing instabilities and high solution gradients, high-order schemes outperform

Table 4.1 Shock interaction with helium bubble in air, Mach $= 1.22$, $t/(R_0/c_{air}) = 3.84$. These results were first reported in the author's earlier work [1]

Method	Error in minimum density (%)	Error in maximum velocity (%)	CPU time (s)
First-Order, 460 Cells	180	36	$1.5e+4$
First-Order, 1840 Cells	90	18	$1.2e+5$
Fifth-Order, 115 Cells	13	15	$8.1e+2$

low-order methods by a huge margin, enabling large-scale shock-bubble interaction computations which would not be possible with low-order methods.

The simple estimate and the above numerical results reveal that increasing the order of accuracy of the spatial discretization, which allows the use of a coarser grid compared to the low-order schemes, can be a more efficient approach to achieving a specific level of accuracy. Of course, increasing the order of accuracy of the spatial scheme (beyond two) also necessitates the corresponding increase in the order of accuracy of the temporal scheme. Therefore, we must present high-order spatial and temporal schemes, especially those of orders three and above. In Chap. 2, we addressed high-order spatial schemes. In this chapter, we focus on high-order temporal schemes.

There are two general approaches for constructing high-order time integration schemes. An intuitive approach consists of using the solution at a sequence of time steps to construct a high-order solution for the next time step. This approach leads to methods dubbed multi-step schemes [2]. Dahlquist showed that for a multi-step scheme to be A-stable, its order of accuracy cannot exceed two [2], which significantly limits the applicability of these schemes for stiff problems encountered in the computations of transport phenomena. (Appendix A reviews the basic schemes, stiffness, and various stability properties.) Alternatively, one may obtain a high-order approximation for the next time step by computing the solution at specific points or nodes situated appropriately between the current and next time steps. This strategy gives rise to the so-called multi-stage methods. The number of stages within the time integration interval (or the number of nodes) and the explicit or implicit computations of the solution at these nodes allow a sufficient number of coefficients that can be optimized to obtain high-order accuracy with desirable stability properties. Runge-Kutta (RK) methods and spectral deferred correction (SDC) schemes are examples of multi-stage schemes. They are high-order schemes that provide effective approaches for the computation of transport phenomena with potentially stiff processes. We discuss them in this chapter.

While we here aim to understand RK and SDC schemes and how they are constructed and implemented, the detailed stability analysis of these schemes is beyond this text's scope. For detailed stability analysis of RK schemes, we refer to the books by Butcher [3] and Hairer and Wanner [4], and manuscripts by Kennedy and Carpenter [5, 6]. Results on the stability analysis of the semi-implicit and multi-implicit analysis of the spectral deferred correction schemes can be found in Refs. [7–9].

4.1 Runge-Kutta Methods

Runge-Kutta methods are examples of one-step, multi-stage schemes because to obtain the solution at time step $n+1$, they require only the solution value at time step n. They, however, compute the solution at time step $n+1$ via a series of derivative evaluations at stages or nodes.

We discuss RK methods in the most general form, deriving implicit schemes before simplifying them to obtain diagonally implicit and explicit versions. We also discuss implicit-explicit (or semi-implicit) forms effective when stiff and nonstiff terms coexist in the underlying ODEs.

4.1.1 Construction of Runge-Kutta Methods

We seek the numerical solution of the nonlinear system of ODEs in the form

$$w'(t) = f(t, w(t)) \tag{4.2}$$

on a temporal grid $\{t_0 = 0, t_1, \ldots t_n, \ldots, t_N = t_{\text{final}}\}$. In Eq. (4.2), we assumed f to be a smooth function of t and $w(t)$. The exact solution of (4.2) from time step t_n to t_{n+1} in the integral form is expressed as

$$w(t_{n+1}) = w(t_n) + \int_{t_n}^{t_{n+1}} f(t, w(t))dt \tag{4.3}$$

For a numerical solution, we must approximate the integral in Eq. (4.3). In a single step scheme, the function $f(t, w(t))$ is approximated over the interval $[t_n, t_{n+1}]$. One possibility that simplifies the computation is the use of the Lagrange interpolation to construct an approximation of $f(t, w(t))$ using the function values at a set of points (nodes), $c_0, \ldots, c_r$, as described in Sect. 2.1 for the spatial approximation. Based on Eq. (2.6), the interpolant of the Lagrange polynomial of degree r is

$$y_r(t) = \sum_{i=0}^{r} f(t_n + c_i \Delta t)\ell_i(t) = f(t, w(t)) + \underbrace{\mathcal{O}(\tau^{r+1})}_{\text{Error}} \quad t \in [t_n, t_{n+1}] \tag{4.4}$$

where τ is a measure of spacing between the set of nodes $c_0, \ldots, c_r$. Note that the truncation error term is of power $r + 1$. Substituting the polynomial interpolation in the exact integral solution (4.3) then yields

$$w(t_{n+1}) = w(t_n) + \int_{t_n}^{t_{n+1}} f(t, w(t))dt = w(t_n) + \int_{t_n}^{t_{n+1}} y_r(t)dt + \underbrace{\mathcal{O}(\tau^{r+2})}_{\text{Error}} \tag{4.5}$$

Note that the error term increases in power by one from $\mathcal{O}(\tau^{r+1})$ to $\mathcal{O}(\tau^{r+2})$. Replacing for y_r from (4.4) yields

$$w(t_{n+1}) = w(t_n) + \int_{t_n}^{t_{n+1}} y_r(t)dt = \int_{t_n}^{t_{n+1}} \sum_{i=0}^{r} f(t + c_i \Delta t)\ell_i(t)dt + \underbrace{\mathcal{O}(\tau^{r+2})}_{\text{Error}} \tag{4.6}$$

4.1 Runge-Kutta Methods

which, upon integration, yields

$$w(t_{n+1}) = w(t_n) + \underbrace{\Delta t \sum_{i=0}^{r} b_i f(t_n + c_i \Delta t, w(t_n + c_i \Delta t))}_{\approx \int_{t_n}^{t_{n+1}} f(w)dt} + \underbrace{\mathcal{O}(\tau^{r+2})}_{\text{Error}} \quad (4.7)$$

where the weight b_i is given as

$$b_i = \int_0^1 \ell_i(t)dt \quad \forall i = 0, \ldots, r \quad (4.8)$$

Equation (4.7) can be used to approximate the solution at time level t_{n+1} provided an estimate of the derivative values or solution values at the nodes, $f(t_n + c_i \Delta t, w(t_n + c_i \Delta t))$ or $w(t_n + c_i \Delta t)$, can be obtained with sufficient accuracy. Given the Δt factor in front of the summation, the required accuracy level for approximation of the derivative values is $\mathcal{O}(\tau^{r+1})$, which retains the accuracy of the approximation of the integral.

One approach is to follow the same integral approach as (4.3), but over the sub-steps (or stages) $[t_{n,0}, t_{n,i}]$ where $t_{n,i} = t_n + c_i \Delta t$

$$w(t_{n,i}) = w(t_n) + \int_{t_n}^{t_{n,i}} f(t, w(t))dt \quad \forall i = 1, \ldots, r-1 \quad (4.9)$$

and use the same interpolant (4.4) to approximate the integral, yielding

$$w(t_{n,i}) = w(t_n) + \Delta t \sum_{j=0}^{r} a_{ij} f(t_{n,j}, w(t + c_j \Delta t)) + \underbrace{\mathcal{O}(\tau^{r+2})}_{\text{Error}}, \quad i = 1, \ldots, r-1 \quad (4.10)$$

where the weight $a_{i,j}$ is given as

$$a_{i,j} = \int_0^{c_i} \ell_j(t)dt \quad i = 1, \ldots, r-1, \; j = 0, \ldots, r \quad (4.11)$$

Note that in Eq. (4.11), the range of i is based on the assumption that the quadrature points include the end points. Counting Eqs. (4.7) and (4.10), we have r equations for r unknowns ($r-1$ stage values, $w_{n,1}, \ldots, w_{n,r-1}$, and w_{n+1}). We may turn Eqs. (4.7) and (4.10) into a numerical integration scheme by dropping the error terms to obtain

$$w_{n,i} = w_n + \Delta t \sum_{j=0}^{r} a_{ij} f(t_{n,j}, w_{n,j}), \quad i = 1, \ldots, r-1 \quad (4.12a)$$

$$w_{n+1} = w_n + \Delta t \sum_{i=0}^{r} b_i f(t_{n,i}, w_{n,i}) \quad (4.12b)$$

We call the scheme given in (4.12) a Runge-Kutta scheme. Since the computations of stage solutions and step (target) solution depend on all stage and target values, the scheme (4.12) is implicit.

Note that the step (target) solution $w(t_{n+1})$ and the stage values $w(t_n + c_i \Delta t)$ are all calculated at least with the order of accuracy of $r+1$, i.e.,

$$|w(t_{n+1}) - w_{n+1}| = \mathcal{O}(\Delta t^{r+2}) \quad |w(t_n + c_i \Delta t) - w_{n,i}| = \mathcal{O}(\Delta t^{r+2}) \quad (4.13)$$

In other words, the order of the scheme and the order of stages are both $r+1$. Note that the global order can be one power lower due to the accumulation of the local truncation error from a time step to another up to the N times to the final integration time of t_f:

$$\text{global error} = \mathcal{O}(N \times (\text{local error})) = \mathcal{O}\left(\frac{t_f}{\Delta t} \Delta t^{r+2}\right) = \mathcal{O}\left(\Delta t^{r+1}\right) \quad (4.14)$$

We can rewrite the scheme (4.12) in a more compact form as

$$w_{n,i} = w_n + \Delta t \sum_{j=0}^{r} d_{ij} f(t_{n,j}, w_{n,j}), \quad i = 1, \ldots, r \quad (4.15)$$

with

$$d_{i,j} = \int_0^{c_i} \ell_j(t) dt \quad i = 1, \ldots, r, \ j = 0, \ldots, r \quad (4.16)$$

However, the two-equation form (4.12) is customary and the nodes c_i and coefficients b_i and a_{ij} are often given in a tabular form as

$$\frac{c \mid \mathcal{A}}{\mid b^T} = \begin{array}{c|ccc} c_0 & a_{00} & \cdots & a_{1,r} \\ \vdots & \vdots & \ddots & \vdots \\ c_r & a_{r,0} & \cdots & a_{r,r} \\ \hline & b_0 & \cdots & b_r \end{array}$$

called Butcher array [10]. In the Butcher array representation, if we assume the interpolation nodes include the end points, $a_{0i} = 0$, and $a_{rj} = b_j$. If one of the endpoints appears in the approximation, the Runge-Kutta scheme is based on Radau quadrature points [11]. For Radau quadrature points, if $c_0 = 0$, $a_{0i} = 0$, and if $c_r = 1$,

4.1 Runge-Kutta Methods

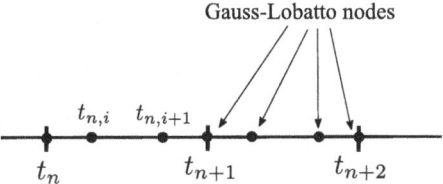

Fig. 4.2 Illustration of the four-node Gauss-Lobatto quadrature in the interval $[t_n, t_{n+1}]$ used in Runge-Kutta and spectral deferred correction time integration schemes

$a_{rj} = b_j$. Otherwise, when both endpoints are not included in the approximation like Gauss quadrature points, $a_{0i} \neq 0$ and $a_{rj} \neq b_j$. Equispaced nodes are not used for high-order schemes due to the issues discussed in Sect. 2.1 related to the Runge phenomenon. The Gauss-Lobatto quadrature nodes are clustered near the endpoints and include the endpoints [11]. They yield $2r + 1$ order scheme with a stage order of $r + 1$ [11]. The $r + 1$-Gauss quadrature yields a $(2r + 2)$th-order scheme with a stage order of $r + 1$ [11]. Figure 4.2 shows the Gauss-Lobatto points on a time interval $[t_n, t_{n+1}]$.

The implicit Runge-Kutta scheme expressed in (4.12) involves solving a (nonlinear) algebraic system. The nonlinearity can come from the derivative function $f(t, w(t))$. If the original ODE arises from the spatial discretization of PDEs using the method of lines (MOL), then the algebraic system can be very large. Given the number of spatial grid points N_x and the number of $PDEs$ N_{PDE}, the size of the system is $r \times N_{PDE} \times N_x$ which can be very large because the spatial grid size is often huge.

To reduce the size of the system, the stage approximation $w_{n,i}$ may use the first $i + 1$ nodes, still resulting in the same overall order provided the following condition

$$\Delta t \left| \frac{\partial f}{\partial w} \right| \ll 1 \quad (4.17)$$

is satisfied along with the order conditions which will be discussed in Sect. 4.1.3. When the condition (4.17) is violated as is the case in stiff problems, i.e.,

$$\Delta t \left| \frac{\partial f}{\partial w} \right| \geq 1 \quad (4.18)$$

the overall order of the scheme drops (the scheme suffers from order reduction as commonly referred to in the literature). We will further discuss the order reduction for systems arising from spatial discretizations of PDEs and stiff problems in Sect. 4.1.3.

The resultant scheme from the use of stage approximations with $i + 1$ nodes is called diagonally implicit Runge-Kutta (DIRK) scheme and has the form

$$w_{n,i} = w_n + \Delta t \sum_{j=0}^{i} a_{ij} f(t_{n,j}, w_{n,j}), \quad i = 1, \ldots, r-1 \tag{4.19a}$$

$$w_{n+1} = w_n + \Delta t \sum_{i=0}^{r} b_i f(t_{n,i}, w_{n,i}) \tag{4.19b}$$

with a Butcher array as

$$\frac{c \mid \mathcal{A}}{\mid b^T} = \begin{array}{c|cccccc} c_0 & a_{00} & & & & \\ c_1 & a_{10} & a_{11} & & & \\ c_2 & a_{20} & a_{21} & a_{22} & & \\ \vdots & \vdots & \vdots & \vdots & \ddots & \ddots \\ c_r & a_{r,0} & a_{r,1} & \cdots & a_{r,r-1} & a_{r,r} \\ \hline & b_1 & b_2 & \cdots & b_{r-1} & b_r \end{array}$$

In a DIRK scheme, the solutions at the nodes are computed sequentially by solving a system of size $N_{PDE} \times N_x$. Hence, the overall scheme involves solving r systems of size $N_{PDE} \times N_x$ each, which is easier and less expensive to solve than the system for the fully implicit scheme with the size $r \times N_{PDE} \times N_x$.

An explicit Runge-Kutta scheme involves strictly lower triangular matrix $\mathcal{A}$ and has the form

$$w_{n,i} = w_n + \Delta t \sum_{j=0}^{i-1} a_{ij} f(t_{n,j}, w_{n,j}), \quad i = 1, \ldots, r-1 \tag{4.20a}$$

$$w_{n+1} = w_n + \Delta t \sum_{i=0}^{r-1} b_i f(t_{n,i}, w_{n,i}) \tag{4.20b}$$

with the Butcher array as

$$\frac{c \mid \mathcal{A}}{\mid b^T} = \begin{array}{c|cccccc} c_0 & 0 & & & & \\ c_1 & a_{10} & 0 & & & \\ c_2 & a_{20} & a_{21} & 0 & & \\ \vdots & \vdots & \vdots & \vdots & \ddots & \ddots \\ c_r & a_{r,0} & a_{r,1} & \cdots & a_{r,r-1} & 0 \\ \hline & b_0 & b_1 & \cdots & b_{r-1} & b_r \end{array}$$

In DIRK and explicit RK schemes, the coefficients $\mathcal{A} = [a_{i,j}]$, b, and the quadrature points c are determined such that the resultant RK scheme yields the highest order of accuracy while possessing some desired stability properties. This is discussed in Sects. 4.1.3 and 4.1.4.

4.1.2 Alternative Form of RK Schemes

The RK scheme (4.12) is based on the solution's value at each stage or node. Alternatively, we may equivalently express the RK scheme based on the (scaled) derivative values at stages or nodes as follows [12, 13]:

$$F_{n,i} = \Delta t \, f(t_{n,j}, w_n + \sum_{j=0}^{r} a_{i,j} F_{n,j}), \quad i = 1, \ldots, r-1 \quad (4.21\text{a})$$

$$w_{n+1} = w_n + \sum_{i=0}^{r} b_i F_{n,i} \quad (4.21\text{b})$$

This form helps extract the stability functions of RK schemes, as the end-of-chapter exercises reinforce.

Consider applying the RK scheme (4.21) to the system of ODEs in the form

$$w'(t) = J(t)w + g(t) \quad (4.22)$$

where $J(t)$ is a $N_x \times N_x$ matrix, and w and g are N_x-dimensional vectors. This system can arise from the spatial discretization of PDEs using the method of lines. Following [12], the matrix form of the scheme (4.21) is expressed as

$$w_{n+1} = R(Z)w_n + \Delta t \, Q(Z)\tilde{g}_n \quad (4.23)$$

$$R(Z) = I + \tilde{b}^T \left(I - Z\tilde{A}\right)^{-1} Z(e \otimes I) \quad (4.24)$$

$$Q(z) = \tilde{b}^T \left(I - Z\tilde{A}\right)^{-1} \quad (4.25)$$

where

$$Z = \Delta t \tilde{J}_n, \quad (4.26)$$

and

$$\tilde{J}_n = J_{n,0} \oplus \cdots \oplus J_{n,r} \quad (4.27)$$

$$\tilde{g}_n = (g_{n,0}, \cdots, g_{n,r})^T \quad (4.28)$$

$$\tilde{A} = A \otimes I \quad \tilde{b}^T = b^T \otimes I \quad (4.29)$$

$$e = (1, \ldots, 1)^T \in \mathbb{R}^{r+1} \quad (4.30)$$

$$J_{n,i} = J(t_n + c_i \Delta t) \quad g_{n,i} = g(t_n + c_i \Delta t) \quad (4.31)$$

The operators $\oplus$ and $\otimes$ signify a direct sum and an outer (Kronecker) product, respectively. The identity matrix I in the outer products (4.27) is $N \times N$, while in other occurrences is of unambiguous appropriate size.

To clarify the above operations involving operators $\otimes$ and $\oplus$, we give examples of these operations for a system of ODEs consisting of two equations. For a two-dimensional vector $g(t) = [g^0(t), g^1(t)]^T$,

$$\tilde{g}_n = (g_{n,0}, \ldots, g_{n,r})^T = \left[(g^0_{n,0}, g^1_{n,0}), \ldots, (g^0_{n,r}, g^1_{n,r})\right]^T \tag{4.32}$$

and for a 2×2 matrix J_n in the form

$$J(t) = \begin{pmatrix} J^{0,0}(t) & J^{0,1}(t) \\ J^{1,0}(t) & J^{1,1}(t) \end{pmatrix}, \tag{4.33}$$

$$\tilde{J}_n = J_{n,0} \oplus \cdots \oplus J_{n,r} = \begin{pmatrix} J^{0,0}_{n,0} & J^{0,1}_{n,0} & & \\ J^{1,0}_{n,0} & J^{1,1}_{n,0} & & 0 \\ & & \ddots & \\ & 0 & & J^{0,0}_{n,r} & J^{0,1}_{n,r} \\ & & & J^{1,0}_{n,r} & J^{1,1}_{n,r} \end{pmatrix} \tag{4.34}$$

To clarify the operation $\otimes$, for a system of two ODEs,

$$\tilde{\boldsymbol{b}}^T = \boldsymbol{b}^T \otimes I = \begin{pmatrix} b_0 & 0 & b_1 & 0 & \cdots & b_r & 0 \\ 0 & b_0 & 0 & b_1 & \cdots & 0 & b_r \end{pmatrix} \tag{4.35}$$

and

$$\tilde{A} = A \otimes I = \begin{pmatrix} a_{0,0} & 0 & \cdots & a_{0,r} & 0 \\ 0 & a_{0,0} & \cdots & 0 & a_{0,r} \\ & & \vdots & & \\ a_{r,0} & 0 & \cdots & a_{r,r} & 0 \\ 0 & a_{r,0} & \cdots & 0 & a_{r,r} \end{pmatrix} \tag{4.36}$$

Finally, $e \otimes I$ is defined as

$$e \otimes I = \begin{pmatrix} 1 & 0 \\ 0 & 1 \\ \vdots & \vdots \\ 1 & 0 \\ 0 & 1 \end{pmatrix} \tag{4.37}$$

4.1.3 Error of Runge-Kutta Schemes

The stage and step exact solutions, $w(t_{n,i})$ and $w(t_{n+1})$ will not necessarily satisfy exactly the RK scheme (4.21) and result in some residuals. Specifically,

4.1 Runge-Kutta Methods

$$F_{n,i}^* = \Delta t \ f\left(t_{n,j}, w(t_n) + \sum_{j=0}^{r} a_{i,j} F_{n,j}^* + \delta_{n,i}\right), \quad i = 1, \ldots, r-1 \quad (4.38a)$$

$$w(t_{n+1}) = w(t_n) + \sum_{i=0}^{r} b_i F_{n,i}^* + \delta_{n+1}. \quad (4.38b)$$

where $\delta_{n,i}$ and δ_{n+1} are stage and step errors. If we now specialize (4.38) to the system of ODE (4.22), we obtain

$$w(t_{n+1}) = R(Z)w(t_n) + \Delta t \, Q(Z)\tilde{g}_n + \tilde{b}^T \left(I - Z\tilde{A}\right)^{-1} Z\tilde{\delta}_n + \delta_{n+1} \quad (4.39)$$

where

$$\tilde{\delta}_n = (\delta_{n,0}, \ldots, \delta_{n,r})^T = \left[(\delta_{n,0}^0, \delta_{n,0}^1), \ldots, (\delta_{n,r}^0, \delta_{n,r}^1)\right]^T \quad (4.40)$$

for the case of a system of two ODEs, for instance.

Denoting the error at time step n and $n+1$ by $\epsilon_{n+1} = w(t_{n+1}) - w_{n+1}$ and $\epsilon_n = w(t_n) - w_n$, upon subtraction of (4.39) from (4.23), we have

$$\epsilon_{n+1} = R(Z)\epsilon_n + \underbrace{\tilde{b}^T \left(I - Z\tilde{A}\right)^{-1} Z\tilde{\delta}_n} + \delta_{n+1} \quad (4.41)$$

$$\text{"accumulated" stage error}$$

If we consider the exact solution in the RK scheme (4.12)

$$w(t_{n,i}) = w(t_n) + \Delta t \sum_{j=0}^{r} a_{ij} f(t_{n,j}, w(t_{n,j})) + \delta_{n,i}, \quad i = 1, \ldots, r \quad (4.42a)$$

$$w(t_{n+1}) = w(t_n) + \Delta t \sum_{i=0}^{r} b_i f(t_{n,i}, w(t_{n,i})) + \delta_{n+1} \quad (4.42b)$$

we can derive an expression for the stage and step errors by expanding the Taylor series of $w(t_{n,i})$, and $w(t_{n+1})$, and the first derivative $w^{(1)}(t_{n,i}) = f(t_{n,i}, w(t_{n,i}))$. Substituting the Taylor expansions

$$w(t_{n,i}) = \sum_{k=0} \frac{\Delta t^k c_i^k}{k!} w^{(k)}(t_n) \quad (4.43)$$

and

$$f(t_{n,j}, w(t_{n,i})) = w^{(1)}(t_{n,i}) = \sum_{k=0} \frac{\Delta t^k c_i^k}{k!} w^{(k+1)}(t_n) \quad (4.44)$$

into (4.42) then yields

$$\delta_n = \sum_{k=1} \frac{\Delta t^k}{(k-1)!} \left[Ac^{k-1} - \frac{1}{k} c^k \right] w^{(k)}(t_n) \tag{4.45}$$

and

$$\delta_{n+1} = \sum_{k=1} \frac{\Delta t^k}{(k-1)!} \left[b^T c^{k-1} - \frac{1}{k} \right] w^{(k)}(t_n) \tag{4.46}$$

where $c^k = [c_0, \ldots, c_r]^T$, and $\delta_n = [\delta_{n,0}, \ldots, \delta_{n,r}]^T$. We have used $c_r = 1$ in (4.46).

Examining Eqs. (4.45) and (4.46) implies that in order to have step order of p, or $\delta_{n+1} = \mathcal{O}(\Delta t^{p+1})$, we need to satisfy the condition

$$b^T c^{k-1} - \frac{1}{k} = 0, \quad k = 1, \ldots, p \tag{4.47}$$

and for the stage order of q, or $\delta_{n,i} = \mathcal{O}(\Delta t^{q+1})$, we need to satisfy

$$\tau^k = Ac^{k-1} - \frac{1}{k} c^k = 0, \quad k = 1, \ldots, q \tag{4.48}$$

Given $\delta_{n+1} = \mathcal{O}(\Delta t^{p+1})$ and $\delta_{n,i} = \mathcal{O}(\Delta t^{q+1})$ by satisfying the order conditions (4.47) and (4.48), we now need to examine the first term in (4.41) and the second term, "accumulated" stage error, to determine the conditions under which the overall step error ϵ_{n+1} is of the order p. The stability of the RK scheme dictates that if $\epsilon_n = \mathcal{O}(\Delta t^{p+1})$, then the first term is at least of the order of $p+1$, i.e.,

$$R(Z)\epsilon_n = \mathcal{O}(\Delta t^{p+1}) \tag{4.49}$$

To understand the impact of the "accumulated" stage error on the overall error, we need to determine the role of term $\left(I - Z\tilde{A} \right)^{-1} Z \otimes I$. To this end, we consider two regimes of nonstiff and stiff cases:

$$|Z| \to 0 \quad \text{nonstiff} \tag{4.50}$$

and

$$|Z| > 1 \quad \text{stiff} \tag{4.51}$$

where the magnitude of a matrix signifies a matrix norm, for instance

$$|B| = \max_{x \neq 0} \frac{|Bx|}{|x|} \tag{4.52}$$

4.1 Runge-Kutta Methods

For the nonstiff case, the Neumann series yields

$$\left(I - Z\tilde{A}\right)^{-1} Z = \left(I + Z\tilde{A} + (Z\tilde{A})^2 + \cdots\right) Z \tag{4.53}$$

Using the identity

$$\tilde{A}Z = Z\tilde{A} \tag{4.54}$$

for the black diagonal matrix Z with identical block diagonal sub-matrices that arise for the constant-coefficient matrix J, we can simplify the Neumann expansion as

$$\left(I - Z\tilde{A}\right)^{-1} Z = \left(I + ZA + (ZA)^2 + \cdots\right) Z = \left(Z + Z^2\tilde{A} + Z^3\tilde{A}^2 + \cdots\right) \tag{4.55}$$

Now, substituting into the "accumulated" stage error term, we have

$$\tilde{b}^T \left(I - Z\tilde{A}\right)^{-1} Z\tilde{\delta}_n = Z\tilde{b}^T \tilde{\delta}_n + Z^2 \tilde{b}^T \tilde{A}\tilde{\delta}_n + \cdots \tag{4.56}$$

Now, satisfying the condition

$$b^T A^j \tau^k = 0 \quad k > 0, \; j + k < p \tag{4.57}$$

results in an "accumulated" stage error of $\mathcal{O}(\Delta t^{p+1})$, and hence, the overall error of $\mathcal{O}(\Delta t^{p+1})$. The definition of τ^k is given in Eq. (4.48).

For the stiff case, on the other hand, the Neumann series yields

$$\left(I - Z\tilde{A}\right)^{-1} Z = -\tilde{A}^{-1} \left(I - \tilde{A}^{-1} Z^{-1}\right)^{-1} \tag{4.58}$$

$$= -\left(\tilde{A}^{-1} + Z^{-1}\tilde{A}^{-1} + Z^{-2}\tilde{A}^{-2} + \cdots\right) \tag{4.59}$$

and the order condition (4.57) no longer results in $\mathcal{O}(\Delta t^{p+1})$ and a reduced order of at least the stage order q is expected for the overall scheme [14].

For the case where J is not a constant matrix in the linearization of the nonlinear system, for instance, the same analysis applies provided Z is formed by a J that is a maximum of J_i in an appropriate norm.

4.1.4 Error Due to Boundary Condition Imposition

Let us now discuss a case of stiffness involving the solution of PDEs using the method of lines and RK schemes. Consider the system of ODEs arising from the spatial discretization of a PDE, for instance, an advection equation with a positive advection speed. When the first-order upwind scheme is used for the spatial discretization and the left boundary value is given as $B_0(t_i) = B(t_i, x = 0)$, then

$$|Z| = \max_{w \neq 0} \frac{|Zw|}{|w|} = \max_{w \neq 0} \left(\frac{\left| \Delta t \frac{\partial w}{\partial x} - \Delta t \frac{\tilde{B}w}{\Delta x} \right|}{|w|} \right) \tag{4.60}$$

where $\tilde{B} = [B_{i,j}]$ is an $M \times M$ matrix with only nonzero $B_{0,0} = B_0(t_i)$.

When spatial derivative $\partial w/\partial x$ is bounded, the first term in the numerator goes to zero for $\Delta t \to 0$. However, the second term is of the order of one when $\Delta x = \mathcal{O}(\Delta t)$, which holds for advection-type problems. Hence,

$$|Z| = \mathcal{O}(B_0(t_i)) \tag{4.61}$$

Now if we consider the modified Neumann series as

$$\left(I - Z\tilde{A} \right)^{-1} Z = \left(I + Z\tilde{A} + (Z\tilde{A})^2 + \cdots - \tilde{B} \right) Z \tag{4.62}$$

where $\tilde{B}$ is a matrix formed by $B_0(t_i)$ and of appropriate size as mentioned above. If the Dirichlet boundary condition is time-independent, $B_0(t_i) = B_0$, then $|Z| \to B_0$, and the first equation in the order condition (4.57) yields an "accumulated" stage error of $\mathcal{O}(\Delta t^{p+1})$, and hence, the overall error of $\mathcal{O}(\Delta t^{p+1})$. However, if the boundary condition is time-dependent, the order condition no longer results in $\mathcal{O}(\Delta t^{p+1})$ and an order reduction is expected.

The order reduction of RK schemes in solving PDEs with time-dependent boundary conditions using the method of lines have been observed and studied [15, 16]. For explicit Runge-Kutta schemes, if the Runge-Kutta boundary solutions, instead of the exact values, are imposed on stage solutions, the original accuracy order are recovered [15]. Recently, other remedies have been proposed restoring the intended order of accuracy of the RK schemes in the presence of time-dependent Dirichlet boundary conditions [14, 17]. However, extensions of these methods to additive RK schemes where both stiff and nonstiff terms are present remain.

4.1.5 Example of Implicit RK Schemes

We now obtain instances of implicit Runge-Kutta methods by using the Radau nodes in the numerical quadrature. When a Radau quadrature of $r + 1$ nodes are used, an implicit RK scheme of order $2r + 1$ results that is both A- and L-stable. For $r = 2$ and $r = 3$, the Radau RK schemes have the following Butcher tables:

4.1 Runge-Kutta Methods

$$
\begin{array}{c|cc}
\frac{1}{3} & \frac{5}{12} & -\frac{1}{12} \\
1 & \frac{3}{4} & \frac{1}{4} \\
\hline
 & \frac{3}{4} & \frac{1}{4}
\end{array}
\qquad
\begin{array}{c|ccc}
0 & \frac{1}{9} & \frac{-1-\sqrt{6}}{18} & \frac{-1+\sqrt{6}}{18} \\
\frac{6-\sqrt{6}}{10} & \frac{1}{9} & \frac{88+7\sqrt{6}}{360} & \frac{88-43\sqrt{6}}{360} \\
\frac{6+\sqrt{6}}{10} & \frac{1}{9} & \frac{88+43\sqrt{6}}{360} & \frac{88-7\sqrt{6}}{360} \\
\hline
 & \frac{1}{9} & \frac{16+\sqrt{6}}{36} & \frac{16-\sqrt{6}}{36}
\end{array}
\qquad (4.63)
$$

4.1.6 Diagonally Implicit Runge-Kutta Schemes

In a DIRK scheme, each stage solution involves the solution of a system. For linear problems, since the coefficients of the system remain identical for all stages, efficient strategies such as LU decomposition or constructing a preconditioner can be reused for each state throughout the time step. For nonlinear problems, one approach to solve the implicit system at each stage is to use a Newton-type scheme. The residual equation for an intermediate stage can be written as

$$
\mathcal{R}_{n,i} = w_{n,i} - w_n + \Delta t \sum_{j=0}^{i} a_{i,j} f(t_{n,i}, w_{n,j}) \qquad (4.64)
$$

The Jacobian for the given internal state approximation is then

$$
J_{n,i} = \frac{\partial \mathcal{R}_{n,i}}{\partial w_{n,i}} = I - \Delta t a_{ii} \frac{\partial f(t_{n,i}, w_{n,i})}{\partial w_{n,i}} \qquad (4.65)
$$

where I is the identity matrix. If the second term in the Jacobian $\frac{\partial f(t_{n,i}, w_{n,i})}{\partial w_{n,i}}$ varies slowly for Δt, then it can be approximated as constant over one or more time steps [18]. Similar to the linear case, if all a_{ii}s are identical, a constant Jacobian for all stages of the RK on one or more times steps results, significantly enhancing the efficiency of the implicit RK scheme. Methods with identical diagonal coefficients are called singly-diagonally implicit RK (SDIRK) schemes [19].

Design of DIRK Schemes

The design of an SDIRK scheme amounts to determining the coefficient matrix $\mathcal{A}$, the coefficient vector $\boldsymbol{b}^T$, and the abscissa vector $\boldsymbol{c}$ under a multitude of constraints. The constraints consist of the satisfaction of the desired order of accuracy p and optimum accuracy in some norm, the satisfaction of stability properties including A-stability, L-stability, and nonlinear (algebraic) stability (relevant for nonlinear problems), and finally the satisfaction of the efficiency requirements. The number of stages r can also be varied to satisfy these requirements. Boom and Zingg provided numerical optimization strategies and derived SDIRK schemes of up to the fifth order of accuracy [18]. Kennedy and Carpenter introduced SDIRK schemes of up to order six that satisfy the A- and L-stability and have small eigenvalues associated with the algebraic stability constraint [6]. As is clear from the recent findings [6, 18], the optimized SDIRK schemes of high-order accuracy $p > 4$ require the number of stages to be larger than the accuracy order, i.e., $r > p$. We refer the reader to the manuscript by Kennedy and Carpenter for more details about the latest SDIRK scheme [6]. Recently, DIRK schemes of orders six to eight have been constructed [20]. Determining all coefficients of the eight-order scheme that is A-stable requires devising highly accurate numerical solutions for a system of 200 equations in over 100 variables [20], indicating that constructing increasingly high-order DIRK schemes involve challenging optimization problems consisting of huge numbers of equations and variables.

4.1.7 Examples of Diagonally Implicit Runge-Kutta Schemes

Reference [21] provides two examples of diagonally implicit methods with a parameter $\gamma > 0$, in the forms

$$
\begin{array}{c|cc}
\gamma & \gamma & \\
1-\gamma & 1-2\gamma & \gamma \\
\hline
 & \frac{1}{2} & \frac{1}{2}
\end{array}
\qquad
\begin{array}{c|ccc}
0 & 0 & & \\
2\gamma & \gamma & \gamma & \\
1 & b_1 & b_2 & \gamma \\
\hline
 & b_1 & b_2 & \gamma
\end{array}
\tag{4.66}
$$

with $b_1 = 3/2 - \gamma - 1/(4\gamma)$, and $b_2 = -1/2 + 1/(4\gamma)$. For $\gamma = 1/2 \pm \sqrt{3}/6$, both methods are third-order accurate, whereas for other γ-values, the methods are second-order accurate.

Both DIRK schemes given in (4.66) share the stability function

$$R(z) = \frac{1 + (1 - 2\gamma)z + (1/2 - 2\gamma + \gamma^2)}{(1 - \gamma z)^2} \tag{4.67}$$

and both methods are A-stable if and only if $\gamma \geq 1/4$; hence, for the two γ-values leading to the third-order accuracy, only $\gamma = 1/2 + \sqrt{3}/6$ yields A-stability. L-stability requires $\gamma = 1 \pm \sqrt{2}/2$ meaning that the L-stable scheme is only second-order accurate. We leave the verification of the stability of these two schemes to an end-of-chapter exercise.

4.1.8 Examples of Explicit Runge-Kutta Schemes

The first-order explicit RK scheme is the explicit Euler scheme. Two examples of second-order explicit RK schemes in the Butcher-array form are

4.1 Runge-Kutta Methods

$$\begin{array}{c|cc} 0 & & \\ 1 & 1 & \\ \hline & \frac{1}{2} & \frac{1}{2} \end{array} \qquad \begin{array}{c|cc} 0 & & \\ \frac{1}{2} & \frac{1}{2} & \\ \hline & 0 & 1 \end{array} \qquad (4.68)$$

The scheme on the left is the explicit trapezoidal rule or the θ-method with $\theta = 1/2$. The method on the right is called the mid-point rule. Typical examples of the third- and fourth-order schemes are

$$\begin{array}{c|ccc} 0 & & & \\ \frac{1}{3} & \frac{1}{3} & & \\ \frac{2}{3} & 0 & \frac{2}{3} & \\ \hline & \frac{1}{4} & 0 & \frac{3}{4} \end{array} \qquad \begin{array}{c|cccc} 0 & & & & \\ \frac{1}{2} & \frac{1}{2} & & & \\ \frac{1}{2} & 0 & \frac{1}{2} & & \\ 1 & 0 & 0 & 1 & \\ \hline & \frac{1}{6} & \frac{1}{3} & \frac{1}{3} & \frac{1}{6} \end{array} \qquad (4.69)$$

Another example of explicit RK schemes which is effective in the time integration of systems arising from spatial discretization of conservation laws using shock-capturing schemes is the third-order total variational diminishing one [22]. It is given in the Butcher array as

$$\begin{array}{c|ccc} 0 & & & \\ 1 & 1 & & \\ \frac{1}{2} & \frac{1}{4} & \frac{1}{4} & \\ \hline & \frac{1}{6} & \frac{1}{6} & \frac{2}{3} \end{array} \qquad (4.70)$$

Similar to the explicit Euler scheme, the explicit Runge-Kutta method, despite its simplicity, is only conditionally stable and lacks the stability properties required for stiff problems.

4.1.9 Removal of the Boundary Condition Error in the Third-Order Runge-Kutta Scheme

As discussed in Sect. 4.1.4, when Runge-Kutta methods are used in solving a system of ODEs arsing from the spatial discretization of PDEs with time-dependent Dirichlet boundary conditions, $u_b = g(t)$, a reduced rate of convergence occurs if the time-dependent boundary conditions are imposed exactly for stage solution updates. To avoid this order reduction, Carpenter et al. showed that if the Runge-Kutta solution of the boundary data are imposed at the stage boundary values, the original orders are recovered [15]. For the third-order RK schemes, this approach is effective for both linear and nonlinear systems. For instance for the third-order TVD RK scheme (4.70), the boundary condition at each stage are imposed based on

$$u_b^n \sim g(t_n) \tag{4.71a}$$

$$u_b^{n,1} \sim g(t_n) + \Delta t g'(t) \tag{4.71b}$$

$$u_b^{n,1} \sim g(t_n) + \frac{1}{2}\Delta t g'(t) + \frac{1}{4}\Delta t^2 g''(t) \tag{4.71c}$$

Similar approach can be used for the fourth-order RK scheme. However, it only restore the fourth-order accuracy for linear PDEs. An end-of-chapter exercise addresses the stage boundary values for the classical fourth-order RK scheme.

To demonstrate the reduced order of accuracy of the RK scheme with the exact boundary conditions and the restoration of the order of accuracy with the RK stage boundary condition, we solved the simple advection equation with speed $a = 1$ for a final time of $t = 1.0$ in a spatial domain of $[0, 1]$. The initial condition is chosen as

$$u(t = 0, x) = \sin(2\pi x) \tag{4.72}$$

with the exact solution of

$$u(t, x) = \sin(2\pi(x - t)) \tag{4.73}$$

On the left boundary, the Dirichlet boundary condition is imposed based on the exact solution. The spatial dicretization is based on a fourth-order spatial derivative scheme. For temporal scheme, the third-order RK scheme with the time step size of $\Delta t = 0.5 \Delta x^{4/3}$ is used. The results are shown in Table 4.2, clearly showing that the exact imposition of the stage boundary conditions leads to a approximately first-order rate of convergence. The imposition of the stage boundary condition using the RK scheme as given in (4.71) yields the theoretical fourth-order rate of convergence.

Table 4.2 Errors and convergence rate for solving simple advection equation with speed one in $[-1, 1]$ for final time of $t = 1.0$ using a fourth-order spatial discretization and the third-order TVD Runge-Kutta scheme (4.70). $\Delta t = 0.5\Delta x^{4/3}$

	Exact stage BC				RK stage BC			
Δx	L_∞ error	Rate	L_1	Rate	L_∞ error	Rate	L_1	Rate
0.1	$3.4e-2$		$1.7e-2$		$3.3e-2$		$1.7e-2$	
0.05	$1.4e-2$	1.3	$8.1e-2$	1.1	$1.7e-3$	4.3	$9.6e-4$	4.1
0.025	$5.9e-3$	1.3	$3.5e-3$	1.2	$1.0e-4$	4.1	$6.0e-5$	4.0
0.0125	$2.5e-3$	1.3	$1.4e-3$	1.3	$6.2e-6$	4.0	$3.8e-6$	4.0
0.00625	$1.0e-3$	1.3	$5.7e-4$	1.3	$3.9e-7$	4.0	$2.4e-7$	4.0

4.1.10 Implicit-Explicit Runge-Kutta Schemes

As discussed in Chap. 1, compressible transport phenomena may involve shock waves. Efficient solutions of problems with shock waves using implicit schemes are challenging as the Jacobian in the Newton scheme is highly ill-conditioned. Therefore, using an explicit time integration for nonstiff terms and an implicit scheme for the stiff terms makes sense, giving rise to implicit-explicit, additive Runge-Kutta schemes [23]. Let us consider a model problem with a nonstiff and a stiff part as

$$w' = f^{[ns]}(w) + f^{[s]}(w) \tag{4.74}$$

where $f^{[ns]}(w)$ and $f^{[s]}(w)$ represent the nonstiff and stiff terms. The additive RK scheme is formulated as

$$w_{n,i} = w_n + \Delta t \sum_{j=0}^{r} a_{ij}^{[E]} f^{[ns]}(w^{n,j}) + \Delta t \sum_{i=1}^{s} a_{ij}^{[I]} f^{[s]}(w_{n,j}), \quad i = 1, \ldots, s, \tag{4.75}$$

$$w_{n+1} = w_n + \Delta t \sum_{i=0}^{r} b_i^{[E]} f^{[ns]}(w_{n,i}) + b_i^{[I]} f^{[s]}(w_{n,i}) \tag{4.76}$$

The coefficient matrices $A^{[E]} = [a_{ij}^{[E]}]$ and $A^{[I]} = [a_{ij}^{[I]}]$, and the coefficient vectors $b^{[E]}$ and $b^{[I]}$ are determined based on the satisfaction of the L-stability for the implicit integrator and the maximum domain stability for the explicit integrator. Kennedy and Carpenter present a third-order scheme with four stages that are L-stable and stiffly accurate [5]. They provide a fourth-order with seven stages and a fifth-order scheme with eight stages with desirable stability, stiff accuracy, and efficiency requirement [23]. It is essential to point out that the explicit and implicit coefficients must be determined in tandem to ensure the stability of the overall additive scheme, and separate stability of the explicit and the implicit schemes is not sufficient [21].

The third-order scheme of Kennedy and Carpenter has $b_i^{[E]} = b_i^{[I]}$ and $c_i^{[E]} = c_i^{[I]}$ and the following coefficients for the explicit part

$$\frac{c \mid A^{[E]}}{\mid b^T} = \begin{array}{c|cccc} 0 & 0 & 0 & 0 & 0 \\ \frac{1767732205903}{2027836641118} & \frac{1767732205903}{2027836641118} & 0 & 0 & 0 \\ \frac{3}{5} & \frac{5535828885825}{10492691773637} & \frac{788022342437}{10882634858940} & 0 & 0 \\ 1 & \frac{6485989280629}{16251701735622} & \frac{-4246266847089}{9704473918619} & \frac{10755448449292}{10357097424841} & 0 \\ \hline b_i & \frac{1471266399579}{7840856788654} & \frac{-4482444167858}{7529755066697} & \frac{11266239266428}{11593286722821} & \frac{1767732205903}{4055673282236} \end{array} \tag{4.77}$$

and for the implicit part

$$
\begin{array}{c|ccccc}
\dfrac{c\ A^{[I]}}{b^T} & 0 & 0 & 0 & 0 & 0 \\
& \dfrac{1767732205903}{2027836641118} & \dfrac{1767732205903}{4055673282236} & \dfrac{1767732205903}{4055673282236} & 0 & 0 \\
& \dfrac{3}{5} & \dfrac{2746238789719}{10658868560708} & \dfrac{-640167445237}{6845629431997} & \dfrac{1767732205903}{4055673282236} & 0 \\
& 1 & \dfrac{1471266399579}{7840856788654} & \dfrac{-4482444167858}{7529755066697} & \dfrac{11266239266428}{11593286722821} & \dfrac{1767732205903}{4055673282236} \\
\hline
b_i & & \dfrac{1471266399579}{7840856788654} & \dfrac{-4482444167858}{7529755066697} & \dfrac{11266239266428}{11593286722821} & \dfrac{1767732205903}{4055673282236}
\end{array}
$$
(4.78)

Appendix A in Ref. [23] lists the coefficients of the fourth- and fifth-order implicit-explicit Runge-Kutta methods.

Additive RK schemes exhibit a reduced order of accuracy for stiff problems [23].

4.2 Spectral Deferred Correction Scheme

Due to challenges involving the construction of very high-order additive Runge-Kutta methods namely large optimization problems, the schemes of orders higher than five that have stability properties suitable for problems with stiff and nonstiff characteristics have yet to be introduced [24, 25]. Another difficulty is that the increasingly higher order RK schemes require the number of stages to be higher than the order of accuracy. Furthermore, constructing RK schemes that allow for different time step sizes (more than two) to integrate processes with widely differing time scales, such as those in compressible fluid dynamics with reaction and phase change, is challenging [8, 26]. These challenges in devising RK schemes motivate the search for an alternative time integrator that allow arbitrary levels of accuracy while are effective in approximating problems with multiple stiff and nonstiff parts. To this end, spectral deferred correction (SDC) schemes are attractive options [7].

4.2.1 Defect Corrections

The deferred correction scheme is just an adoption of the general defect correction approach designed to simplify the solution of difficult problems [27], which in the case here is the numerical integration of ODEs. Let us denote the original, difficult-to-solve problem by the operator $\mathcal{L}^2(w)$ as

$$\mathcal{L}^2(w) = 0 \quad \text{(original problem)} \tag{4.79}$$

Instead of directly solving (4.79), a simplified version of the problem is considered denoted by the operator $\mathcal{L}^1(w)$ as

$$\mathcal{L}^1(w) = 0 \quad \text{(simplified problem)} \tag{4.80}$$

4.2 Spectral Deferred Correction Scheme

Since the solution to the simplified problem is expected to have a low accuracy, it must be modified to yield an improved accuracy. Given an approximation to the solution of the original problem $\tilde{w}$, the difference between the original operator and the simplified version, namely the defect,

$$\text{defect} \equiv \mathcal{L}^1(\tilde{w}) - \mathcal{L}^2(\tilde{w}) \tag{4.81}$$

can be used to modify the simplified problem

$$\mathcal{L}^1(w) - [\text{defect}] = 0 \tag{4.82}$$

or

$$\mathcal{L}^1(w) - \left[\mathcal{L}^1(\tilde{w}) - \mathcal{L}^2(\tilde{w})\right] = 0 \quad \text{(correction problem)} \tag{4.83}$$

Solving the correction (or corrected) problem yields an improved solution to the original problem, w. Naturally, the process can be repeated for a higher level of accuracy. An iteration of the repeated application of the defect correction process is symbolically expressed as

$$\mathcal{L}^1(w^{(k)}) - \left[\mathcal{L}^1(w^{(k-1)}) - \mathcal{L}^2(w^{(k-1)})\right] = 0 \quad \text{(correction iteration)} \tag{4.84}$$

where $w^{(k)}$ is the improved solution at the kth iteration starting from the initial provisional solution $w^{(0)}$.

4.2.2 From Defect Correction to Deferred Correction

We now specialize the defect correction iteration (4.84) to solving the system (4.12), arising from the implicit RK scheme, which is rewritten in a single equation form in (4.15). Specifically, the $\mathcal{L}^2(w)$ operator is written as

$$\mathcal{L}^2(w_{n,1}, \ldots, w_{n,r}) = \begin{pmatrix} w_{n,1} - w_{n,0} - c_1 \Delta t \sum_{j=0}^{r} d_{1j} f(t_{n,j}, w_{n,j}) \\ \vdots \\ w_{n,r} - w_{n,0} - c_r \Delta t \sum_{j=0}^{r} d_{rj} f(t_{n,j}, w_{n,j}) \end{pmatrix} \tag{4.85}$$

where $w_{n,0} = w_n$ and $w_{n,r} = w_{n+1}$. For a simplified operator, the simplest choice is the first-order forward Euler approximation of the integral in (4.3) as

$$\mathcal{L}^1(w_{n,1}, \ldots, w_{n,r}) = \begin{pmatrix} w_{n,1} - w_{n,0} - c_1 \Delta t f(t_n, w_{n,0}) \\ \vdots \\ w_{n,r} - w_{n,0} - c_r \Delta t f(t_n, w_{n,0}) \end{pmatrix} \qquad (4.86)$$

The use of Eq. (4.84) yields one iteration of the correction equation as

$$\begin{pmatrix} w_{n,1}^{(k)} - w_{n,0}^{(k)} - c_1 \Delta t f(t_n, w_{n,0}^{(k)}) \\ \vdots \\ w_{n,r}^{(k)} - w_{n,0}^{(k)} - c_r \Delta t f(t_n, w_{n,0}^{(k)}) \end{pmatrix} - \left[\begin{pmatrix} w_{n,1}^{(k-1)} - w_{n,0}^{(k-1)} - c_1 \Delta t f(t_n, w_{n,0}^{(k-1)}) \\ \vdots \\ w_{n,r}^{(k-1)} - w_{n,0}^{(k-1)} - c_r \Delta t f(t_n, w_{n,0}^{(k-1)}) \end{pmatrix} \right.$$
$$\left. - \begin{pmatrix} w_{n,1}^{(k-1)} - w_{n,0}^{(k-1)} - c_1 \Delta t \sum_{j=0}^{r} d_{1j} f(t_{n,j}, w_{n,j}^{(k-1)}) \\ \vdots \\ w_{n,r}^{(k-1)} - w_{n,0}^{(k-1)} - c_r \Delta t \sum_{j=0}^{r} d_{rj} f(t_{n,j}, w_{n,j}^{(k-1)}) \end{pmatrix} \right] = 0 \qquad (4.87)$$

After simplification, one iteration of the correction equation becomes

$$\begin{pmatrix} w_{n,1}^{(k)} \\ \vdots \\ w_{n,r}^{(k)} \end{pmatrix} = \begin{pmatrix} w_{n,0}^{(k-1)} + c_1 \Delta t \sum_{j=0}^{r} d_{1j} f(t_{n,j}, w_{n,j}^{(k-1)}) \\ \vdots \\ w_{n,0}^{(k-1)} + c_r \Delta t \sum_{j=0}^{r} d_{rj} f(t_{n,j}, w_{n,j}^{(k-1)}) \end{pmatrix} \qquad (4.88)$$

where $w_{n,0}^{(k)} = w_n$ and $w_{n,r}^{(k)} = w_{n+1}^{(k)}$. The first iteration of (4.88) requires the values of $w_{n,j}^{(0)}$ which is often approximated with a first-order accuracy by taking $w_{n,j}^{(0)} = w_n$. Since the quadrature rule is of the order of $r + 1$, the scheme (4.88) adds one order of accuracy to the approximate solution after every iteration thanks to the factor Δt, and after r iterations, the overall order of accuracy is $r + 1$. Further iteration does not, however, add to the accuracy as the quadrature rule is only $(r + 1)$th order accurate. Therefore, to achieve an accuracy order $r + 1$, which is the accuracy of the integral evaluation, r correction iterations are required.

Unlike the original problem involving solving a system (4.12), the scheme (4.88) involves r vector updates in each of its iterations. Hence, it is an explicit scheme.

The scheme (4.88) is called a deferred correction scheme, and if particular nodes such as Gauss or Gauss-Lobatto nodes are used in the integration, it is called a spectral deferred correction scheme. It is the simplest explicit spectral deferred correction scheme.

4.2 Spectral Deferred Correction Scheme

Alternatively, the integral in Eq. (4.3) is approximated using the first-order backward Euler scheme

$$\mathcal{L}^1(w_{n,1}, \ldots, w_{n,r}) = \begin{pmatrix} w_{n,1} - w_{n,0} - c_1 \Delta t f(t_n, w_{n,1}) \\ \vdots \\ w_{n,r} - w_{n,0} - c_r \Delta t f(t_n, w_{n,r}) \end{pmatrix} \quad (4.89)$$

Now substituting the $\mathcal{L}^1$ and $\mathcal{L}^2$ operators, Eqs. (4.89) and (4.85), into the correction equation (4.84) and simplifying, we obtain

$$\begin{pmatrix} w_{n,1}^{(k)} \\ \vdots \\ w_{n,r}^{(k)} \end{pmatrix} = \begin{pmatrix} w_{n,0}^{(k-1)} + c_1 \Delta t \left[f(t_{n,1}, w_{n,1}^{(k)}) - f(t_{n,1}, w_{n,1}^{(k-1)}) \right] + c_1 \Delta t \sum_{j=0}^{r} d_{1j} f(t_{n,j}, w_{n,j}^{(k-1)}) \\ \vdots \\ w_{n,0}^{(k-1)} + c_r \Delta t \left[f(t_{n,r}, w_{n,r}^{(k)}) - f(t_{n,r}, w_{n,r}^{(k-1)}) \right] + c_r \Delta t \sum_{j=0}^{r} d_{rj} f(t_{n,j}, w_{n,j}^{(k-1)}) \end{pmatrix}$$

(4.90)

The scheme (4.90) is an implicit scheme, where every stage solution requires solving one nonlinear system of size $N_{\text{PDE}} \times N_x$. The complexity of $N_{\text{PDE}} \times N_x$ arises because an application of the method of lines in the discretization of the PDEs governing the compressible flows yields a system of ODEs of dimensions $N_{\text{PDE}} \times N_x$, where N_x is the number of spatial grid points. Similar to the explicit SDC scheme, every iteration adds one order of accuracy to the approximate solution, starting from the first-order accuracy of $w_{n,j}^{(0)} = w_n$.

When $f(t, w(t))$ consists of stiff and nonstiff parts that are separable as

$$f(t, w(t)) = \underbrace{f^s(t, w(t))}_{\text{stiff}} + \underbrace{f^{ns}(t, w(t))}_{\text{nonstiff}} \quad (4.91)$$

the $\mathcal{L}^1$ operator of the stiff part is an implicit scheme, such as the backward Euler, and that of the nonstiff part is an explicit scheme, such as the forward Euler. The results of the backward and forward Euler schemes for the stiff and nonstiff terms yield the following SDC scheme:

$$\begin{pmatrix} w_{n,1}^{(k)} \\ \vdots \\ w_{n,r}^{(k)} \end{pmatrix} = \begin{pmatrix} w_{n,0}^{(k-1)} + c_1 \Delta t \left[f^s(t_{n,1}, w_{n,1}^{(k)}) - f^s(t_{n,1}, w_{n,1}^{(k-1)}) \right] + c_1 \Delta t \sum_{j=0}^{r} d_{1j} f(t_{n,j}, w_{n,j}^{(k-1)}) \\ \vdots \\ w_{n,0}^{(k-1)} + c_r \Delta t \left[f^s(t_{n,r}, w_{n,r}^{(k)}) - f^s(t_{n,r}, w_{n,r}^{(k-1)}) \right] + c_r \Delta t \sum_{j=0}^{r} d_{rj} f(t_{n,j}, w_{n,j}^{(k-1)}) \end{pmatrix}$$

(4.92)

For the three-point Gauss-Lobatto quadrature, the coefficients $d_i = [d_{i0}, d_{i1}, d_{i2}]$ (for $i = 1$, and 2) are

$$d_1 = \{\tfrac{5}{24}, \tfrac{8}{24}, -\tfrac{1}{24}\} \quad (4.93a)$$
$$d_2 = \{\tfrac{1}{6}, \tfrac{2}{3}, \tfrac{1}{6}\} \quad (4.93b)$$

Finally, the following provisional solution values are used to start and perform the iterations:

$$w_{n,0}^{(k)} = w_n \tag{4.94}$$

$$w_{n,i}^{(0)} = w_n \tag{4.95}$$

Remark 4.6 The version of the SDC scheme presented here is based on [28, 29] and draws similarity with the RK scheme in that the stage integrals are carried out over $[t_n, t_{n,i}]$. The SDC scheme (4.88) can be expressed as an RK scheme; Ref. [29] provides examples for an explicit SDC scheme. An alternative formulation of the SDC scheme is possible where the integrals are performed over $[t_{n,i-1}, t_{n,i}]$ instead of over $[t_n, t_{n,i}]$ [7–9]. For instance, in the alternative formulation, an explicit SDC scheme is obtained by choosing the following $\mathcal{L}^2$ and $\mathcal{L}^1$ operators:

$$\mathcal{L}^2(w_{n,1}, \ldots, w_{n,r}) = \begin{pmatrix} w_{n,1} - w_{n,0} - c_1 \Delta t \sum_{j=0}^{r} d'_{1j} f(t_{n,j}, w_{n,j}) \\ \vdots \\ w_{n,r} - w_{n,r-1} - (c_r - c_{r-1}) \Delta t \sum_{j=0}^{r} d'_{rj} f(t_{n,j}, w_{n,j}) \end{pmatrix} \tag{4.96}$$

and

$$\mathcal{L}^1(w_{n,1}, \ldots, w_{n,r}) = \begin{pmatrix} w_{n,1} - w_{n,0} - c_1 \Delta t f(t_n, w_{n,0}) \\ \vdots \\ w_{n,r} - w_{n,r-1} - (c_r - c_{r-1}) \Delta t f(t_n, w_{n,r-1}) \end{pmatrix} \tag{4.97}$$

Using the defect iteration $\mathcal{L}^1(w^{(k)}) - [\mathcal{L}^1(w^{(k-1)}) - \mathcal{L}^2(w^{(k-1)})] = 0$, we obtain the following expression for the solution at stage m (for $m = 1, \ldots, m = r$):

$$w_{n,m}^{(k)} = w_{n,m-1}^{(k)} + (c_m - c_{m-1}) \Delta t \left[f(t_{n,m}, w_{n,m-1}^{(k)}) - f(t_{n,r}, w_{n,m-1}^{(k-1)}) \right] + (c_m - c_{m-1}) \Delta t \sum_{j=0}^{r} d'_{mj} f(t_{n,j}, w_{n,j}^{(k-1)}) \tag{4.98}$$

Note that Scheme (4.98) has the same order of accuracy as Scheme (4.88). However, the scheme (4.88) allows the parallel computations of the stage values because each stage value computation is independent of others [29, 30], while the scheme (4.98) compute the stage values in sequence, and it is not amenable to the parallel computations. Moreover, in the scheme (4.88), in the last correction iteration ($k = k_f$), only the correction equation for the last stage value, $w_{n,r}^{(k_f)}$ is needed, and other stage values, $w_{n,j}^{(k_f)}$ ($j < r$), are not required, resulting in a reduced number of function evaluations, and, in turn, the reduced cost.

4.2.3 Arbitrarily High-Order SDC Method

As discussed above, an SDC method consists of two main parts: the quadrature and the correction iteration. The quadrature is the numerical approximation using a set of quadrature points over an interval spanned by each consecutive quadrature point. The number and the type of quadrature points determine the order of accuracy of the integration. The correction iteration often involves the first-order Euler scheme due to its simplicity, and a sufficient number of iterations ensures that the overall accuracy coincides with the accuracy of the quadrature. We now describe how to carry out the quadrature with an arbitrarily high order of accuracy using the Gauss-Lobatto quadrature points.

Gauss-Lobatto Quadrature

Assume the total of $r+1$ Gauss-Lobatto quadrature points x_i defined over $[-1, 1]$, and x_0 and x_r are the left and right endpoints. (The nodes c_i are determined using $c_i = (x_i + 1)/2$.) Then, the quadrature weights $W_{m+1,i}$ ($\forall\, m \in [0, r]$)

$$W_{m+1,i} = \frac{1}{(x_i^2 - 1) P_r''(x_i)} \int_{x_0}^{x_{m+1}} \frac{(x^2 - 1) P_r'(x)}{x - x_i} dx \quad 0 < i < M \quad (4.99\text{a})$$

$$W_{m+1,0} = \frac{1}{2 P_r'(-1)} \int_{x_0}^{x_{m+1}} (1 - x) P_r'(x) dx \quad i = 0 \quad (4.99\text{b})$$

$$W_{m+1,r} = \frac{1}{2 P_r'(1)} \int_{x_0}^{x_{m+1}} (x + 1) P_r'(x) dx \quad i = r \quad (4.99\text{c})$$

Here, $P_r'(x)$ and $P_r''(x)$ are the first and second derivatives of the Legendre polynomial $P_r(x)$ that is of degree r and defined over $[-1, 1]$ [11]. Expressions given in Eq. (4.99) are based on derivations in Hildebrand's numerical analysis book [31]. The internal Gauss-Lobatto quadrature points are zeros of $P_r'(x)$. Hence, to perform the integrals above, we must know the Legendre polynomial and the Gauss-Lobatto quadrature points. Finally, $d_{m+1,i}$ in Eq (4.92) are

$$d_{m,i} = \frac{W_{m+1,i}}{2} \quad (4.100)$$

For $r = 2$, $x_0 = -1$, $x_1 = 0$ and $x_2 = 1$ and the quadrature weights d_1 and d_2 are given in (4.93). For $r = 3$, the quadrature nodes are $x_0 = -1$, $x_1 = -\frac{1}{\sqrt{5}}$, $x_2 = \frac{1}{\sqrt{5}}$, and $x_3 = 1$, and the quadrature weights d_1, d_2, and d_3 calculated based on (4.99) are

$$d_{1,0} = \frac{1}{4800}\sqrt{5}(3\sqrt{5} + 40) + \frac{17}{192} \tag{4.101a}$$

$$d_{1,1} = \frac{1}{120}\sqrt{5}(5\sqrt{5} - 1) \tag{4.101b}$$

$$d_{1,2} = \frac{1}{120}\sqrt{5}(5\sqrt{5} - 13) \tag{4.101c}$$

$$d_{1,3} = -\frac{1}{4800}\sqrt{5}(3\sqrt{5} - 40) - \frac{1}{192} \tag{4.101d}$$

$$d_{2,0} = \frac{1}{4800}\sqrt{5}(3\sqrt{5} - 40) + \frac{17}{192} \tag{4.102a}$$

$$d_{2,1} = \frac{1}{120}\sqrt{5}(5\sqrt{5} - 1) + \frac{7}{60}\sqrt{5} \tag{4.102b}$$

$$d_{2,2} = \frac{1}{120}\sqrt{5}(5\sqrt{5} - 13) + \frac{7}{60}\sqrt{5} \tag{4.102c}$$

$$d_{2,3} = -\frac{1}{4800}\sqrt{5}(3\sqrt{5} + 40) - \frac{1}{192} \tag{4.102d}$$

$$d_{3,0} = \frac{1}{12} \tag{4.103a}$$

$$d_{3,1} = \frac{1}{120}\sqrt{5}(5\sqrt{5} - 14) + \frac{7}{60}\sqrt{5} \tag{4.103b}$$

$$d_{3,2} = \frac{1}{120}\sqrt{5}(5\sqrt{5} - 14) + \frac{7}{60}\sqrt{5} \tag{4.103c}$$

$$d_{3,3} = \frac{1}{12} \tag{4.103d}$$

For the list of nodes and weights for higher order Gauss-Lobatto quadratures refer to Ref. [32].

4.2.4 Multi-implicit SDC Schemes

In compressible fluid dynamics with advective, diffusive, and reactive (or phase change) processes, there are often more than two temporal scales, namely the time scale for the advection, the time scale for the diffusion, and the time scale for reaction. (An example of a problem with three-time scales is compressible two-phase flow with heat conduction and finite rate of phase change which is discussed in Sect. 9.3.) Advective processes often have the slowest time scale and feature strongly nonlinear phenomena such as shock waves. An explicit time integration of the advective

4.2 Spectral Deferred Correction Scheme

terms in the governing equations is often adopted because implicit solution schemes are difficult to optimize for high-Mach number flows. The diffusive fluxes feature weaker nonlinearity and are often the second fastest process with significant stiffness, demanding an implicit time integration strategy with smaller time step sizes than the advective terms. The reactive processes are local and nonlinear and often possess the fastest time scale, also requiring an implicit time integration with the smallest time step sizes compared to the advection and diffusion.

To tackle the multi-scale nature of the compressible flows, we adopt a multi-implicit scheme, where the integration of advection, diffusion, and reaction are separated, each handled with its specialized numerical solution strategy. The SDC framework allows high-order splitting strategies [9], which are challenging to accomplish using other schemes such as additive RKs.

Let us consider a model of the compressible flow with advective, diffusive, and reactive terms as

$$w'(t) = f_A(t, w(t)) + f_D(t, w(t)) + f_R(t, w(t)) \tag{4.104}$$

An integration of (4.104) based on the splitting approach requires three integration operations to be carried out in sequence as

$$w_A(t + \Delta t) = w(t) + \int_t^{t+\Delta t} f_A(\tau, w(\tau)) d\tau \tag{4.105a}$$

$$w_D(t + \Delta t) = w_A(t + \Delta t) + \int_t^{t+\Delta t} f_D(\tau, w_A(\tau)) d\tau \tag{4.105b}$$

$$w(t + \Delta t) = w_D(t + \Delta t) + \int_t^{t+\Delta t} f_R(\tau, w_D(\tau)) d\tau \tag{4.105c}$$

It is not difficult to show that this splitting scheme is at most second-order accurate locally and first-order globally [33], since the advection, diffusion, and reaction all act simultaneously and order of action matters. Comparing the splitting scheme (4.105) with the non-split integration

$$w(t + \Delta t) = w(t) + \int_t^{t+\Delta t} [f_A(\tau, w(\tau)) + f_D(\tau, w(\tau)) + f_R(\tau, w(\tau))] d\tau \tag{4.106}$$

we notice the differences as

$$\int_t^{t+\Delta t} [f_D(\tau, w(\tau)) - f_D(\tau, w_A(\tau))] d\tau \approx \frac{\partial f_D}{\partial w} \frac{\partial w}{\partial t} \Delta t^2 = \mathcal{O}(\Delta t^2) \tag{4.107a}$$

$$\int_t^{t+\Delta t} [f_R(\tau, w(\tau)) - f_R(\tau, w_D(\tau))] d\tau \approx \frac{\partial f_R}{\partial w} \frac{\partial w}{\partial t} \Delta t^2 = \mathcal{O}(\Delta t^2) \tag{4.107b}$$

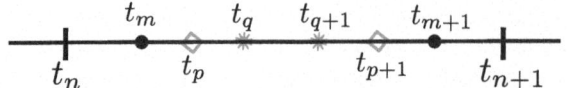

Fig. 4.3 Illustration of the Gauss-Lobatto quadrature nodes for the advection, diffusion, and reaction updates in the interval $[t_n, t_{n+1}]$ used in multi-implicit SDC (MISDC) scheme

Intuitively, the reason for the low-order accuracy of the splitting approach is the lack of coupling from one integration to the other. The multi-implicit SDC (MISDC) scheme provides sufficient coupling between the sequential integration operations. In MISDC, to allow smaller time steps for diffusion and yet smaller for reaction, a sub-time interval $[t_m, t_{m+1}]$ arising from the quadrature nodes for the advective term integration is divided into N_D time steps such that $t_{m,0} < t_{m,1} < \cdots < t_{m,p} < \cdots < t_{m,N_D} = t_{m+1}$ for the diffusion steps. Then, the sub-interval $[t_{n,p}, t_{n,p+1}]$ is further divided into N_R smaller intervals such that $t_{m,p,0} < t_{m,p,0} < \cdots < t_{m,p,q} < \cdots < t_{m,p,N_R} = t_{m,p+1}$. In the following, where there is no ambiguity, we use t_p for representing $t_{m,p}$ and t_q for $t_{m,p,q}$ for notational simplicity. Also, let $\Delta t_p \equiv t_{p+1} - t_p$, and $\Delta t_q \equiv t_{q+1} - t_q$. Figure 4.3 shows an example of the sub-division of $[t_m, t_{m+1}]$. In the implementation, t_m, t_p, and t_q are Gauss-Lobatto nodes in the intervals $[t_n, t_{n+1}]$, $[t_m, t_{m+1}]$, and $[t_p, t_{p+1}]$, respectively. Following [9], the application of the splitting approach to the SDC scheme with an explicit treatment of the advection and implicit treatments of diffusion and reaction yields the following update equations for the diffusion and reaction processes

$$w_{D,n,m,p}^{(k)} = u_{D,n,m,0}^{(k)} + c_p \Delta t_m \left[f_A(w_{A,n,m}^k) - f_A(w_{A,n,m}^{k-1}) + f_D(u_{D,n,m,p}^k) - f_D(u_{D,n,m,p}^{k-1}) \right]$$
$$+ c_p \Delta t_m \sum_{j=0}^{j=r} d_{p,j} \left[f_A(w_{n,m,p}^{(k-1)}) + f_D(w_{n,m,p}^{(k-1)}) + f_R(w_{n,m,p}^{(k-1)}) \right] \quad (4.108)$$

$$w_{n,m,p,q}^{(k)} = w_{n,m,p,0}^{(k-1)} + c_q \Delta t_p \left[f_A(w_{A,n,m}^{(k)}) - f_A(w_{A,n,m}^{(k-1)}) + f_D(w_{D,n,m,p}^{(k)}) - f_D(w_{D,n,m,p}^{(k-1)}) \right.$$
$$\left. + f_R(w_{n,m,p,q}^{(k)}) - f_R(w_{n,m,p,q}^{(k-1)}) \right]$$
$$+ c_q \Delta t_m \sum_{j=0}^{j=r} d_{q,j} \left[f_A(w_{n,m,p,q}^{(k-1)}) + f_D(w_{n,m,p,q}^{(k-1)}) + f_R(w_{n,m,p,q}^{(k-1)}) \right]$$
$$(4.109)$$

Since the advection term is treated explicitly, no separate update equation is required for $w_A^{(k)}$.

Algorithm for MISDC
Improving the accuracy of w^k by one requires the following step:

$$
\begin{aligned}
&\text{For } m = 0, \ldots, N_{A-1} \\
&\quad \text{For } p = 0, \ldots, N_{D-1} \\
&\quad\quad \text{Solve (4.108) for } w_{D,p}^{(k)}; \text{ Compute } f_D(w_{D,p}^{(k)}) \\
&\quad\quad \text{For } q = 0, \ldots, N_{R-1} \\
&\quad\quad\quad \text{Solve (4.108) for } w_q^{(k)}; \text{ Compute } F_R(u_q^{(k)}) \\
&\quad\quad \text{End} \\
&\quad\quad \text{Update } f_D(w_p^{(k)}) \\
&\quad \text{End} \\
&\quad \text{Update } f_A(w_{m+1}^{(k)}) \\
&\text{End}
\end{aligned} \quad (4.110)
$$

The initial provisional solution is obtained following the steps:

$$
\begin{aligned}
&\text{For } m = 0, \ldots, N_{A-1} \\
&\quad w_{A_{m+1}}^0 = w_m^{(0)} + \Delta t_m f_{A_m}(w_m^{(0)}) \\
&\quad w_{D,p}^{(0)} = w_p^{(0)} + \Delta t_p \left[f_A(w_m^{(0)}) + f_D(w_{D,p+1}^{(0)}) \right]; \\
&\quad \text{Compute } f_D(w_{D,p}^{(0)}) \\
&\quad \text{For } q = 0, \ldots, N_{R-1} \\
&\quad\quad w_q^{(0)} = w_q^{(0)} + \Delta t_q \left[f_A(w_m^{(0)}) + f_D(w_{D,p+1}^{(0)}) + f_R(w, q^{(0)}) \right] \\
&\quad \text{End} \\
&\text{End} \\
&\quad \text{Update } f_A(w_{m+1}^{(0)}) \\
&\text{End}
\end{aligned} \quad (4.111)
$$

4.3 Final Comments

Implicit-explicit, additive Runge-Kutta schemes, and spectral deferred correction schemes are high-order temporal schemes providing effective integration strategies for the highly stiff compressible transport phenomena [26, 34, 35]. The applications of these schemes to problems involving shock waves and phase change or extremely stiff reactions in highly compressible media are ongoing research.

4.4 Exercises

1. Consider a linear scalar ODE in the form

$$w' = \lambda w. \tag{4.112}$$

When an RK scheme is written as

$$w_{n+1} = R(z)w_n \tag{4.113}$$

with $z = \lambda \Delta t$, $R(z)$ is called the stability function of the RK scheme.] **(a)** Using the derivative form of RK scheme (4.21), show that the stability function $R(z)$ has the form $R(z) = 1 + z\boldsymbol{b}^T (I - z\mathcal{A})^{-1} \boldsymbol{e}$, where $\boldsymbol{e}(1, \ldots, 1)^T$, an $(r+1)$-dimensional vector. **(b)** For the fourth-order explicit RK scheme given in (4.69), determine $R(z)$. Hint: the inverse of a $n \times n$ lower triangular matrix $T = I - N$, where N is a strictly lower triangular matrix, is $T^{-1} = I + \sum_{j=1}^{n-1} N^j$. **(c)** For λ as a complex number, plot the stability region of the fourth-order scheme on the complex plane.

2. Consider a non-homogeneous linear scalar ODE in the form

$$w' = \lambda w + g(t). \tag{4.114}$$

(a) Using the derivative form of RK scheme (4.21), show that the RK scheme can be expressed in a matrix-vector form as $w_{n+1} = R(z)w_n + Q(z)g_n$ where $R(z)$ is as given in the previous exercise, $Q(z) = \boldsymbol{b}^T (I - z\mathcal{A})^{-1} \boldsymbol{e}$, and $g_n = \Delta t(g(t_n + c_1 \Delta t), \ldots, g(t_n + c_r \Delta t))^T$. **(a)** For the fourth-order explicit RK scheme given in (4.69), derive an explicit expression for $Q(z)$. Hint: the inverse of a $n \times n$ lower triangular matrix $T = I - N$, where N is a strictly lower triangular matrix, is $T^{-1} = I + \sum_{j=1}^{n-1} N^j$.

3. Derive the stability function (4.67) for schemes given by the Butcher arrays (4.66).
4. Analyze the A-stability and L-stability of the DIRK schemes (4.66).
5. Consider the ODE system arising from the spatial discretization of the advection-diffusion equation as

$$u'(t) = F_A(t, u(t)) + F_D(t, u(t))$$

where F_A and F_D are the advection and diffusion operators. Also, for time integration of the above system, consider a splitting scheme that sequentially, rather than simultaneously, integrates the effect of advection and diffusion as

$$u_A(t + \Delta t) = u(t) + \int_t^{t+\Delta t} F_A(\tau, u_A(\tau)) d\tau \tag{4.115}$$

$$u(t + \Delta t) = u_A(t) + \int_t^{t+\Delta t} F_D(\tau, u(\tau)) d\tau \tag{4.116}$$

Show that if F_A and F_D do not commute, the above splitting scheme is locally second-order and globally first-order only [33].

6. Provide an RK Butcher tableau for a second and third-order SDC scheme.
7. For the third and fourth-order Runge-Kutta schemes given in (4.69), determine the stage boundary values that results in theoretical convergence rates when solving the ODEs arising from spatial discretization of PDEs using time-dependent Dirichlet boundary condition. Follow similar approach to the stage boundary values in the third-order TVD Runge-Kutta scheme given in (4.71)
8. Consider hemoglobin unbinding and binding with oxygen in the lung's alveoli as a two-way reaction. The reaction is symbolically shown as $W_1 \xrightarrow{k_1} W_2 \xrightarrow{k_2} W_1$ with reaction rates of k_1 and k_2, representing the hemoglobin-oxygen unbinding and binding. Hence, it is reasonable to assume that $k_2 \gg k_1$. Using mass action law, this two-way reaction can be modeled using the following system of ODEs

$$w_1'(t) = -k_1 w_1(t) + k_2 w_2(t) \tag{4.117a}$$
$$w_2'(2) = k_1 w_1(t) - k_2 w_2(t) \tag{4.117b}$$

where $w_1(t)$ and $w_2(t)$ represent the bonded and unbonded hemoglobin and oxygen, respectively. (a) Write this system of ODEs in a matrix-vector form and identify the matrix entries. (b) Derive the exact solution to the problem using the eigendecomposition of the matrix, $A = U \Lambda U^{-1}$ and transforming the equations into the eigenspace with $v = U^{-1} w$. (c) Write out a third-order explicit, a third-order implicit, and a third-order diagonally implicit Runge-Kutta scheme for the time integration of the system. (d) Write a computer code for the integration of the system using Euler's explicit method with $w_1(0) = 0.2$ and $w_2(0) = 0.8$. For the final time of $T = 1$, obtain the approximate concentrations for $\tau = 1/50, 1/100, \ldots 1/5000$ and three pairs of the reaction rates $(k_1, k_2) = (1, 10), (1, 100)$ and $(1, 1000)$. Plot and discuss the data. (e) Determine the order of accuracy of the explicit and implicit schemes as the stiffness varies.
9. Consider Burgers' equation with a nonlinear reaction term in the form

$$u_t + u u_x = \frac{\delta(1-\gamma)}{2} u_{xx} + \frac{2(2\gamma - 1)}{\delta} u(u-1)^2 \tag{4.118}$$

with the initial condition

$$u(t=0, x) = \frac{1}{2} - \frac{1}{2} \tanh\left(\frac{x}{\delta}\right) \tag{4.119}$$

and the boundary conditions

$$u(t, -2) = 1 \qquad u(t, 2) = 0 \tag{4.120}$$

(a) Show that the exact solution to the above Burgers' equation is $u(t,x) = \frac{1}{2} - \frac{1}{2}\tanh\left(\frac{x-\gamma t}{\delta}\right)$. (b) Discretize the Burgers equation in space using appropriate spatial schemes for the advective and diffusive terms. (c) Write a computer code for a third-order semi-implicit additive Runge-Kutta method for the time integration of the resulting system of ODEs. (d) Program a multi-implicit spectral deferred time integration of order 3. (e) For varying values of δ that control the stiffness of the equation, verify the order of both temporal schemes and compare the performance of both schemes in terms of accuracy and efficiency when the equation is integrated into the final time $t = 0.5$.

References

1. Shahbazi, K.: Robust second-order scheme for multi-phase flow computations. J. Comput. Phys. **339**, 163–178 (2017)
2. Dahlquist, G.: A special stability problem for linear multistep methods. BIT Numer. Math. **3**, 27–43 (1963)
3. Runge–Kutta Methods, Chap. 3, pp. 143–331. Wiley (2016)
4. Hairer, E., Wanner, G.: Solving Ordinary Differential Equations II: Stiff and Differential - Algebraic Problems. Springer Series in Computational Mathematics., Springer, Berlin Heidelberg (2013)
5. Kennedy, C.A., Carpenter, M.H.: Additive Runge–Kutta schemes for convection–diffusion–reaction equations. Appl. Numer. Math. **44**(1), 139–181 (2003)
6. Kennedy, C.A., Carpenter, M.H.: Diagonally implicit Runge-Kutta methods for stiff odes. Appl. Numer. Math. **146**, 221–244 (2019)
7. Dutt, A., Greengard, L., Rokhlin, V.: Spectral deferred correction methods for ordinary differential equations. BIT Numer. Math. **40**(2), 241–266 (2000)
8. Minion, M.L.: Semi-implicit spectral deferred correction methods for ordinary differential equations. Commun. Math. Sci. **1**(3), 471–500 (2003)
9. Bourlioux, A., Layton, A.T., Minion, M.L.: High-order multi-implicit spectral deferred correction methods for problems of reactive flow. J. Comput. Phys. **189**(2), 651–675 (2003)
10. Butcher, J.C.: Coefficients for the study of Runge-Kutta integration processes. J. Aust. Math. Soc. **3**, 185–201 (1963)
11. Abramowitz, M., Stegun, I.A.: Handbook of Mathematical Functions with Formulas Graphs, and Mathematical Tables. Dover Publications (1983)
12. Alexander, R.K.: Stability of Runge-Kutta methods for stiff ordinary differential equations. SIAM J. Numer. Anal. **31**(4), 1147–1168 (1994)
13. Burrage, K., Hundsdorfer, W.H., Verwer, J.G.: A study of b-convergence of Runge-Kutta methods. Computing **36**(1), 17–34 (1986)
14. Ketcheson, D.I., Seibold, B., Shirokoff, D., Zhou, D.: Dirk schemes with high weak stage order. In: Sherwin, S.J., Moxey, D., Peiró, J., Vincent, P.E., Schwab, C. (eds.) Spectral and High Order Methods for Partial Differential Equations ICOSAHOM 2018, pp. 453–463. Springer International Publishing, Cham (2020)
15. Carpenter, M.H., Gottlieb, D., Abarbanel, S., Don, W.-S.: The theoretical accuracy of Runge–Kutta time discretizations for the initial boundary value problem: a study of the boundary error. SIAM J. Sci. Comput. **16**(6), 1241–1252 (1995)
16. Abarbanel, S., Gottlieb, D., Carpenter, M.H.: On the removal of boundary errors caused by Runge–Kutta integration of nonlinear partial differential equations. SIAM J. Sci. Comput. **17**(3), 777–782 (1996)
17. Biswas, A., Ketcheson, D.I., Seibold, B., Shirokoff, D.: Design of dirk schemes with high weak stage order. Commun. Appl. Math. Comput. Sci. **18**(1), 1–28 (2023)

References

18. Boom, P.D., Zingg, D.W.: Optimization of high-order diagonally-implicit Runge–Kutta methods. J. Comput. Phys. **371**, 168–191 (2018)
19. Hairer, E., Wanner, G.: Solving Ordinary Differential Equations II. Stiff and Differential-Algebraic Problems, Springer, Berlin (1996)
20. Alamri, Y., Ketcheson, D.I.: Very high-order a-stable stiffly accurate diagonally implicit Runge-Kutta methods with error estimators (2023)
21. Hundsdorfer, W., Verwer, J.: Numerical Solution of Time-Dependent Advection-Diffusion-Reaction Equations, 1st edn. Springer, Berlin, Heidelberg (2003)
22. Shu, C.-W., Osher, S.: Efficient implementation of essentially non-oscillatory shock-capturing schemes, ii. J. Comput. Phys. **83**(1), 32–78 (1989)
23. Kennedy, C.A., Carpenter, M.H.: Higher-order additive Runge-Kutta schemes for ordinary differential equations. Appl. Numer. Math. **136**, 183–205 (2019)
24. Gottlieb, S., Ketcheson, D.I., Shu, C.-W.: High order strong stability preserving time discretizations. J. Sci. Comput. **38**(3), 251–289 (2009)
25. Guo, R., Xia, Y., Yan, X.: Semi-implicit spectral deferred correction methods for highly nonlinear partial differential equations. J. Comput. Phys. **338**, 269–284 (2017)
26. Layton, A.T., Minion, M.L.: Conservative multi-implicit spectral deferred correction methods for reacting gas dynamics. J. Comput. Phys. **194**(2), 697–715 (2004)
27. Böhmer, K., Hemker, P.W., Stetter, H.J.: The Defect Correction Approach, pp. 1–32. Springer Vienna, Vienna (1984)
28. Abgrall, R.: High order schemes for hyperbolic problems using globally continuous approximation and avoiding mass matrices. J. Sci. Comput. **73**(2), 461–494 (2017)
29. Abgrall, R., Le Mélédo, É., Öffner, P., Torlo, D.: Relaxation deferred correction methods and their applications to residual distribution schemes. SMAI J. Comput. Math. **8**, 125–160 (2022)
30. Ketcheson, D.I., Waheed, U.B.: A comparison of high-order explicit Runge-Kutta, extrapolation, and deferred correction methods in serial and parallel. Commun. Appl. Math. Comput. Sci. **9**(2), 175–200 (2014). Publisher Copyright: 2014 Mathematical Sciences Publishers
31. Hildebrand, F.B.: Introduction to Numerical Analysis. Tata McGraw-Hill Publishing Company Ltd., New Delhi (1956)
32. Jay, L.O.: Lobatto Methods, pp. 817–826. Springer, Berlin, Heidelberg (2015)
33. Strang, G.: On the construction and comparison of difference schemes. SIAM J. Numer. Anal. **5**(3), 506–517 (1968)
34. Muscat, L., Puigt, G., Montagnac, M., Brenner, P.: A coupled implicit-explicit time integration method for compressible unsteady flows. J. Comput. Phys. **398**, 108883 (2019)
35. Zingale, M., Katz, M.P., Bell, J.B., Minion, M.L., Nonaka, A.J., Zhang, W.: Improved coupling of hydrodynamics and nuclear reactions via spectral deferred corrections. Astrophys. J. **886**(2), 105 (2019)

Chapter 5
High-Order Nonlinear Schemes for Nonlinear Conservation Laws

As discussed in Chap. 1, the nonlinear conservation laws support shock waves as their solution. As a result, in approximating shock waves and other discontinuities, appropriate spatial discretizations are needed to avoid spurious oscillations near discontinuities that do not diminish with grid refinement. To remedy the spurious oscillations, we have introduced in Chap. 2 nonlinear schemes based on weighted essentially nonoscillatory (WENO) and targeted essentially nonoscillatory (TENO) schemes that are high-order accurate and offer simplicity and generality.

A further difficulty in discretizing nonlinear conservation laws is the application of proper upwinding which is essential for stability. Since the characteristic speeds are not unidirectional in nonlinear conservation laws, waves travel in different directions; hence, the upwinding is not straightforward. An approach overcoming this difficulty is the Lax-Friedrichs splitting [1], in which the nonlinear flux splits into two parts: one tends to yield a positive advection speed, and the other a negative advection speed. The flux part with a positive advection speed is then approximated with a left-biased scheme, while the flux with a negative speed is with a right-biased one. The first-order Lax-Friedrichs (LF) scheme is provably nonoscillatory (monotone) [2]. However, the extension of the LF scheme to high orders requires nonlinear reconstructions based on WENO or TENO schemes to overcome spurious oscillations in the presence of sharp gradients and discontinuities. The resultant high-order method is highly effective in solving nonlinear conservation laws.

Finally, in addition to the requirement of high-order accuracy with no spurious oscillations near discontinuities, the numerical solution of the conservation laws must produce approximations that fall inside the admissible set spanned by the exact solution function. Thus, the numerical solution must always be larger than the global minimum value of the solution and smaller than the global maximum. Such numerical schemes are called maximum-principle preserving schemes, and if the solution has a lower bound of zero and an upper bound of infinity, the schemes are called positivity-preserving schemes. Failure to satisfy the maximum-principle conditions can be detrimental to computations as it may yield unphysical imaginary characteristic speeds resulting in a computational overflow.

We begin this chapter with the first-order LF scheme for scalar nonlinear conservation laws before presenting high-order LF-type methods. We then introduce high-order LF-type methods for a system of conservation laws, including Euler's equations of gas dynamics. Since LF splitting relies on the characteristic speeds, it is essential that the scheme yields real values for the sound speed or positive values for the square of the speed. We will present the analysis of the monotonicity of the first-order LF scheme, which yields maximum-principle preservation of the scheme and the positivity of the square of the sound speed, before presenting algorithms or modifications to the base high-order LF schemes that make the scheme maximum-principle preserving and, in the case of the Euler system, make the square of the sound speed positive.

We have deliberately focused on the LF schemes and left out discussions of other methods for flux approximations. The reasons for this are at least twofold. First, the LF schemes are simple, straightforward, and robust, and increasing the approximation order yields solutions comparable to other flux approximation schemes. Second, as our discussion of the discretization of the two-fluid model and diffusive fluxes in Chaps. 6 and 7 will reveal, the LF schemes are extendable to the diffusive fluxes in the compressible Navier-Stokes equations and the two-fluid models.

5.1 First-Order Lax-Friedrichs Scheme

Consider a nonlinear scalar conservation law in the form

$$\frac{\partial u}{\partial t} + \frac{\partial f(u)}{\partial x} = 0 \tag{5.1}$$

where $f \equiv f(u)$ is a nonlinear flux. The conservative finite difference scheme for this conservation law has the form

$$u_j^{n+1} = u_j^n - \frac{\tau}{h}(\hat{f}_{j+\frac{1}{2}}^n - \hat{f}_{j-\frac{1}{2}}^n) \tag{5.2}$$

where $\hat{f}_{j+\frac{1}{2}}$ and $\hat{f}_{j-\frac{1}{2}}$ are numerical fluxes that approximate the exact fluxes $f_{j+\frac{1}{2}}^n = f(u(t_n, x_{j+\frac{1}{2}}))$ and $f_{j-\frac{1}{2}}^n = f(u(t_n, x_{j-\frac{1}{2}}))$. For stability, the numerical scheme must be upwind-biased.

To this end, we use the LF scheme to split the flux into two fluxes as

$$f = \frac{1}{2}(f + \alpha u) + \frac{1}{2}(f - \alpha u) \tag{5.3}$$

5.1 First-Order Lax-Friedrichs Scheme

where α is the maximum of the absolute value of the Jacobian of the flux

$$\alpha = \max_x \left| \frac{df(u(x))}{du} \right| \tag{5.4}$$

Substituting the splitting back into the nonlinear conservation law yields

$$\frac{\partial u}{\partial t} + \frac{1}{2}\frac{\partial (f+\alpha u)}{\partial x} + \frac{1}{2}\frac{\partial (f-\alpha u)}{\partial x} = 0 \tag{5.5}$$

which its quasi-conservative form yields two terms with positive and negative characteristic speeds as

$$\frac{\partial u}{\partial t} + \underbrace{\frac{1}{2}\left(\frac{\partial f}{\partial u}+\alpha\right)}_{\geq 0}\frac{\partial u}{\partial x} + \underbrace{\frac{1}{2}\left(\frac{\partial f}{\partial u}-\alpha\right)}_{\leq 0}\frac{\partial u}{\partial x} = 0 \tag{5.6}$$

thanks to the definition of α, Eq. (5.4). Hence, we may denote

$$f^+ = \frac{1}{2}(f+\alpha u) \quad \text{and} \quad f^- = \frac{1}{2}(f-\alpha u) \tag{5.7}$$

and refer to the LF splitting as a splitting into a positive flux and a negative flux. If the maximum is carried out over the entire x interval, the scheme is referred to as the global LF scheme. If the maximum is carried out over a local stencil, the scheme is referred to as the local LF scheme.

Using (5.3), the exact flux at mid-points $f^n_{j+\frac{1}{2}} = f(u(t_n, x_{j+\frac{1}{2}}))$ and $f^n_{j-\frac{1}{2}} = f(u(t_n, x_{j-\frac{1}{2}}))$ are expressed as

$$f^n_{j+\frac{1}{2}} = \frac{1}{2}(f^n_{j+\frac{1}{2}} + \alpha u^n_{j+\frac{1}{2}}) + \frac{1}{2}(f^n_{j+\frac{1}{2}} - \alpha u^n_{j+\frac{1}{2}}) \tag{5.8a}$$

$$f^n_{j-\frac{1}{2}} = \frac{1}{2}(f^n_{j-\frac{1}{2}} + \alpha u^n_{j-\frac{1}{2}}) + \frac{1}{2}(f^n_{j-\frac{1}{2}} - \alpha u^n_{j-\frac{1}{2}}) \tag{5.8b}$$

The positive and negative fluxes are then approximated using the first-order left and right schemes, respectively, as

$$f^n_{j+\frac{1}{2}} \approx \hat{f}^n_{j+\frac{1}{2}} = \underbrace{\frac{1}{2}(f^n_j + \alpha u^n_j)}_{\text{left scheme}} + \underbrace{\frac{1}{2}(f^n_{j+1} - \alpha u^n_{j+1})}_{\text{right scheme}} \tag{5.9a}$$

$$f^n_{j-\frac{1}{2}} \approx \hat{f}^n_{j-\frac{1}{2}} = \frac{1}{2}(f^n_{j-1} + \alpha u^n_{j-1}) + \frac{1}{2}(f^n_j - \alpha u^n_j) \tag{5.9b}$$

Substitution into the conservative form of the finite difference scheme (5.2) then yields the first-order LF finite difference scheme as

$$u_j^{n+1} = u_j^n - \frac{\tau}{2h}\left[(f_j^n + \alpha u_j^n + f_{j+1}^n - \alpha u_{j+1}^n) - (f_{j-1}^n + \alpha u_{j-1}^n + f_j^n - \alpha u_j^n)\right] \tag{5.10}$$

5.1.1 Monotonicity and Maximum Principle-Preserving

Following Ref. [2], we rearrange the LF finite difference scheme for the scalar conservation law, expressed in Eq. (5.10), to obtain

$$u_j^{n+1} = u_j^n - \frac{\tau}{2h}\left[2\alpha u_j^n + (f_{j+1}^n - \alpha u_{j+1}^n) - (f_{j-1}^n + \alpha u_{j-1}^n)\right] \tag{5.11}$$

$$u_j^{n+1} = \left(1 - \frac{\tau\alpha}{h}\right)u_j^n + \frac{\tau}{2h}(\alpha u_{j+1}^n - f_{j+1}^n) + \frac{\tau}{2h}(\alpha u_{j-1}^n + f_{j-1}^n) \tag{5.12}$$

$$u_j^{n+1} = \mathcal{H}(u_{j-1}, u_j, u_{j+1}) \tag{5.13}$$

From the expression for stabilization parameter α, Eq. (5.4), and the stability condition for the explicit Euler's scheme, $\tau \leq \frac{h}{\alpha}$ with α being the global maximum (maximum overall x), it follows that $\mathcal{H}(a, b, c)$ is an increasing function of all three arguments. Let us unpack this. The first term on the right-hand side of Eq. (5.12) is an increasing function of u_j^n because

$$\left(1 - \frac{\tau\alpha}{h}\right) > 0 \implies \frac{d}{du_j^n}\left[\left(1 - \frac{\tau\alpha}{h}\right)u_j^n\right] > 0 \tag{5.14}$$

The second and third terms on the right-hand side of Eq. (5.12) are also an increasing function of u_{j+1}^n because

$$\alpha = \max_x \left|\frac{df(u(x))}{du}\right| \implies \frac{d}{du_{j+1}^n}\left[\frac{\tau}{2h}(\alpha u_{j+1}^n - f_{j+1}^n)\right] \geq 0 \tag{5.15}$$

Finally, the third term can be shown to be an increasing function of u_{j-1}^n.

Since an increasing function is a particular form of a monotone function, the LF scheme is a monotone scheme. The scheme is also consistent, i.e.,

$$\mathcal{H}(a, a, a) = a \tag{5.16}$$

The monotonicity is an essential property; it simply implies that if the distribution of u is initially increasing or decreasing in x, at a later time, the distribution is also increasing or decreasing, respectively, i.e., no new local minimum or maximum

5.1 First-Order Lax-Friedrichs Scheme

is generated. Hence, given a monotone initial distribution, monotone $\mathcal{H}$ yields a monotone distribution of u over time.

From the monotonicity and the consistency, we deduce that if the solution at time level n falls in $[m, M]$, or $m \leq u_{j-1}^n, u_j^n, u_{j+1}^n \leq M$, then

$$m = H(m, m, m) \leq u_j^{n+1} = \mathcal{H}(u_{j-1}^n, u_j^n, u_{j+1}^n) \leq H(M, M, M) = M \quad (5.17)$$

The above expression reveals that no overshoot or undershoot will be generated using the global Lax-Friedrichs scheme. In other words, the global LF method is free from spurious oscillation and is maximum-principle preserving. As a result, it always generates a solution that is within the bounds of the initial exact solution, which is verified by setting m and M as the initial solution's global minimum and maximum. The scheme keeps the solution's positivity when $m = 0$. Therefore, despite its low accuracy, the first-order LF scheme has a special place in the numerical solution of nonlinear conservation laws. For instance, high-order methods are often compared to the first-order LF scheme in terms of their non-oscillatory or maximum-principle preservation properties.

5.1.2 Dissipation and Stability of the Lax-Friedrichs Scheme

The spatial approximation to the conservation law using the first-order LF splitting yields

$$\left.\frac{\partial u}{\partial t}\right|_{x_j} = -\frac{1}{2h}(f_{j+1}^n - f_{j-1}^n) + \frac{\alpha}{2h}(u_{j-1}^n - 2u_j^n + u_{j+1}^n) \quad (5.18)$$

The first term on the right-hand side is a second-order approximation of the first derivative,

$$\frac{1}{2h}(f_{j+1}^n - f_{j-1}^n) = \left.\frac{\partial u}{\partial x}\right|_{x_j} + \mathcal{O}(h^2) \quad (5.19)$$

and the second term is a second-order approximation of the second derivative,

$$\frac{1}{h^2}(u_{j-1}^n - 2u_j^n + u_{j+1}^n) = \left.\frac{\partial^2 u}{\partial x^2}\right|_{x_j} + \mathcal{O}(h^2) \quad (5.20)$$

After substitution of (5.19) and (5.20) into (5.18), the LF scheme satisfies

$$\left.\frac{\partial u_j}{\partial t}\right|_{x_j} = \left.\frac{\partial u}{\partial x}\right|_{x_j} + \frac{\alpha h}{2}\left.\frac{\partial^2 u}{\partial x^2}\right|_{x_j} + \mathcal{O}(h^2) \quad (5.21)$$

Therefore, the LF scheme is an $\mathcal{O}(h^2)$ approximation of the original conservation law modified with the inclusion of a diffusive term. Also, the diffusive term is scaled by $\alpha h/2$, indicating the numerical dissipation introduced when one uses the LF splitting. Since the coefficient of the numerical diffusion is $\alpha h/2$, the term $\pm\alpha u$ in the LF flux is called the stabilization term, and α is the stabilization parameter. The higher the α, the higher the rate of numerical dissipation. Furthermore, the numerical dissipation diminishes with the $\mathcal{O}(h)$ rate; in the limit of vanishing grid spacing, the added numerical dissipation approaches zero, revealing the consistency of the LF scheme.

Remark 5.1 For nonlinear conservation laws, numerical dissipation plays a more critical role in ensuring the integrity of the numerical solution compared to the linear equations. The central approximation of the first derivative, the second term in Eq. (5.18), is a dispersive approximation of a non-dispersive model. In the absence of any dissipation, the numerical dispersion causes the separation of the solution modes over time as they are propagated at different speeds. This dispersion of different modes is exacerbated by the generation of higher and higher modes as a result of nonlinearity. For the case of gas dynamic equations, the spurious oscillations can easily lead to negative density, in turn, resulting in an unbounded increase in the internal energy and pressure, eventually leading to the computational overflow.

Fortunately, the numerical diffusion inherent in the LF scheme effectively dissipates these higher modes, and the solution's integrity is retained.

Remark 5.2 As demonstrated in Appendix A, the Fourier analysis reveals that the numerical dissipation effectively smooths out the large gradients. This means that any discontinuity is smeared out as a result of numerical diffusion. For nonlinear discontinuities, this is not an issue as the characteristics internal to the sharp gradients are convergent, and this convergence of characteristics counters the smearing of the numerical dissipation. As a result, the discontinuity is captured within a fixed number of grid points. On the contrary, linear discontinuities such as material interfaces smear out over time. Hence, high-order numerical schemes introducing significantly lower dissipation are preferable; their discussion concerns us next.

5.2 High-Order Lax-Friedrichs Scheme for Scalar Conservation Law

A high-order Lax-Friedrichs scheme is based on the high-order spatial derivative computation Eq. (2.40) as discussed in the previous chapter. For conciseness, we first write the LF splitting of the nonlinear flux $f(u)$ as

$$f(u) = \frac{1}{2}(f^+ + f^-) \qquad (5.22)$$

with

$$f^+ = f + \alpha u \quad f^- = f - \alpha u \qquad (5.23)$$

5.2 High-Order Lax-Friedrichs Scheme for Scalar Conservation Law

Then, the high-order derivative approximation (2.40) using the LF splitting has the following form

$$\left.\frac{\partial f(u)}{\partial x}\right|_{x_i} \approx \frac{1}{h}\sum_{j=1}^{r+1} d_j \left[\frac{1}{2}\left(f^+_{i+j-\frac{1}{2}} + f^-_{i+j-\frac{1}{2}}\right) - \frac{1}{2}\left(f^+_{i-j+\frac{1}{2}} + f^-_{i-j+\frac{1}{2}}\right)\right] \quad (5.24)$$

where coefficient d_j are given in Table 2.2. Alternatively, using Eq. (2.42), the high-order derivative approximation with the LF splitting is expressed as

$$\left.\frac{\partial f(u)}{\partial x}\right|_{x_i} \approx \frac{1}{h}\left\{d'_1\left[\frac{1}{2}\left(f^+_{i+\frac{1}{2}} + f^-_{i+\frac{1}{2}}\right) - \frac{1}{2}\left(f^+_{i-\frac{1}{2}} + f^-_{i-\frac{1}{2}}\right)\right] + \sum_{j=2}^{p} d'_j \left[f(x_{i+j-1}) - f(x_{i-j+1})\right]\right\}$$
(5.25)

where coefficient d_j are given in Table 2.3. In Eq. (5.24) or (5.25), $f^+_{i+j-\frac{1}{2}}$ and $f^-_{i+j+\frac{1}{2}}$ can be obtained from a left and right WENO approximation of the mid-point values of the solution u, or directly from the WENO approximations of the mid-point values of fluxes $f(u) \pm \alpha u$. Although, in some cases, both approaches are acceptable, in some challenging problems, the reconstruction of the solution or some functions of the solution is preferable. For instance, in the case of Euler equations of gas dynamics, the reconstruction of the primitive variables yields solutions with smaller oscillations and a higher degree of positiveness. Furthermore, as will be discussed in the next chapter, for two-fluid systems, it is essential to use the WENO reconstruction of the primitive variables. Thus, we herein adopt the reconstruction of the solution or primitive variable solution to determine the positive and negative fluxes given as

$$f^+_{i+j-\frac{1}{2}} = f(u^+_{r,\text{WENO},i+j-\frac{1}{2}}) + \alpha u^+_{r,\text{WENO},i+j-\frac{1}{2}} \quad (5.26a)$$

$$f^-_{i+j-\frac{1}{2}} = f(u^-_{r,\text{WENO},i+j-\frac{1}{2}}) - \alpha u^-_{r,\text{WENO},i+j-\frac{1}{2}} \quad (5.26b)$$

Instead of WENO reconstructions, one may use TENO reconstructions, and doing so yields the positive and negative fluxes in the form

$$f^+_{i+j-\frac{1}{2}} = f(u^+_{r,\text{TENO},i+j-\frac{1}{2}}) + \alpha u^+_{r,\text{TENO},i+j-\frac{1}{2}} \quad (5.27a)$$

$$f^-_{i+j-\frac{1}{2}} = f(u^-_{r,\text{TENO},i+j-\frac{1}{2}}) - \alpha u^-_{r,\text{TENO},i+j-\frac{1}{2}} \quad (5.27b)$$

In Eqs. (5.26) and (5.27), the superscript "+" or "−" on u signifies a left- or right-biased spatial approximation.

5.2.1 Conservative Form of the Discretized Scalar Nonlinear Conservation Law

We now concern ourselves with the conservative form of the discretized equation after performing the high-order spatial and temporal schemes, as the conservation property is essential for capturing shock waves. The conservative form of the high-order spatial derivative approximation has been derived and presented in Sect. 2.2.4 for spatial discretization only. Here, we present the conservative discretized equation in the context of a high-order explicit Runge-Kutta scheme.

The method of lines starts with the discretization of the spatial derivative based on the derivation in Sect. 2.2.4 leads to

$$\frac{\partial u_i}{\partial t} = -\frac{1}{h}\left(\hat{F}_{i+\frac{1}{2}}\{u_i\} - \hat{F}_{i-\frac{1}{2}}\{u_i\}\right) \tag{5.28}$$

The notation $\hat{F}_{i+1/2}\{u_i\}$ means the flux $f_{i+1/2}$ depends on a subset of solution vector $\{u_i\}$ determined by the effective stencil of the spatial approximation. Now, using the recipe for the high-order RK scheme given in Sect. 4.1, we can derive the conservative form of the fully discretized equation for one time step in the form

$$u_i^{n+1} = u_i^n - \frac{\tau}{h}\left(\hat{F}^{RK}_{i+\frac{1}{2}} - \hat{F}^{RK}_{i-\frac{1}{2}}\right) \tag{5.29}$$

where

$$\hat{F}^{RK}_{i\pm\frac{1}{2}} = \sum_{q=1}^{s} b_q F_q^{\pm} \tag{5.30}$$

and using (4.20)

$$F_1^{\pm} = \hat{F}^n_{i\pm\frac{1}{2}}\{u_i\} \tag{5.31a}$$

$$F_2^{\pm} = \hat{F}^n_{i\pm\frac{1}{2}}\left\{u_i - \frac{\tau}{h}[\alpha_{21}(F_1^+ - F_1^-)]\right\} \tag{5.31b}$$

$$F_3^{\pm} = \hat{F}^n_{i\pm\frac{1}{2}}\left\{u_i - \frac{\tau}{h}[\alpha_{31}(F_1^+ - F_1^-) + \alpha_{32}(F_2^+ - F_2^-)]\right\} \tag{5.31c}$$

$$\vdots$$

$$F_i^{\pm} = \hat{F}^n_{i\pm\frac{1}{2}}\left\{u_i - \frac{\tau}{h}\sum_{j=1}^{i-1}[\alpha_{ij}(F_j^+ - F_j^-)]\right\} \tag{5.31d}$$

The coefficients b_q and α_{ij} are listed up to the fourth order in Sect. 4.1.

5.2.2 Maximum-Principle Preservation of High-Order Schemes

A WENO or TENO scheme used for the high-order discretization of the conservation laws provides "essentially" non-oscillatory solutions, meaning that after the reconstruction, some relatively small oscillations that vanish with grid refinement may be present near discontinuities and sharp features. In many applications, these oscillations have negligible impact on the overall solution; however, for certain scenarios, they can pose problems. For example, as demonstrated in Ref. [3], simulations that involve very high-Mach shock waves, or small density/pressure can quickly develop negative density and or pressure values leading to blowup. Therefore, we need an algorithm to ensure the positivity of the high-order scheme. Knowing that the first-order LF scheme is maximum-principle preserving, an effective approach that is extendable to higher space dimensions, a system of conservation laws, and even two-fluid systems, is to limit the high-order fluxes toward the first-order LF fluxes.

Specifically, the first-order LF solution at time step $n + 1$, u_i^{n+1}, using the high-order solution at time step n, u_i^n, is expressed as

$$u_i^{n+1} = u_i^n - \frac{\tau}{h}\left(\hat{f}_{i+\frac{1}{2}}^n - \hat{f}_{i-\frac{1}{2}}^n\right) \tag{5.32}$$

and the high-order spatial and temporal solution, U_i^{n+1}, in the conservative form as

$$U_i^{n+1} = u_i^n - \frac{\tau}{h}\left(\hat{F}_{i+\frac{1}{2}}^n - \hat{F}_{i-\frac{1}{2}}^n\right) \tag{5.33}$$

Then, the maximum-principle preserving algorithm seeks the largest value of parameters $0 \leq \theta_{i+\frac{1}{2}} \leq 1$ and $0 \leq \theta_{i-\frac{1}{2}} \leq 1$ such that the modified scheme in the form

$$\tilde{U}_i^{n+1} = u_i^n - \frac{\tau}{h}\left(\tilde{F}_{i+\frac{1}{2}}^n - \tilde{F}_{i-\frac{1}{2}}^n\right) \tag{5.34}$$

with

$$\tilde{F}_{i+\frac{1}{2}}^n = \hat{f}_{i+\frac{1}{2}}^n + \theta_{i+\frac{1}{2}}\left(\hat{F}_{i+\frac{1}{2}}^n - \hat{f}_{i+\frac{1}{2}}^n\right) \tag{5.35a}$$

$$\tilde{F}_{i-\frac{1}{2}}^n = \hat{f}_{i-\frac{1}{2}}^n + \theta_{i-\frac{1}{2}}\left(\hat{F}_{i-\frac{1}{2}}^n - \hat{f}_{i-\frac{1}{2}}^n\right) \tag{5.35b}$$

yields the modified solution $\tilde{U}_i^{n+1}$ that falls in the physical bound of the solution given the maximum-principle preservation of solution u_i^n, namely,

$$m \leq u_i^n \leq M \implies m \leq \tilde{U}_i^{n+1} \leq M \tag{5.36}$$

In Eq. (5.35), $\hat{f}^n_{i+\frac{1}{2}}$ and $\hat{f}^n_{i-\frac{1}{2}}$ are the first-order fluxes and in Eq. (5.36), m and M are the global minima and maxima of the exact solution.

Clearly, the smaller the limiting parameter θ, the closer the modified flux to the first-order one. Also, the case of $\theta = 0$ yields the first-order flux, and $\theta = 1$ corresponds to no limiting and keeps the high-order flux unchanged. The goal is to maintain the formal high-order accuracy of the scheme while enforcing maximum-principle preservation.

Let us demonstrate how to determine the limiting parameter for enforcing the lower bound:

$$m \leq \tilde{U}^{n+1}_i = u^n_i - \frac{\tau}{h}\left(\tilde{F}^n_{i+\frac{1}{2}} - \tilde{F}^n_{i-\frac{1}{2}}\right) \tag{5.37}$$

During the intermediate step of the derivation of the limiting parameter, let us denote the limiting parameter of the left and right fluxes for the cell i arising from enforcing the lower bound by $\Lambda^m_{+,i}$ and $\Lambda^m_{-,i}$, respectively. Substituting the expression of the modified flux (5.35) into the above expression yields

$$-m + \left[u^n_i - \lambda(\hat{f}_{i+\frac{1}{2}} - \hat{f}_{i-\frac{1}{2}})\right] - (\lambda \Lambda^m_{+,i} F_+) + (\lambda \Lambda^m_{+,i} F_-) \geq 0 \tag{5.38}$$

where $F_+ = \hat{F}^{RK}_{i+\frac{1}{2}} - \hat{f}_{i+\frac{1}{2}}$ and $F_- = \hat{F}^{RK}_{i-\frac{1}{2}} - \hat{f}_{i-\frac{1}{2}}$ and $\lambda = \tau/h$. The term $\left[u^n_i - \lambda(\hat{f}_{i+\frac{1}{2}} - \hat{f}_{i-\frac{1}{2}})\right]$ is simply the first-order solution at times step $n+1$ and since it is maximum principle preserving, we have

$$d^m_i = m - \left[u^n_i - \lambda(\hat{f}_{i+\frac{1}{2}} - \hat{f}_{i-\frac{1}{2}})\right] \leq 0 \tag{5.39}$$

Substituting this definition back into (5.38) gives

$$(-\lambda \Lambda^m_{+,i} F_+) + (\lambda \Lambda^m_{-,i} F_-) - d^m_i \geq 0 \tag{5.40}$$

Noting that the third term $-d^m_i$ is always greater than zero, we can consider the four possible cases for the other two terms and identify the required flux limiter values.

1. $F_+ \leq 0$ and $F_- \geq 0$

   Here both the first and second terms in (5.40) are greater than zero, so no limiting is required. Thus, we can assign the limiters as

   $$\Lambda^m_{+,i} = 1$$
   $$\Lambda^m_{-,i} = 1.$$

2. $F_+ > 0$ and $F_- \geq 0$

   The first term in (5.40) is less than zero, while the second is not. We must thus solve for the $\Lambda_{+,i}$ value that guarantees the inequality for any $\Lambda_{-,i}$. For this case, we can assign the limiters as

5.2 High-Order Lax-Friedrichs Scheme for Scalar Conservation Law

$$\Lambda^m_{-,i} = 1$$

$$\Lambda^m_{+,i} = \begin{cases} 1 & \text{if } \frac{d^m_i}{-\lambda F_+ - \epsilon} > 1 \\ \frac{d^m_i}{-\lambda F_+ - \epsilon} & \text{otherwise} \end{cases}$$

3. $F_- < 0$ and $F_+ \leq 0$

   The first term in (5.40) is greater than zero while the second is less than zero. Solving the corresponding limiters gives

$$\Lambda^m_{-,i} = \begin{cases} 1 & \text{if } \frac{d^m_i}{\lambda F_- - \epsilon} > 0 \\ \frac{d^m_i}{\lambda F_- - \epsilon} & \text{otherwise} \end{cases}$$

$$\Lambda^m_{+,i} = 1.$$

4. $F_- < 0$ and $F_+ > 0$

   Finally, we have the case where both the first and second terms in (5.40) are less than zero. We can take $\Lambda_{+,i} = \Lambda_{-,i}$ and then solve directly for either to obtain the following limiters

$$\Lambda^m_{-,i} = \begin{cases} 1 & \text{if } \frac{d^m_i}{-\lambda F_+ + \lambda F_- - \epsilon} \geq 1 \\ \frac{d^m_i}{-\lambda F_+ + \lambda F_- - \epsilon} & \text{otherwise} \end{cases}$$

$$\Lambda^m_{+,i} = \begin{cases} 1 & \text{if } \frac{d^m_i}{-\lambda F_+ + \lambda F_- - \epsilon} \geq 1 \\ \frac{d^m_i}{-\lambda F_+ + \lambda F_- - \epsilon} & \text{otherwise}. \end{cases}$$

The parameter ϵ is a small number near machine zero to avoid division by zero. After repeating a similar algorithm to find the limiting parameters $\Lambda^M_{i,\pm}$ that enforce the upper bound, we select the following limiting parameters for each cell i:

$$\begin{cases} \theta_{i+\frac{1}{2}} = \min \left\{ \Lambda^m_{-,i+1}, \Lambda^M_{-,i+1}, \Lambda^m_{+,i}, \Lambda^M_{+,i} \right\} \\ \theta_{i-\frac{1}{2}} = \min \left\{ \Lambda^m_{-,i}, \Lambda^M_{-,i}, \Lambda^m_{+,i-1}, \Lambda^M_{+,i-1} \right\} \end{cases} \quad (5.41)$$

This last step ensures that two adjacent cells have the same limiting parameters, and, hence, the fluxes are continuous at the intercell boundaries and the conservation property of the scheme remains untouched. In practice, this lower bound is enforced using $m' = \min_i \{u^{n+1}_i, m + \epsilon\}$ where u^{n+1}_i is the first-order solution and ϵ is a small number near machine epsilon. Similarly, the upper bound is enforced using $M' = \max_i \{u^{n+1}_i, M - \epsilon\}$.

This approach is inspired by the flux-limiting approach introduced in Ref. [4] for one-dimensional conservation law solved using the flux-based WENO schemes [5]. The extensions to two-dimensional scalar conservation law and one- and two-dimensional Euler systems have also been introduced [6–8].

5.3 Lax-Friedrichs Scheme for System of Conservation Laws

There are two main differences between the LF scheme for a scalar equation and the LF scheme for a system of equations. These differences are in the definition of the stabilization parameter and the definition of the upwinding scheme. The stabilization parameter α for the system of nonlinear conservation law is the maximum of the absolute value of the eigenvalues of the Jacobian matrix. If the system has m equations for solution vector $\bar{u}$, the Jacobian matrix is

$$[J_k] = \left[\frac{\partial f_{k,i}(\bar{u})}{\partial u_j}\right] \quad \forall k = 1, \cdots, n, i = 1, \cdots, m, j = 1, \cdots, m \quad (5.42)$$

where n is the total number of grid points. The eigenvalues of the Jacobian matrix are denoted as $\lambda_{k,j}$ and they depend on the solution vector $\bar{u}$. Then,

$$\alpha = \max_{k=1}^{n} \max_{j=1}^{m} |\lambda_{k,j}(\bar{u})| \quad (5.43)$$

Here, the stabilization is defined over all spatial grid points. Note that similar to the scalar case, the maximum is obtained over a local stencil or over all grid points giving rise to the local and global LF schemes, respectively.

For the Euler equation of gas dynamics, the eigenvalues of the Jacobian matrix are

$$\lambda_{k,1} = v - c \quad \lambda_{k,2} = v \quad \lambda_{k,3} = v + c \quad (5.44)$$

Therefore, in the context of the Euler equations, the stabilization parameter is defined as

$$\alpha = \max_{k=1}^{n}(|v_k| + c_k) \quad (5.45)$$

Moreover, the direction of the wave propagation in the case of a system of equations can only be interpreted as the direction of characteristics. Therefore, for proper upwinding, the WENO reconstruction must take place on the characteristic variables. Now the question is what set of variables should be chosen for transformation to the characteristic space. Should we use conservative variables or primitive variables? As discussed above, primitive variables are more advantageous for satisfying the non-oscillatory and positivity of certain properties, and, thus, are more robust. As long as the eigenvectors of the system can be computed easily and with non-significant computational cost, the reconstruction should be performed on the transformation of the primitive variables to the characteristic space. However, if the eigenvectors are computationally expensive, then the WENO (or TENO) reconstruction can be performed directly on the primitive variables. Specifically, given the vector of primitive variables $\bar{w}$ and the matrices of the eigendecomposition $[Q]$ and $[Q^{-1}]$, we define the WENO reconstruction as

5.3 Lax-Friedrichs Scheme for System of Conservation Laws

$$\bar{w}_{\text{WENO}} = [Q] \, \mathcal{N}_{\text{WENO}}([Q^{-1}]\bar{w}) \tag{5.46}$$

where $\mathcal{N}_{\text{WENO}}(\bar{u})$ is a WENO reconstruction applied to vector $\bar{u}$. A detailed discussion of the characteristic implementation of the WENO or TENO schemes has been given in Sect. 2.3.4.1.

For the Euler equations of the gas dynamics, with $\bar{w} = (\rho, v, p)^T$ as primitive variable vector, and $\bar{u} = [\rho, \rho v, E]^T$ as a conservative variable vector, the spatial derivative approximation of the mass flux using Scheme 1 for the high-order derivative is expressed as

$$\left.\frac{\partial(\rho v)}{\partial x}\right|_{x_i} \approx \frac{1}{h} \sum_{j=1}^{r+1} d_j \left\{ \frac{1}{2}\left[(\rho v)^+_{i+j-\frac{1}{2}} + (\rho v)^-_{i+j-\frac{1}{2}}\right] - \frac{1}{2}\left[(\rho v)^+_{i-j+\frac{1}{2}} + (\rho v)^-_{i-j+\frac{1}{2}}\right] \right\} \tag{5.47}$$

where d_js are listed in Table 2.2. A similar expression can be used for the derivative approximation using Scheme 2. For details consult with Sect. 2.2.3. The positive and negative fluxes in Eq. (5.47) are obtained as

$$(\rho v)^+_{i+j-\frac{1}{2}} = \rho^+_{\text{WENO},i+j-\frac{1}{2}} \, v^+_{\text{WENO},i+j-\frac{1}{2}} + \alpha \, \rho^+_{\text{WENO},i+j-\frac{1}{2}} \tag{5.48a}$$

$$(\rho v)^-_{i+j-\frac{1}{2}} = \rho^-_{\text{WENO},i+j-\frac{1}{2}} \, v^-_{\text{WENO},i+j-\frac{1}{2}} - \alpha \, \rho^-_{\text{WENO},i+j-\frac{1}{2}} \tag{5.48b}$$

where the superscript "+" or "−" on a primitive variable signifies a left- or right-biased spatial approximation. The stabilization parameter is given in (5.45). Also, the columns of the matrix Q are eigenvectors of the quasi-linear form of the Euler equations with respect to the primitive variables ρ, v, and p.

The derivative of momentum and energy fluxes are computed analogously to that of the mass flux given in Eqs. (5.47) and (5.48).

Similar expressions apply if instead of WENO reconstructions, TENO reconstructions are used.

5.3.1 Numerical Example of Comparison of Approximation Orders for Ultrasound Pulse Propagation

We now compare the performance of WENO schemes of different approximation orders in solving the Euler equations of gas dynamics as presented in the author's article [9]. The propagation of a short ultrasound pulse in a $[-10, 10]$ medium of air is considered [9]. The pulse enters the domain from the left side with a sinusoidal velocity profile of $u = A \sin(2\pi f)$ where $A = 0.058$ and $f = 3$ are the amplitude and frequency of the wave. Given this velocity at the left boundary, density, and pressure are imposed using the characteristic method and isentropic relations for a period of $T = 1$, which corresponds to three wavelengths. Specifically, at each time step, knowing the properties at the second left-most grid point, u_2, c_2, p_2, ρ_2, and the boundary velocity, u_1, the boundary pressure and density are determined as

$$-u_1 + \frac{2c_1}{\gamma - 1} = -u_2 + \frac{2c_2}{\gamma - 1} \quad \text{along} \quad \frac{dx}{dt} = u - c \quad (5.49)$$

This implies

$$c_1 = c_2 + (u_1 - u_2)(\gamma - 1)/2, \quad (5.50a)$$

$$\rho_1 = \rho_2 (c_1/c_2)^{2/(\gamma-1)} \quad (5.50b)$$

$$p_1 = p_2 (c_1/c_2)^{2\gamma/(\gamma-1)} \quad (5.50c)$$

The Euler equations of gas dynamics are solved with the third-, fifth-, and ninth-order accurate WENO scheme. The fourth-order Runge-Kutta is used for the integration in time. The results for all orders are reported with a CFL condition of 0.5, as no noticeable differences were observed with smaller CFL conditions. The computed pressures are shown in Fig. 5.1. As seen the pressure profile deforms due to nonlinear shock formation. The reduced pressure amplitude and uneven wavelength are due to the action of shock formation, which results in a reduction of total kinetic energy and an increase of total internal energy. Figure 5.2 shows the total energy, internal and kinetic energy illustrating this phenomenon where the total energy is almost constant, and the kinetic energy drops over time as internal energy rises.

Comparison of the computed pressure profiles obtained with different orders of $n = 3, 5$ and 9 and the CPU times for each computation are shown in Table 5.1. The L_1 error values at two resolutions of $N = 1000$ and $N = 2000$ and the corresponding

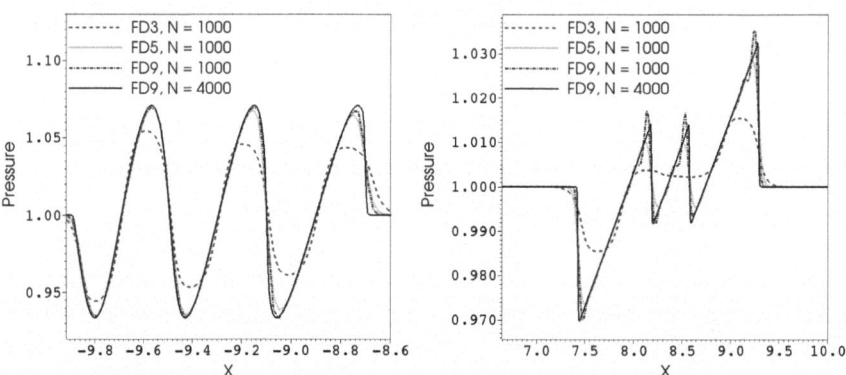

Fig. 5.1 Computed pressure for an ultrasound pulse entering a $[-10, 10]$ domain of air from left side for $N = 1000$ grid points and finite difference orders of third-order (FD3), fifth-order (FD5) and ninth-order (FD9) and for $N = 4000$ with FD9 solution; left $T = 1.0$ and right $T = 16.0$ clearly depicting the nonlinear deformation of an initial sinusoidal wave with compression/expansion of wave length and pressure amplitude attenuation. These results were first appeared in the author's earlier work [9]

5.3 Lax-Friedrichs Scheme for System of Conservation Laws

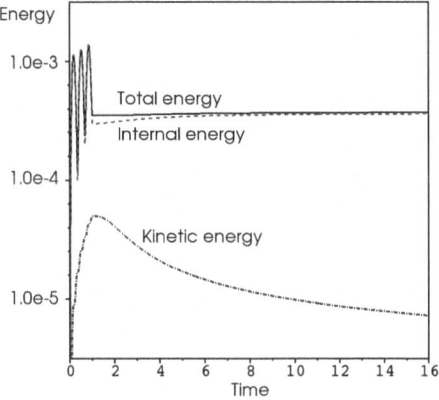

Fig. 5.2 The kinetic, internal, and total energy of air computed over time for an ultrasound pulse entering a $[-10, 10]$ domain of air from the left side for $N = 4000$ grid points and approximation order of $n = 5$, demonstrating the decline of the kinetic energy and the rise of internal energy which are the result of the nonlinear shock formation. These results were first reported in the author's earlier work [9]

Table 5.1 Errors in pressure and CPU times for computations of an ultrasound pulse propagating from the left in the $[-10, 10]$ domain of air divided into two grids of $N = 1000$ and $N = 2000$ cells. Data for finite difference methods of third order (FD3), fifth order (FD5), and ninth order (FD9) are shown. All CPU times are normalized by the CPU time for FD3 on the courser grid of $N = 1000$. These results were first reported in the author's earlier work [9]

	FD order = 3		FD order = 5		FD order = 9	
	L_1 Error (%)	CPU Time	L_1 Error (%)	CPU Time	L_1 Error (%)	CPU Time
N = 1000	52	1	13	1.4	11	2.5
N = 2000	22	4	5.0	5.6	3.5	10.0

CPU times are shown. The L_1 error is computed using $N = 10000$ of the fifth-order method as an "exact" solution. From the table, the third-order scheme results in a solution with 22% errors in 4s while the fifth and ninth-order methods yield solutions with 13% and 11% errors in 1.5s and 2.5s, respectively. In the rubric of error level per CPU time, the fifth- and ninth-order methods outperform the third-order methods for this test problem.

5.3.2 Positivity of Pressure in First-Order Scheme

Positivity of density in the first-order LF scheme follows the proof presented in Sect. 5.1.1. We now need to discuss the positivity of the pressure. The proof of the positivity of the pressure at time step $n + 1$ relies on the positivity of the pressure at time step n and the concavity of the pressure as a function of conservation variable vectors.

Denoting the conservative variable vector in the Euler system by $\bar{u}$, we first note that the pressure is a concave function, i.e., for $0 \leq \theta \leq 1$ and any two conservative variable vector $\bar{u}_1$ and $\bar{u}_2$,

$$p[\theta \bar{u}_1 + (1-\theta)\bar{u}_2] \geq \theta p(\bar{u}_1) + (1-\theta)p(\bar{u}_2) \tag{5.51}$$

(The proof is left to an exercise.) Expressing the first-order LF scheme for the Euler system as

$$\bar{u}_j^{n+1} = \left(1 - \frac{\tau\alpha}{h}\right)\bar{u}_j^n + \frac{\tau\alpha}{2h}\left[\bar{u}_j^n - \frac{1}{\alpha}\bar{f}(\bar{u}_{j+1}^n)\right] + \frac{\tau\alpha}{2h}\left[\bar{u}_{j-1}^n + \frac{1}{\alpha}\bar{f}(\bar{u}_{j-1}^n)\right] \tag{5.52}$$

Since $\bar{u}_j^{n+1}$ is a convex combination of the three vectors $\bar{u}_j^n$, $\bar{u}_j^n - \frac{1}{\alpha}\bar{f}(\bar{u}_{j+1}^n)$ and $\bar{u}_{j-1}^n + \frac{1}{\alpha}\bar{f}(\bar{u}_{j-1}^n)$, and at time step n the solution assumed to have positive density, it is sufficient to show $\bar{u}_j^n - \frac{1}{\alpha}\bar{f}(\bar{u}_{j+1}^n)$ and $\bar{u}_{j-1}^n + \frac{1}{\alpha}\bar{f}(\bar{u}_{j-1}^n)$ yield positive pressures. Dropping subscripts and superscripts for simplicity and writing the internal energy per unit volume, $\tilde{e} = \rho e$, since $p = (\gamma - 1)\tilde{e}$, it is sufficient to show the positivity of $(\tilde{e}\bar{u} \pm \frac{1}{\alpha}\bar{f}(\bar{u}))$. To this end, we first note that the components of $\bar{u} \pm \frac{1}{\alpha}\mathbf{f}(\mathbf{u})$ are

$$\bar{u} \pm \frac{1}{\alpha}\bar{f}(\bar{u}) = \left[\rho \pm \rho v/\alpha, \rho v \pm (\rho v^2 + p)/\alpha, E \pm [(E+p)v]/\alpha\right]^T \tag{5.53}$$

Then,

$$\tilde{e}\left(\bar{u} \pm \frac{1}{\alpha}\bar{f}(\bar{u})\right) = \left(1 \pm \frac{v}{\alpha}\right)E \pm \frac{vp}{\alpha} - \frac{1}{2}\frac{[(1 \pm v/\alpha)\rho v \pm p/\alpha]^2}{(1 \pm v/\alpha)\rho} \tag{5.54a}$$

$$= \left[1 - \frac{p^2}{2(\alpha \pm v)^2 \rho^2 e}\right]\left(1 - \frac{v}{\alpha}\right)\rho e \tag{5.54b}$$

Clearly, for the positivity of $\tilde{e}(\bar{u} \pm \frac{1}{a_0}\bar{f}(\bar{u}))$, it is sufficient to have

$$\left[1 - \frac{p^2}{2(\alpha \pm v)^2 \rho^2 e}\right] \geq 1 \tag{5.55}$$

Using the relation $p = (\gamma - 1)\rho e$ and the speed of sound for the ideal gas $c = \sqrt{\gamma p/\rho}$, we obtain the condition on α for which the positivity is satisfied, namely

$$\alpha \geq |v| + \frac{p}{\sqrt{\rho^2 e}} = |v| + \left[\frac{(\gamma-1)p}{\rho}\right]^{1/2} = |v| + c\left(\frac{\gamma-1}{\gamma}\right)^{1/2} \tag{5.56}$$

The α from this requirement is smaller than the α needed for the stability of the LF scheme, $\alpha = |v| + c$, resulting in the positivity of the pressure for the first-order LF scheme without any penalty on the CFL condition.

5.3.3 Positivity Enforcement of Pressure in High-Order Scheme

The enforcement of the positivity of pressure relies heavily on the concavity of the pressure function as given in Eq. (5.51). Furthermore, since the conservative variables are linearly dependent on the limiting vector $\bar{\theta} = (\theta_1, \theta_2)$, then for two limiting vectors $\bar{\theta}_1$ and $\bar{\theta}_1$, we also have

$$p(\eta\bar{\theta}_1 + (1-\eta)\bar{\theta}_2) \geq \eta p(\bar{\theta}_1) + (1-\eta) p(\bar{\theta}_2) \qquad (5.57)$$

As a result of this concavity property, the admissible set (the set of $\bar{\theta}$ that yields positive pressure) is convex, proved by showing that the line connecting two points within the set, lies entirely within the set, i.e., all points on the line yield positive pressure. The convexity of the admissible set also means there is no hole in the set, and any subset of the admissible set consists of points that yield positive pressure.

Taking the limiting parameters as zero, we recover the first-order scheme where its pressure is positive based on the proof in Sect. 5.3.2. Therefore, using the zero limiting parameters $\bar{\Lambda}_0 = (0,0)$ and limiting parameters obtained from enforcing the positivity of the density, $\bar{\Lambda} = (\Lambda_{+,i}, \Lambda_{-,i})$ for each cell I_i, obtained prior to enforcing the intercell continuity of the limiting (5.41), a set spanning a quadrilateral is formed with the four vertices as

$$S_\rho = \{\bar{\Lambda}_0 = (0,0), \bar{\Lambda}_1 = (\Lambda_{+,i}, 0), \bar{\Lambda}_2 = (0, \Lambda_{-,i}), \bar{\Lambda}_3 = (\Lambda_{+,i}, \Lambda_{-,i})\} \qquad (5.58)$$

Now, we can take the three nonzero limiting parameters $\bar{\Lambda}_1$, $\bar{\Lambda}_2$, and $\bar{\Lambda}_3$ and examine if any one yields a positive pressure. The pressure results from

$$p = (\gamma - 1)\left(E - \frac{m}{\rho}\right) \qquad (5.59)$$

where the total energy E, the momentum m, and the density ρ are computed using the modified conservative variables using the limiting parameters $\bar{\Lambda}_j$ as

$$\tilde{U}_i^{n+1} = \bar{u}_i^n - \frac{\tau}{h}\left(\tilde{\bar{F}}_{i+\frac{1}{2}}^n - \tilde{\bar{F}}_{i-\frac{1}{2}}^n\right) \qquad (5.60)$$

with

$$\tilde{F}^n_{i+\frac{1}{2}} = \hat{f}^n_{i+\frac{1}{2}} + \Lambda_{j,1}\left(\hat{F}^n_{i+\frac{1}{2}} - \hat{f}^n_{i+\frac{1}{2}}\right) \qquad (5.61a)$$

$$\tilde{F}^n_{i-\frac{1}{2}} = \hat{f}^n_{i-\frac{1}{2}} + \Lambda_{j,2}\left(\hat{F}^n_{i-\frac{1}{2}} - \hat{f}^n_{i-\frac{1}{2}}\right) \qquad (5.61b)$$

where $\bar{u}^n_i$ is the vector of conservative variables at time step n, and $\hat{f}^n_{i\pm 1/2}$ and $\hat{F}^n_{i\pm 1/2}$ are the first and high-order LF fluxes.

The limiting parameters $\bar{\Lambda}$s giving positive pressures are stored unchanged as $\bar{L}_j = \bar{\Lambda}_j$; the ones yielding negative pressures are scaled by a to-be-determined constant r, $\bar{L}_j = r\bar{\Lambda}_j$. Then, r is determined such that the pressure evaluated using $\bar{L}_j$ yields positive value, i.e.,

$$p(\bar{L}) = p(r\bar{\Lambda}) = \epsilon \implies p(r\bar{\Lambda}) - \epsilon = 0 \qquad (5.62)$$

where ϵ is a positive tolerance and $0 \leq r \leq 1$. The obtained three limiting vectors $\bar{L}_1$, $\bar{L}_2$, and $\bar{L}_3$ and the origin $\bar{\Lambda}_0$ span a subset of S_p, denoted by S_p,

$$S_p = \{\bar{\Lambda}_0 = (0,0), \bar{L}_1, \bar{L}_2, \bar{L}_3\} \qquad (5.63)$$

that yields positive pressure. To yield the highest accuracy, we need to choose the highest values of limiting parameters. Therefore, the vertex of the quadrilateral with the largest limiting values should be selected. However, in practice, due to floating point precision a small deviation from this vertex can produce a point outside of the admissible set and yield a negative pressure. To remedy this, we can form a rectangular subset of the quadrilateral S_p with the following vertices

$$S'_p = \{[0,0], [\min(L_{1,1}, L_{3,1}), 0], [0, \min(L_{2,2}, L_{3,2})], [\min(L_{1,1}, L_{3,1}), \min(L_{2,2}, L_{3,2})]\} \qquad (5.64)$$

which ensures the positivity of the pressure with the highest level of accuracy by choosing the following limiting vector

$$\bar{L}^p = [\min(L_{1,1}, L_{3,1}) - \epsilon, \min(L_{2,2}, L_{3,2}) - \epsilon] \qquad (5.65)$$

where ϵ is a small tolerance.

Now, the obtained limiting vector for each cell, $\bar{L}^p_i$, can be used to determine limiting parameters that ensure continuity of the intercell fluxes similar to (5.41) as (Fig. 5.3)

$$\theta_{i+\frac{1}{2}} = \min\left\{L^p_{i,1}, L^p_{i+1,2}\right\} \qquad (5.66a)$$

$$\theta_{i-\frac{1}{2}} = \min\left\{L^p_{i-1,1}, L^p_{i,2}\right\} \qquad (5.66b)$$

5.4 Numerical Examples for Positivity Preservation

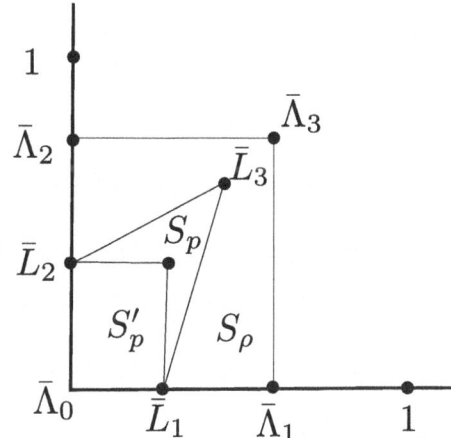

Fig. 5.3 The sets spanned by the limiting parameters $\theta_{i+1/2}$ and $\theta_{i-1/2}$ for positivity of density (S_ρ) and positivity of pressure (S_p and S'_p) given in Eqs. (5.58), (5.63), and (5.64)

Finally, using the above limiting parameters $\theta_{i\pm1/2}$, the modified conservative variables are determined using (5.60) with

$$\tilde{F}^n_{i+\frac{1}{2}} = \hat{\tilde{f}}^n_{i+\frac{1}{2}} + \theta_{i+\frac{1}{2}} \left(\hat{\tilde{F}}^n_{i+\frac{1}{2}} - \hat{\tilde{f}}^n_{i+\frac{1}{2}} \right) \tag{5.67a}$$

$$\tilde{F}^n_{i-\frac{1}{2}} = \hat{\tilde{f}}^n_{i-\frac{1}{2}} + \theta_{i-\frac{1}{2}} \left(\hat{\tilde{F}}^n_{i-\frac{1}{2}} - \hat{\tilde{f}}^n_{i-\frac{1}{2}} \right) \tag{5.67b}$$

5.3.3.1 Determining the Scaling for the Limiter

To determine r for scaling the limiting parameters obtained from positivity enforcement of the density such that the pressure becomes positive, the nonlinear equation (5.62) needs to be solved. The root finding method we found particularly effective was the Brent method [10]. As with any root finding scheme, the Brent method is an iterative scheme, and care must be exercised in the choices of tolerances. Reference [3] gives some effective strategies for the successful solution of (5.62) using the Brent method.

5.4 Numerical Examples for Positivity Preservation

Several one- and two-dimensional test problems have been considered to verify the accuracy and performance of the high-order positivity-preserving WENO finite difference scheme described in the previous section. For all test problems, a standard fourth-order Runge-Kutta time integration was used. In all cases, the CFL number was set to 0.5, yielding stable and positive solutions; however, for convergence studies, we used smaller CFL numbers to avoid contamination of the solution by temporal errors. Five problems are presented: Two problems related to scalar conservation laws and three problems related to the Euler system [3].

5.4.1 Linear Advection with Continuous Initial Data

We first consider the linear advection equation

$$\frac{\partial u}{\partial t} + \frac{\partial u}{\partial x} = 0 \tag{5.68}$$

with a smooth initial solution $u(x, 0) = \sin^2(\pi x - \sin(\pi x)/\pi)$ over the domain $x \in [-1, 1]$. Periodic boundary conditions are enforced, and the maximum time is set to $t_{\max} = 20$ corresponding to 10 full advections of the property u through the computational domain. The exact solution is

$$u(x, t) = \sin^2\left(\pi(x - t) - \frac{\sin(\pi(x - t))}{\pi}\right) \tag{5.69}$$

Note that to avoid polluting the spatial error with temporal error, we use a small CFL number (0.05) for this example. The results are tabulated in Table 5.2, where the L_∞

Table 5.2 Convergence comparison for scalar, linear advection with and without maximum-principle preserving limiters with CFL = 0.05, demonstrating that formal high order of accuracy is maintained while maximum-principle preserving is enforced

Order	N	Without limiters				With limiters			
		Error	min(u)	max(u)	Rate	Error	min(u)	max(u)	Rate
5	20	8.152e-02	−6.084e-03	1.001		9.317e-02	8.800e-17	1.000	
	40	2.211e-03	−9.728e-05	1.000	5.204	2.552e-03	9.926e-17	1.000	5.190
	80	6.361e-05	−2.966e-06	1.000	5.119	7.466e-05	9.992e-17	1.000	5.095
	160	1.998e-06	−9.281e-08	1.000	4.992	2.345e-06	9.999e-17	1.000	4.993
7	20	2.772e-02	−3.329e-04	1.001		2.947e-02	9.610e-17	1.000	
	40	2.402e-04	−1.377e-06	1.000	6.851	3.251e-04	9.987e-17	1.000	6.503
	80	1.479e-06	−3.349e-09	1.000	7.343	1.492e-06	9.995e-17	1.000	7.768
	160	9.797e-09	−2.290e-11	1.000	7.238	9.883e-09	9.999e-17	1.000	7.238
9	20	8.687e-03	−5.520e-05	1.003		1.126e-02	1.033e-05	1.000	
	40	1.224e-05	5.464e-09	1.000	9.471	1.224e-05	5.464e-09	1.000	9.846
	80	2.845e-08	5.543e-12	1.000	8.749	2.845e-08	5.543e-12	1.000	8.749
	160	9.901e-11	5.346e-15	1.000	8.166	9.901e-11	5.346e-15	1.000	8.166
11	20	3.903e-03	−7.674e-05	1.002		6.955e-03	3.860e-05	1.000	
	40	1.465e-06	2.018e-10	1.000	11.4	1.465e-06	2.018e-10	1.000	12.2
	80	1.650e-09	2.776e-14	1.000	9.794	1.650e-09	2.776e-14	1.000	9.794

5.4 Numerical Examples for Positivity Preservation

error norms for two cases, with and without the positivity limiters, are given along with the global minimum and maximum values obtained throughout the numerical solution.

The effectiveness and accuracy of the positivity scheme are demonstrated in Table 5.2. In all simulations, the solutions' upper and lower bounds are preserved with no observable impact upon accuracy. In other words, optimal high-order accuracy and maximum-principle property are preserved using our proposed scheme.

5.4.2 Inviscid Burgers Equation

Next, we consider the inviscid Burgers equation

$$\frac{\partial u}{\partial t} + \frac{\partial}{\partial x}\left(\frac{u^2}{2}\right) = 0 \tag{5.70}$$

on the domain $x \in [-1, 1]$ with $t_{max} = 0.2$ and the initial condition $u(x, 0) = \frac{1+\sin(\pi x)}{2}$. For $t_{max} = 0.2$, the solution remains smooth as the time for shock formation is $t_s = 2/\pi$. The exact solution prior to the shock formation is obtained by solving the nonlinear implicit solution: $u = (1 + \sin(\pi(x - ut)))/2$. Once again, we use a small CFL number to avoid contaminating the convergence rate with temporal error. Table 5.3 presents the L_∞ error norms and the minimum and maximum values.

Table 5.3 Convergence comparison for the inviscid Burgers equation with and without maximum-principle preserving scheme with CFL $= 0.05$, showing the formal order of accuracy is maintained while maximum-principle preserving is enforced

Order	N	Without limiters				With lmiters			
		Error	min(u)	max(u)	Rate	Error	min(u)	max(u)	Rate
5	40	1.62e-05	−1.39e-06	1.00		1.72e-05	9.97e-17	1.00	
	80	5.41e-07	−4.61e-08	1.00	4.90	5.63e-07	6.61e-10	1.00	4.93
	160	1.70e-08	−1.37e-09	1.00	4.99	2.11e-08	1.06e-11	1.00	4.74
	320	5.34e-10	−4.05e-11	1.00	4.99	9.36e-10	1.67e-13	1.00	4.49
7	40	2.37e-06	−1.90e-08	1.00		2.38e-06	9.98e-17	1.00	
	80	2.24e-08	−1.28e-10	1.00	6.72	2.25e-08	9.99e-17	1.00	6.73
	160	1.82e-10	−9.70e-13	1.00	6.95	1.84e-10	8.28e-17	1.00	6.93
	320	1.49e-12	−7.51e-15	1.00	6.94	1.67e-12	8.10e-19	1.00	6.78
9	40	5.55e-07	−1.36e-10	1.00		5.55e-07	9.99e-17	1.00	
	80	1.77e-09	−2.34e-13	1.00	8.30	1.77e-09	5.85e-18	1.00	8.30
	160	3.97e-12	−4.34e-16	1.00	8.80	4.13e-12	1.75e-19	1.00	8.74
11	20	3.90e-03	−7.67e-05	1.00		6.95e-03	3.86e-05	1.00	
	40	1.47e-06	2.02e-10	1.00	11.4	1.47e-06	2.02e-10	1.00	12.2
	80	1.65e-09	2.78e-14	1.00	9.79	1.65e-09	2.78e-14	1.00	9.79

Note that for the fine meshes ($N \geq 160$), the higher order schemes give error values of the same order as the error associated with double machine precision; these values are meaningless; hence, we do not report them.

These results confirm the high-order accuracy and maximum-principle preserving quality of the scheme for the nonlinear case. For all orders and grid sizes, the scheme without limiters obtained values below zero while the scheme with limiters always remained positive.

5.4.3 Euler System Interacting Blast Waves

Here, we simulate the collision of a right-moving Mach 199 shock wave with a left-moving Mach 63 shock wave. This is a modified form of the interacting blast wave problem discussed in [11, 12]. Specifically, we have moved the two initial blast waves closer to each other to reduce the effects of numerical diffusion and preserve the sharpness of the shock discontinuities at the collision, making it significantly more challenging to compute. Owing to the large pressure ratios across the shocks and the very low minimum pressure values, this problem is designed to verify the preservation of pressure positivity. The domain is [0, 1], and the specific heat ratio is selected to correspond to air ($\gamma = 1.4$). The initial data are

$$(\rho, v, p) = \begin{cases} (1, 0, 1000) & 0 \leq x < 0.6 \\ (1, 0, \frac{1}{100}) & 0.6 \leq x < 0.7 \\ (1, 0, 100) & 0.7 \leq x \leq 1. \end{cases} \quad (5.71)$$

Figure 5.4 shows eleventh-order pressure solutions at various time steps. Panel (a) shows the two shock waves approaching each other and refraction waves propagating in the opposite direction. Panel (b), the two shock waves collide to generate very high pressure. Note how few grid points are contained within this jump; for the eleventh-order solution, this feature is accurately captured by about five grid points, demonstrating the high-order schemes' ability to resolve extremely fine features. In contrast, the first-order scheme required about ten times the grid resolution to approach the sharpness of the high-order solutions (this is further explored in Fig. 5.6). Panel (c) shows the left and rightward traveling shocks propagating after the collision into the air previously compressed by the wave traveling in the opposite direction. Finally, panel (d) shows the rightward traveling transmitted wave overtaking the initial rightward traveling expansion wave.

The seventh-order density solutions are shown in Fig. 5.5 for the same time steps as in Fig. 5.4. The first panel shows the two shock waves traveling toward each other, and the second shows the sudden density jump immediately following their collision.

The third and fourth panels show the continued evolution of the complex post-shock density profile. Note the sharp features present throughout the simulation; as

5.4 Numerical Examples for Positivity Preservation 149

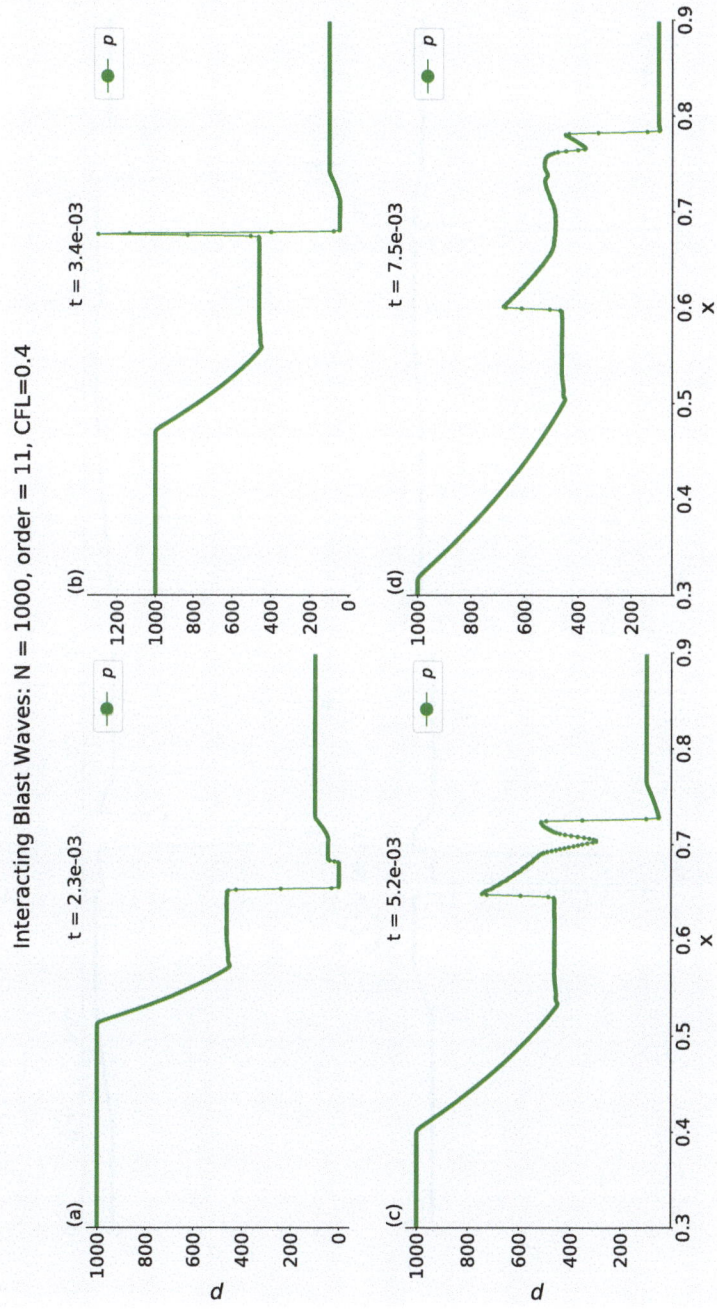

Fig. 5.4 Pressure solutions for the interacting blast waves problem

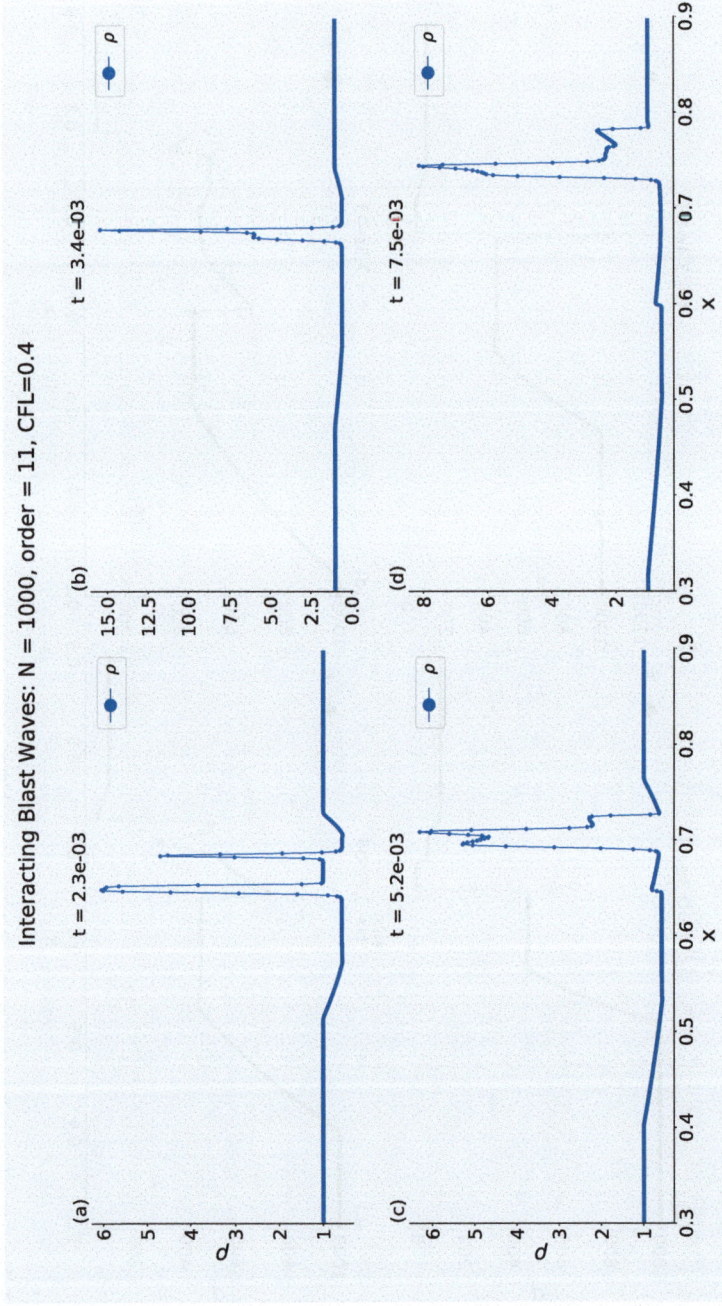

Fig. 5.5 Density solutions for the interacting blast waves problem

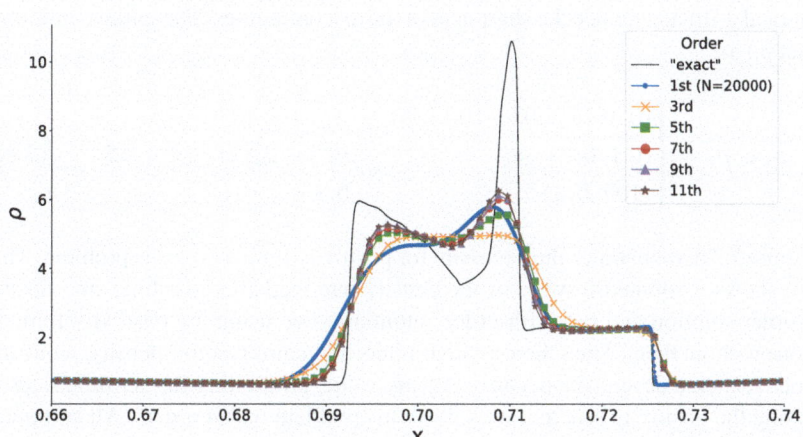

Fig. 5.6 Comparison of high-order density solutions for interacting blast waves problem

evidenced in Fig. 5.6, the high-order schemes are able to effectively capture these profiles, while the lower order schemes require significant grid refinement to approach comparable sharpness.

A comparison of the third-, fifth-, seventh-, ninth-, and eleventh-order density solutions at $t = 5.25 \times 10^{-3}$ is given in Fig. 5.6. The solid black line is the "exact" converged solution calculated using the seventh order scheme with 10^4 grid points. The third-order solution fails to capture any sharp features. The fifth-order solution fares better, but it is noticeably less sharp than the higher orders, which are nearly indistinguishable at this scaling. However, near the peak, it is clear that the eleventh-order solution is indeed the sharpest. This problem demonstrates the positivity-preserving quality of the scheme in the presence of very large pressure jumps (of magnitude 10^5) from low-pressure regions ($\min\{p\} = 0.01$). Furthermore, it was found that for all high-order schemes, the positivity limiters were required to prevent the simulation from crashing due to negative pressure, thus demonstrating the necessity of the positivity algorithm.

5.4.4 Euler System Shock, Interface Problem

Next, we consider the problem of a strong shock wave ($M = 8.96$) interacting with a density interface. This problem is designed to test the ability of the algorithm to preserve the positivity of density in the extreme case of a huge density jump. The shock is initially traveling rightward through helium ($\gamma = 1.67$) and is placed two grid points away from the density interface. This is done to make the problem more challenging by forcing the shock/interface interaction to occur before the effects of numerical diffusion reduce the sharpness of the discontinuities. The initial conditions are set up as

$$(\rho, v, p) = \begin{cases} (0.384, 27.086, 100.176) & 0 \leq x < (0.5 - 2\Delta x) \\ (0.1, 0, 1) & (0.5 - 2\Delta x) \leq x \leq 0.5 \\ (100, 0, 1) & 0.5 < x \leq 1 \end{cases} \quad (5.72)$$

Figure 5.7 demonstrates the necessity for positivity limiters for this problem. This figure shows a zoomed-in view of the density interface after one-time step for the first-order solution and two fifth-order solutions—one using the positivity limiters and one without them. The scheme with limiters maintains positive density, while the standard scheme develops a negative density value, prohibiting further time steps.

Using the positivity scheme, we solved this problem for all orders. All computations were successful CFL number as large as 0.5. Solutions at $t = 0.01$ are presented in Figs. 5.8, 5.9, and 5.10; the black lines indicate the grid converged "exact" solution calculated using the seventh order scheme with 10^4 points, and the dashed lines give the initial conditions.

In the pressure trace, we observe the incident shock wave traveling rightward through the high-density region near $x \approx 0.55$. Additionally, a reflected wave is propagating leftward into the initial high-pressure region near $x \approx 0.35$. The density jump observed immediately trailing the rightward traveling shock results from the shock wave leading to the advection of the initial density/temperature interface. Note the increasing sharpness of this feature with approximation order.

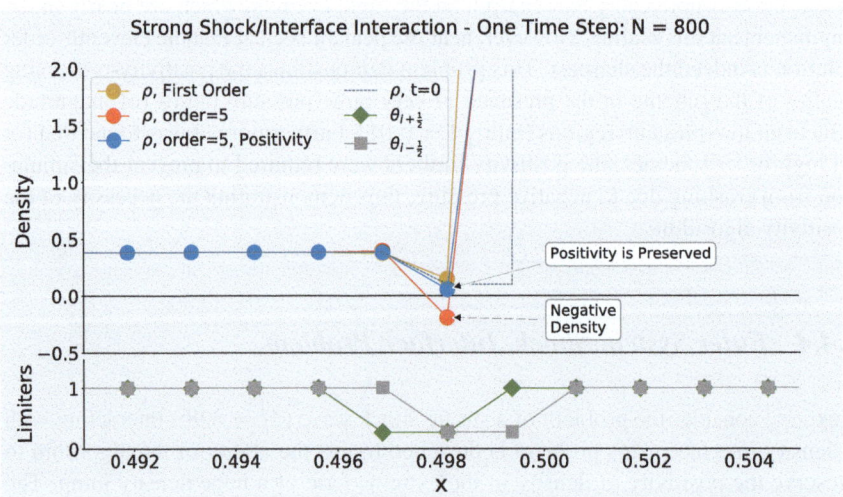

Fig. 5.7 Demonstration of positivity-preserving scheme for strong shock/interface interaction

5.4 Numerical Examples for Positivity Preservation

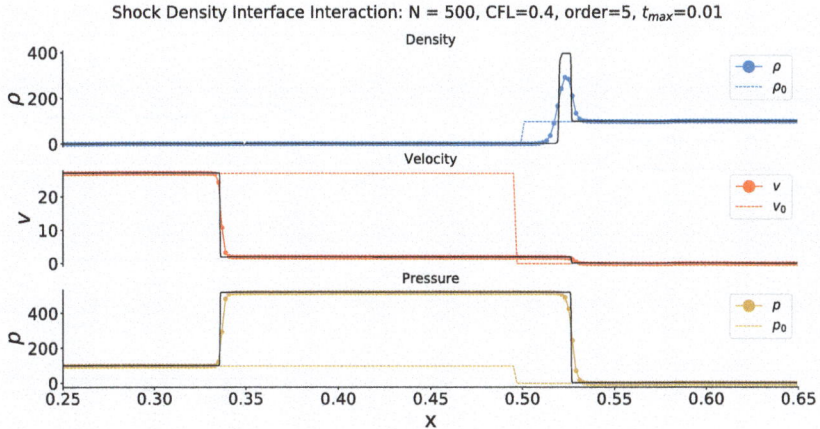

Fig. 5.8 Fifth-order solution for shock air/helium interface problem

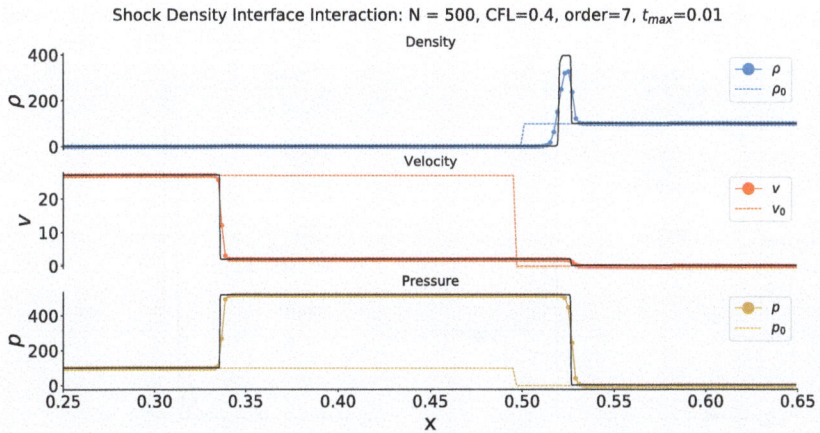

Fig. 5.9 Seventh-order solution for shock air/helium interface problem

To confirm the benefits of high-order schemes, Fig. 5.11 compares high-order solutions to a first-order solution of comparable accuracy. The number of grid points and CPU time for each scheme are shown in the legend. The fifth-order accurate scheme outperforms the first-order scheme as it delivers a similar level of accuracy to the first-order scheme at an approximately 35-fold reduction in CPU time. No changes in the efficiency data resulted at lower CFL numbers.

This test problem demonstrates the algorithm's ability to maintain density positivity under extreme conditions successfully. Additionally, the significantly greater efficiency of high-order schemes in the presence of sharp features is demonstrated. As in the last test problem, the limiters were required for all orders to solve this problem.

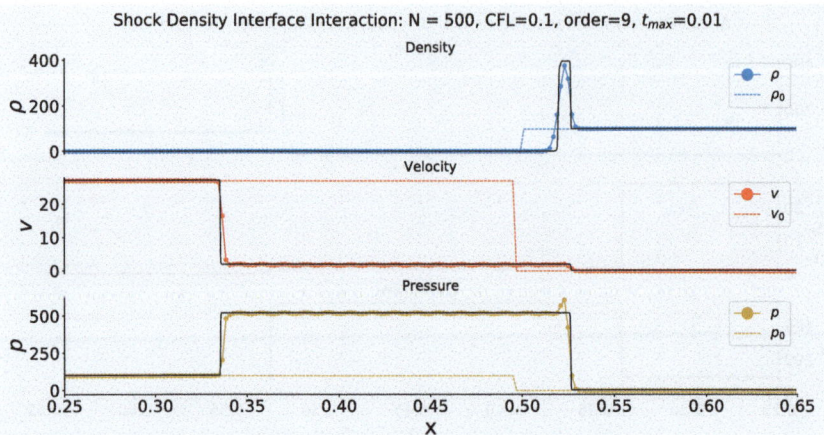

Fig. 5.10 Ninth-order solution for shock air/helium interface problem

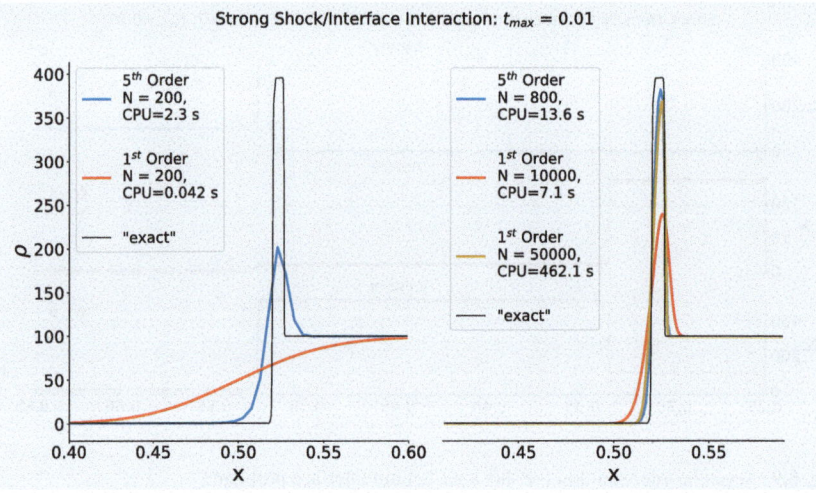

Fig. 5.11 Comparison of high- and low-order solutions for shock interface problem

5.4.5 2D Isentropic Vortex

We now verify the accuracy of the high-order positivity-preserving scheme for two-dimensional stationary isentropic vortex where the field variables are defined on the domain $[-10, 10] \times [-10, 10]$ as

5.4 Numerical Examples for Positivity Preservation

$$v_x = -\frac{\epsilon}{2\pi} \exp\left(\frac{1-r^2}{2}\right) y \qquad (5.73)$$

$$v_y = \frac{\epsilon}{2\pi} \exp\left(\frac{1-r^2}{2}\right) x \qquad (5.74)$$

$$\rho = \left[1 - \frac{(\gamma-1)\epsilon^2}{8\gamma\pi^2} \exp(1-r^2)\right]^{\frac{1}{\gamma-1}} \qquad (5.75)$$

$$p = \rho^\gamma - \Pi \qquad (5.76)$$

where $r^2 = x^2 + y^2$, and ϵ is the vortex strength parameter. The domain size is made sufficiently large to ensure the solution coincides with the undisturbed base field at the boundaries. The $\epsilon = 5$ with $\Pi = 0$ case is commonly used in the literature [13, 14] to verify the performance of high-order schemes. Since taking $\epsilon = 9$ generates low pressure ($p_{min} = 6 \times 10^{-3}$) and density ($\rho_{min} = 2 \times 10^{-2}$) values, this case has been used to confirm positivity preservation [8, 13, 15]. The fluid parameters are chosen as ($\gamma = 1.4$, $\Pi = 0.1$) to coincide with [13]. Periodic boundary conditions are enforced.

The results from the $\epsilon = 5$ case are presented in Table 5.4. Here, the L_1 and L_∞ errors for the density and pressure are shown, confirming the high-order convergence of our scheme. It is important to note that in order to observe the expected convergence

Table 5.4 Convergence study for stationary isentropic vortex with $\epsilon = 5$ and $t = 0.01$. These results were first appeared in the author's earlier work [3]

Order	Resolution	Density				Pressure			
		L_∞ Error	L_∞ Rate	L_1 Error	L_1 Rate	L_∞ Error	L_∞ Rate	L_1 Error	L_1 Rate
5	40 × 40	2.30e-03	3.26	2.59e-03	3.68	1.08e-02	3.19	2.36e-05	4.29
	80 × 80	7.95e-05	4.86	1.13e-04	4.52	5.33e-04	4.34	6.77e-07	5.12
	160 × 160	2.01e-06	5.30	3.26e-06	5.11	1.37e-05	5.28	1.73e-08	5.29
	320 × 320	3.28e-08	5.94	5.45e-08	5.90	1.31e-07	6.70	3.64e-10	5.58
7	40 × 40	8.47e-04	4.11	1.58e-03	3.92	6.41e-03	3.48	1.13e-05	4.98
	80 × 80	2.16e-05	5.29	4.29e-05	5.20	2.11e-04	4.93	1.51e-07	6.22
	160 × 160	2.41e-07	6.49	3.58e-07	6.91	1.65e-06	6.99	1.16e-09	7.02
	320 × 320	1.09e-09	7.79	1.42e-09	7.97	3.56e-09	8.86	6.29e-12	7.53
9	40 × 40	5.35e-04	4.33	1.15e-03	4.05	4.56e-03	3.64	7.12e-06	5.29
	80 × 80	1.06e-05	5.66	1.61e-05	6.17	9.05e-05	5.66	5.28e-08	7.07
	160 × 160	3.29e-08	8.33	3.66e-08	8.78	1.95e-07	8.86	1.22e-10	8.76
	320 × 320	3.69e-11	9.80	4.03e-11	9.83	1.07e-10	10.8	1.71e-13	9.47
11	40 × 40	3.92e-04	4.43	9.24e-04	4.15	3.58e-03	3.75	5.17e-06	5.42
	80 × 80	4.90e-06	6.32	6.10e-06	7.24	3.64e-05	6.62	2.24e-08	7.85
	160 × 160	4.18e-09	10.2	5.35e-09	10.2	3.23e-08	10.1	1.62e-11	10.4
	320 × 320	1.11e-12	11.9	1.61e-12	11.7	4.27e-12	12.9	6.14e-15	11.4

Table 5.5 Convergence study for stationary isentropic vortex with $\epsilon = 9$ and $t = 0.01$. These results were first appeared in the author's earlier work [3]

Order	Resolution	Density				Pressure			
		L_∞ Error	L_∞ Rate	L_1 Error	L_1 Rate	L_∞ Error	L_∞ Rate	L_1 Error	L_1 Rate
5	40×40	1.03e-02	3.06	1.31e-02	2.78	6.75e-02	2.58	9.79e-05	4.04
	80×80	5.41e-04	4.25	5.94e-04	4.47	3.06e-03	4.46	2.97e-06	5.04
	160×160	1.17e-05	5.53	1.60e-05	5.21	7.60e-05	5.33	7.07e-08	5.39
	320×320	1.28e-07	6.52	1.62e-07	6.63	4.63e-07	7.36	1.12e-09	5.99
7	40×40	7.67e-03	3.27	9.55e-03	3.12	5.47e-02	2.75	7.54e-05	4.29
	80×80	1.52e-04	5.66	2.67e-04	5.16	1.27e-03	5.43	8.84e-07	6.41
	160×160	1.49e-06	6.66	2.47e-06	6.76	1.35e-05	6.56	6.68e-09	7.05
	320×320	5.64e-09	8.05	7.59e-09	8.35	2.08e-08	9.34	2.76e-11	7.92
9	40×40	6.02e-03	3.50	7.51e-03	3.43	4.63e-02	2.92	6.17e-05	4.48
	80×80	7.00e-05	6.43	1.71e-04	5.46	9.59e-04	5.59	3.41e-07	7.50
	160×160	5.40e-07	7.02	5.24e-07	8.34	3.03e-06	8.31	1.04e-09	8.36
	320×320	8.75e-10	9.27	4.33e-10	10.2	1.16e-09	11.3	1.19e-12	9.76
11	40×40	4.90e-03	3.72	6.19e-03	3.70	4.06e-02	3.07	5.38e-05	4.61
	80×80	4.22e-05	6.86	1.11e-04	5.80	7.03e-04	5.85	1.96e-07	8.10
	160×160	2.23e-07	7.56	1.22e-07	9.82	8.13e-07	9.76	2.39e-10	9.68
	320×320	1.34e-10	10.7	4.43e-11	11.4	1.11e-10	12.8	8.24e-14	11.5

rate at any order of approximation, we had to sample the L_1 and L_∞ errors of the numerical solution at each grid refinement over a sufficiently large set of grid points, which we took to be the finest grid in the refinement sequence. For this error calculation, the solution from a coarse grid was obtained on the finest level by interpolation with the same accuracy order as the approximation. Failure to sample the error at any grid level over the finest grid level yields artificially low convergence rates and masks the true high-order convergence.

Table 5.5 shows the convergence results for $\epsilon = 9$. In contrast to the $\epsilon = 5$ case, the positivity limiters were required to solve this problem for all orders. Additionally, we observed that the flux-limiting algorithm for pressure outlined in Ref. [8] failed to maintain positivity beyond a few time steps. On the other hand, our scheme with detailed steps in Sect. 5.2.2 performs flawlessly. Using our method, we were able to run the isentropic vortex problem with $\epsilon = 10$ for longer than $t = 14$ without issue; however, using the scheme in [8], the simulation obtained negative pressure values within a few time steps, even with grid refinement and CFL reduction.

5.5 Exercises

1. Describe the impact of higher values of the stabilization parameter of the LF scheme on the quality of the solution and the time step size in an explicit temporal scheme.
2. For the scalar nonlinear conservation law with the flux $f(u) = \frac{u^2}{2}$, show that f^+ and f^- of the Lax-Friedrichs scheme are always positive and negative (more precisely, f^+ and f^- yield positive and negative speeds, respectively.
3. Compute the stabilization parameter of the Lax-Friedrichs (LF) scheme applied to the equations governing a compressible medium. Discuss both local and global LF schemes.
4. Show that the first-order Lax-Friedrichs scheme applied to the advection equation with constant speed is identical to the first-order upwind scheme.
5. Prove the concavity of the pressure function in the Euler equations of gas dynamics.
6. Consider the inviscid Burgers equation on a one-dimensional spatial domain. (**a**) Choose two initial conditions: The first is $u_0(x) = x + 1$ and the second is any smooth initial solution such that it will develop a shock wave at a finite time. Determine the time to the appearance of the shock wave. (**b**) Determine the exact solutions for both initial solutions at any (t, x). (**c**) Propose a high-order numerical scheme for both time and space for the Burgers equation. For the temporal scheme, you may use a third-order or a second-order Runge-Kutta scheme. For spatial schemes use three schemes: a linear third-order upwind scheme, the nonlinear third-order WENO upwind scheme (upwinding through the Lax-Friedrichs (LF) splitting scheme), as well as a fourth-order central scheme. Give the construction of your schemes and the discretized equations. (**d**) Code these schemes. (**e**) Verify the correctness of your implementations of the three schemes of linear upwind, nonlinear, and the central scheme by comparing against the exact solution from the first initial solution and against the exact solution from the second initial condition for a time before shock formation. Carry out the verification both qualitatively (plotting the numerical and exact solutions) and quantitatively (measuring the error). Also, study the impact of the stabilization parameter of the LF scheme on the accuracy and stability of the scheme. (**f**) Demonstrate the behavior of each scheme, linear upwind, nonlinear upwind, and the central schemes for the second initial condition for a time after the shock has formed. Demonstrate why you need the nonlinear WENO upwind scheme.

References

1. Lax, P.D.: Weak solutions of nonlinear hyperbolic equations and their numerical computation. Commun. Pure Appl. Math. **7**(1), 159–193 (1954)
2. Zhang, X., Shu, C.-W.: Maximum-principle-satisfying and positivity-preserving high-order schemes for conservation laws: survey and new developments. Proc. Roy. Soc. A: Math. Phys. Eng. Sci. **467**(2134), 2752–2776 (2011)
3. Boe, D., Shahbazi, K.: A positivity preserving high-order finite difference method for compressible two-fluid flows. Num. Methods Partial Differ. Equ. **39**(6), 4087–4125 (2023)
4. Zhengfu, X.: Parametrized maximum principle preserving flux limiters for high order schemes solving hyperbolic conservation laws: One-dimensional scalar problem. Math. Comput. **83**, 22213–2238 (2014)
5. Jiang, G.-S., Shu, C.-W.: Efficient implementation of weighted eno schemes. J. Comput. Phys. **126**, 202–228 (1996)
6. Liang, C., Zhengfu, X.: Parametrized maximum principle preserving flux limiters for high order schemes solving multi-dimensional scalar hyperbolic conservation laws. J. Sci. Comput. **58**, 41–60 (2014)
7. Hu, X.Y., Adams, N.A., Shu, C.-W.: Positivity-preserving method for high-order conservative schemes solving compressible Euler equations. J. Comput. Phys. **242**, 169–180 (2013)
8. Xiong, T., Qiu, J.-M., Zhengfu, X.: Parametrized positivity preserving flux limiters for high order finite difference weno scheme solving compressible euler equations. J. Sci. Comput. **67**, 1066–1088 (2016)
9. Shahbazi, K.: High-order finite difference scheme for compressible multi-component flow computations. Comput. Fluids **190**, 425–439 (2019)
10. Brent, R.P.: An algorithm with guaranteed convergence for finding the a zero of a function. Comput. J. **14**, 422–425 (1971)
11. Woodward, P., Colella, P.: The numerical simulation of two-dimensional fluid flow with strong shocks. J. Comput. Phys. **54**(1), 115–173 (1984)
12. Gerolymos, G.A., Sénéchal, D., Vallet, I.: Very-high-order weno schemes. J. Comput. Phys. **228**, 8481–8524 (2009)
13. Shahbazi, K.: Robust second-order scheme for multi-phase flow computations. J. Comput. Phys. **339**, 163–178 (2017)
14. Coralic, V., Colonius, T.: Finite-volume weno scheme for viscous compressible multicomponent flows. J. Comput. Phys. **274**, 95–121 (2014)
15. Shahbazi, K.: Positivity preservation of a first-order scheme for a quasi-conservative compressible two-material model. SIAM J. Sci. Comput. **43**(4), B1029–B1055 (2021)

Chapter 6
Schemes for Compressible Two-Fluid Model

In the previous chapter, the low and high-order schemes were introduced and analyzed for conservation laws including compressible single-fluid systems. In this chapter, we turn our attention to two-fluid models. The focus is on the simplest model with no heat transfer and phase change and consisting of two immiscible fluids. The two fluids can be two different gases or a gas and a liquid. A fluid-solid system, under the condition of extreme pressures where the elasticity of solids does not play a significant role, the solid behaves as a fluid, may also be covered with this model.

Despite these simplifications, the application areas of this model are broad and include problems of shock-induced interfacial instabilities, Richtmyer-Meshkov instabilities, shock-induced bubble collapse for a single or cluster of bubbles, and detonation and blast waves.

We here extend, as much as possible, the methodology developed for the single-fluid systems in the previous chapter to two-fluid systems. Hence, for the shock and interfacial discontinuities, as in the single-fluid case, the resultant scheme is of capturing or Eulerian nature, not a tracking or Lagrangian nature. As discussed in Remark 5.2, due to converging characteristics, the nonlinear discontinuities are captured within a fixed number of grid points (approximately seven grid points) independent of the propagation time. However, due to the lack of inherent sharpening mechanisms such as converging characteristics, the linear discontinuities such as interfacial discontinuities, tend to continuously smear out over time as a result of the dissipative nature of the capturing scheme. It is, therefore, desirable to design numerical schemes with minimum dissipation. With significantly lower dissipation, the high-order schemes are highly effective in the computation of compressible two-fluid systems as shown in Ref. [1].

The departing point in this chapter is the discussion of what system of equations should be used as a model of the compressible two-fluid systems. As first discussed in Ref. [2], the model must be in a particular form such that its numerical solution yields non-oscillatory capturing of a simple advection of material interfaces. Here, a thorough analysis of this requirement and the derived model based on Ref. [1] is presented. The first-order discretization scheme for the model is then presented followed by a detailed discussion of the hyperbolicity preserving property of the first-order scheme recently discovered [3]. The high-order finite difference extension of the scheme is also proposed. The strategies of positivity preserving are also introduced based on the recent work [4]. High-order finite volume schemes for the two-fluid schemes [5–7] are not discussed as their multi-dimensional version are significantly more complicated and costly compared to the finite difference scheme. Moreover, the finite volume schemes for the two-fluid model are formally second-order accurate in regions near discontinuities. For clarity, we present the model and its analysis in a one-space dimension.

6.1 Compressible Two-Fluid Model

In this section, we discuss the consistency of the compressible two-fluid model. An appropriate equation of state and its parameters that yield a physically consistent model is first discussed before presenting appropriate advection models required for the oscillation-free capturing of fluid interfaces.

6.1.1 Equation of State

The first step in modeling the compressible two-fluid system is to represent the relation between the internal properties of the fluids with an equation of state (EoS). The ideal gas law taking into account only the atomic or molecular agitation is only applicable to gases with small molecular sizes and very weak inter-molecular bonds, which is not valid for liquids or solids. An equation of state with one level higher complexity that incorporates the finite inter-molecular bonds is the so-called stiffened-gas equation of states [8]. The stiffened-gas EoS is often given as a relation between pressure(p), density (ρ), and the internal energy (e) in the form

$$p = (\gamma - 1)\rho e - \gamma \pi \tag{6.1}$$

The first term on the right-hand side models the contribution of molecular (atomic) agitations to pressure, and is scaled by the parameter γ which can be obtained using Hugoniot relations and experimental data for shock speed versus particle speed [9]. For gases, γ coincides with the physical property of the specific heat ratio (e.g., $\gamma = 1.4$ for air) while for liquids, it is not and merely plays the role of a parameter yielding the best fit to experimental data. For instance, for liquid water, depending on the shock Mach number, γ can be chosen in a range of $4 - 7$ [10]. The second term, having the same dimension as pressure, represents the inter-molecular bonds,

6.1 Compressible Two-Fluid Model

and is scaled by parameter π which can also be determined similarly. For ideal gases $\pi = 0$, for liquid water, however, it is very large and is of the order of $4000 - 6000$ bars.

Remark 9 In the stiffened-gas EoS, the parameter γ must always be chosen larger than one, i.e.,

$$\gamma > 1 \tag{6.2}$$

This requirement is needed for the model to be consistent with the physics of the sound waves. Specifically, the constraint of $\gamma > 1$ is required in order for the limit of sound speed to be zero as density goes to zero, i.e.,

$$\lim_{\rho \to 0} c^2 = \lim_{\rho \to 0} \left(\frac{\partial p}{\partial \rho} \right)_s = 0 \tag{6.3}$$

To find an expression for the sound speed in a fluid with the stiffened-gas EoS, we first take the differential of Eq. (6.1). Then, dividing by $d\rho$ and invoking the Gibbs first property relation for an isentropic process

$$\cancel{Tds} = de - \frac{pd\rho}{\rho^2} \tag{6.4}$$

yield the expression for the sound speed in the form

$$c^2 = \lim_{\rho \to 0} \left(\frac{\partial p}{\partial \rho} \right)_s = \frac{\gamma(p+\pi)}{\rho} \tag{6.5}$$

Similarly, using the Gibbs property relation for an isentropic process, we have

$$(p + \pi) = (p_0 + \pi) \left(\frac{\rho}{\rho_0} \right)^\gamma \tag{6.6}$$

where p_0 and ρ_0 are reference pressure and density, respectively. Using Eqs. (6.5) and (6.6), the limit of the speed of sound is obtained as

$$\lim_{\rho \to 0} c^2 = \lim_{\rho \to 0} \left(\frac{\partial p}{\partial \rho} \right)_s = \lim_{\rho \to 0} \frac{\gamma(p+\pi)}{\rho} = \lim_{\rho \to 0} \frac{\gamma(p_0+\pi)(\rho)^{\gamma-1}}{\rho_0^\gamma} \tag{6.7}$$

This limit is only zero if $\gamma > 1$. Hence, the stiffened-gas EoS is physically consistent only if the first order parameter γ satisfies the condition of $\gamma > 1$.

For ideal gases, $\gamma > 1$ always, as it is the specific heat ratio. However, for liquids, γ is obtained from experimental data fitting; hence, it is important to ensure a value larger than one.

Remark 10 The expression for the sound speed in a stiffened-gas EoS (6.5) reveals that for the case of a liquid where $\Pi > 0$, the pressure can become negative while the model remains consistent, i.e., the speed of sound remains real valued, as is consistent with the physical observations. For instance, high-intensity ultrasound propagation in liquids or biological media involves waves with tensile tails that are characterized by negative pressures.

6.1.2 Two-Fluid Model

Now that we have decided on an appropriate EoS, we need to have a way of determining what fluid is residing at any location during the evolution of the two-fluid system, to assign the proper values of parameters γ and π. For this, we need to capture the interface using an advection equation for an interface identifying function (interface ordering function). One relatively straightforward approach is to use two advection equations for the two parameters of EoS. The analysis which will follow, however, reveals that a particular function of each parameter of EoS needs to be advected not the parameters itself, for the model to be consistent with the underlying physics of a simple advection, i.e., for a simple advection (advection with a uniform speed and pressure) of an interface with a jump in material parameters, the model must preserve uniform velocity and pressure across the interface. Denoting the particular (for now undetermined) functions of the fluid parameters Γ and Π, we have the advection equations as

$$\frac{\partial \Gamma}{\partial t} + v\frac{\partial \Gamma}{\partial x} = 0 \tag{6.8a}$$

$$\frac{\partial \Pi}{\partial t} + v\frac{\partial \Pi}{\partial x} = 0 \tag{6.8b}$$

Supplementing these equations with the mass, momentum, and energy equations leads to the model of the compressible two-fluid system in the form

$$\frac{\partial \rho}{\partial t} + \frac{\partial (\rho v)}{\partial x} = 0 \tag{6.9a}$$

$$\frac{\partial (\rho v)}{\partial t} + \frac{\partial (\rho v^2 + p)}{\partial x} = 0 \tag{6.9b}$$

$$\frac{\partial [\rho(e + \frac{v^2}{2})]}{\partial t} + \frac{\partial [(\rho e + \frac{\rho v^2}{2} + p)v]}{\partial x} = 0 \tag{6.9c}$$

$$\frac{\partial \Gamma}{\partial t} + v\frac{\partial \Gamma}{\partial x} = 0 \tag{6.9d}$$

$$\frac{\partial \Pi}{\partial t} + v\frac{\partial \Pi}{\partial x} = 0 \tag{6.9e}$$

6.1 Compressible Two-Fluid Model

This system consists of three conservation laws and two advection equations; hence, it is quasi-conservative. We may combine the advection equations (6.9d) and (6.9e) with the mass equation (6.9a) to obtain the fully conservative system in the form

$$\frac{\partial \rho}{\partial t} + \frac{\partial (\rho v)}{\partial x} = 0 \tag{6.10a}$$

$$\frac{\partial (\rho v)}{\partial t} + \frac{\partial (\rho v^2 + p)}{\partial x} = 0 \tag{6.10b}$$

$$\frac{\partial [\rho(e + \frac{v^2}{2})]}{\partial t} + \frac{\partial [(\rho e + \frac{\rho v^2}{2} + p)v]}{\partial x} = 0 \tag{6.10c}$$

$$\frac{\partial (\rho \Gamma)}{\partial t} + \frac{\partial (\rho v \Gamma)}{\partial x} = 0 \tag{6.10d}$$

$$\frac{\partial (\rho \Pi)}{\partial t} + \frac{\partial (\rho v \Pi)}{\partial x} = 0 \tag{6.10e}$$

Although the latter system due to its full conservation form seems to be more amenable to the numerical approximation using the Lax-Friedrichs scheme introduced in the previous chapter, the following analysis shows that the former, quasi-conservative, form is the preferred choice as it enables oscillation-free capturing of two-fluid interfaces (under a proper definition of Γ and Π), which is verified by the test problem of a simple advection of an interface.

From the analysis in the next section, it will become clear that special functions for Γ and Π must be used to have consistency for the simple advection of a two-fluid interface. The functions are

$$\Gamma \equiv \frac{1}{\gamma - 1}, \tag{6.11a}$$

$$\Pi \equiv \frac{\gamma \Pi}{\gamma - 1}. \tag{6.11b}$$

With these definitions, pressure can be expressed as

$$p = \frac{E - \frac{1}{2}\rho v^2}{\Gamma} - \frac{\Pi}{\Gamma} \tag{6.12}$$

Or, the total energy E can be written as

$$E = \Gamma p + \Pi + \frac{1}{2}\rho v^2 \tag{6.13}$$

6.1.3 Simple Advection of an Interface

Let us analyze the solution of the above two models (6.9) and (6.10) under the initial condition of a simple advection of an interface located at x_0 as

$$v(x < x_0, t = 0) = v(x > x_0, t = 0) \tag{6.14a}$$
$$p(x < x_0, t = 0) = p(x > x_0, t = 0) \tag{6.14b}$$
$$\rho_1 \equiv \rho(x < x_0, t = 0) \neq \rho_2 \equiv \rho(x > x_0, t = 0) \tag{6.14c}$$
$$\gamma_1 \equiv \gamma(x < x_0, t = 0) \neq \gamma_2 \equiv \gamma(x > x_0, t = 0) \tag{6.14d}$$
$$\pi_1 \equiv \pi(x < x_0, t = 0) \neq \pi_2 \equiv \pi(x > x_0, t = 0) \tag{6.14e}$$

Given the above initial solution, we need to determine the solution at cell i at the next time step, which we show by superscript prime, i.e., v', p', ρ', Γ', and Π'. For clarity, we will not specify the spatial discretization but denote the approximation of the spatial derivative of the flux f at cell i as

$$\left.\frac{\partial f}{\partial x}\right|_i \approx \frac{\partial^{\text{LF}} f(R\{\bar{u}\})}{\partial x} \tag{6.15}$$

where $R\{\bar{u}\}$ stands for the WENO or TENO reconstruction of $\bar{u}$. Moreover, for the constant a, b, and c, consider the flux in the form $af(cu) + b$ where $f(cu)$ is interpreted as the flux of the variable cu. The derivative of the flux $af(cu) + b$ using the global Lax-Friedrichs scheme may be expressed as

$$\frac{\partial [af(cu)+b]}{\partial x} \approx \frac{\partial^{\text{LF}}[af(cu)+b]}{\partial x} = a\left(\frac{f_2 - f_1}{2h}\right) - \frac{h\alpha c}{2}\left(\frac{R\{u_2\} - 2R\{u\} + R\{u_1\}}{h^2}\right) \tag{6.16}$$

where no subscript is used for the solution at x_i, while subscripts 1 and 2 are used to signify quantities at x_{i-1} and x_{i+1}, respectively. To avoid long expressions, we use the following notations

$$\frac{\delta g}{\delta x} = \frac{R\{g_2\} - R\{g_1\}}{2h} \quad \frac{\delta^2 g}{\delta x^2} = -\frac{h\alpha}{2}\left(\frac{R\{g_2\} - 2R\{g\} + R\{g_1\}}{h^2}\right) \tag{6.17}$$

and

$$\frac{\delta fg}{\delta x} = \frac{R\{f_2\}R\{g_2\} - R\{f_1\}R\{g_1\}}{2h} \tag{6.18}$$

6.1.4 Lack of Consistency of the Conservative Model with the Simple Advection

The one-time step integration of the mass equation with the time step size τ along with the uniformity of the velocity and the use of Eq. (6.18) yields the following expression for the density ρ' for either of the systems (6.9) or (6.10):

$$\left.\begin{array}{c}\rho' = \rho - \tau \frac{\partial^{\mathrm{LF}}[R\{v\}R\{\rho\}]}{\partial x} \\ v = \mathrm{const.}\end{array}\right\} \quad \Longrightarrow \quad \rho' = \rho - \tau v \frac{\delta \rho}{\delta x} - \tau \frac{\delta^2 \rho}{\delta x^2} \qquad (6.19)$$

Alternatively, for a discontinuous initial solution,

$$\rho' = \rho - \tau(v \pm \alpha) \frac{\delta \rho}{\delta x} \qquad (6.20)$$

The one-time step integration of the momentum equation with constant velocity and pressure leads to

$$(\rho v)' = \rho v - \tau \frac{\partial^{\mathrm{LF}}[R\{\rho\}R^2\{v\} + R\{p\}]}{\partial x} \qquad (6.21)$$

$$= v\left[\rho - \tau v \frac{\delta \rho}{\delta x} - \tau \frac{\delta^2 \rho}{\delta x^2}\right] \qquad (6.22)$$

Then, using Eqs. (6.19) and (6.22) yields

$$v' \equiv \frac{(\rho v)'}{\rho'} = v \qquad (6.23)$$

Similarly, the total energy after the first time step E' can be written as

$$E' = E - \tau \frac{\partial^{\mathrm{LF}}(E + p)v}{\partial x} \qquad (6.24\mathrm{a})$$

$$= E - \tau \frac{\partial^{\mathrm{LF}}}{\partial x}\left([(R\{\Gamma\} + 1)R\{p\} + R\{\Pi\} + R\{\rho\}R^2\{v\}/2]R\{v\}\right) \qquad (6.24\mathrm{b})$$

$$= E - \tau p v \frac{\delta \Gamma}{\delta x} - \tau v \frac{\delta \Pi}{\delta x} - \frac{\tau v^3}{2}\frac{\delta \rho}{\delta x} - \tau p \frac{\delta^2 \Gamma}{\delta x^2} - \tau \frac{\delta^2 \Pi}{\delta x^2} - \frac{\tau v^2}{2}\frac{\delta^2 \rho}{\delta x^2} \qquad (6.24\mathrm{c})$$

Or, for a discontinuous initial solution,

$$E' = E - \tau p(v \pm \alpha) \frac{\delta \Gamma}{\delta x} - \tau(v \pm \alpha) \frac{\delta \Pi}{\delta x} - \frac{\tau v^2(v \pm \alpha)}{2}\frac{\delta \rho}{\delta x} \qquad (6.25)$$

Now, *for the conservative model, we will show that for $\Pi = 0$, the pressure equilibrium is not maintained.* To obtain an expression for the pressure at the next time step, we need an expression for Γ'. To this end, we have

$$\frac{(\rho\Gamma)'}{\rho'} = \frac{(\rho\Gamma) - \tau \frac{\partial^{LF}(R\{v\}R\{\rho\}R\{\Gamma\})}{\partial x}}{\rho - \tau \frac{\partial^{LF}(R\{v\}R\{\rho\})}{\partial x}} \tag{6.26a}$$

$$= \frac{(\rho\Gamma_1) - \tau(v \pm \alpha)\frac{\delta\rho\Gamma}{\delta x}}{\rho - \tau(v \pm \alpha)\frac{\delta\rho}{\partial x}} \tag{6.26b}$$

$$= \frac{\rho_2\Gamma_2 - \frac{\tau(v\pm\alpha)}{2h}(\rho_2\Gamma_2 - \rho_1\Gamma_1)}{\rho_2 - \frac{\tau(v\pm\alpha)}{2h}(\rho_2 - \rho_1)} \tag{6.26c}$$

$$= \frac{\Gamma_2 - \frac{\tau(v\pm\alpha)}{2h}(\Gamma_2 - \frac{\rho_1}{\rho_2}\Gamma_1)}{1 - \frac{\tau(v\pm\alpha)}{2h}(1 - \frac{\rho_1}{\rho_2})} \tag{6.26d}$$

Now, denoting $\rho_1/\rho_2 = 1 - \Delta\rho/\rho_2$ yield

$$\frac{(\rho\Gamma)'}{\rho'} = \frac{\Gamma_2 - \frac{\tau(v\pm\alpha)}{2h}(\Gamma_2 - \Gamma_1) - \frac{\tau(v\pm\alpha)}{2h}\frac{\Delta\rho}{\rho_2}\Gamma_1}{1 - \frac{\tau(v\pm\alpha)}{2h}\frac{\Delta\rho}{\rho_2}} \tag{6.27}$$

Finally,

$$p' = \frac{E' - \frac{1}{2}\rho'v^2}{\frac{(\rho\Gamma)'}{\rho'}} \tag{6.28a}$$

$$= \frac{E - \tau p(v \pm \alpha)\frac{\delta\Gamma}{\delta x} - \frac{\tau v^2(v\pm\alpha)}{2}\frac{\delta\rho}{\delta x} - \frac{1}{2}\rho'v^2}{\frac{(\rho\Gamma)'}{\rho'}} \tag{6.28b}$$

$$= \frac{E - \frac{p\tau(v\pm\alpha)}{2h}(\Gamma_2 - \Gamma_1) - \frac{1}{2}\rho v^2}{\frac{(\rho\Gamma)'}{\rho'}} \tag{6.28c}$$

$$= \frac{p\Gamma_2 + \frac{1}{2}\rho v^2 - \frac{p\tau(v\pm\alpha)}{2h}(\Gamma_2 - \Gamma_1) - \frac{1}{2}\rho v^2}{\frac{(\rho\Gamma)'}{\rho'}} \tag{6.28d}$$

$$= \frac{p\left[\Gamma_2 - \frac{\tau(v\pm\alpha)}{2h}(\Gamma_2 - \Gamma_1)\right]\left[1 - \frac{\tau(v\pm\alpha)}{2h}\frac{\Delta\rho}{\rho_2}\right]}{\Gamma_2 - \frac{\tau(v\pm\alpha)}{2h}(\Gamma_2 - \Gamma_1) - \frac{\tau(v\pm\alpha)}{2h}\frac{\Delta\rho}{\rho_2}\Gamma_1} \tag{6.28e}$$

$$= p\left[1 + \frac{(B-1)A}{D}\right] \tag{6.28f}$$

6.1 Compressible Two-Fluid Model

where

$$A = \frac{\tau(v \pm \alpha)}{2h} \frac{\Delta \rho}{\rho_2} \qquad (6.29a)$$

$$B = \frac{\Gamma_1}{\Gamma_2 - \frac{\tau(v \pm \alpha)}{2h}(\Gamma_2 - \Gamma_1)} \qquad (6.29b)$$

$$D = 1 - BA \qquad (6.29c)$$

Clearly, for a τ satisfying the CFL condition ($\tau \leq h/\alpha$), p' differs from p with a finite value regardless of the grid resolution; or

$$p' = p + \mathcal{O}(p) \qquad (6.30)$$

In the case of a large jump in the density across the fluid interface, the maximum error in pressure is

$$\frac{p' - p}{p} = \mathcal{O}\left(\frac{\Gamma_1}{\Gamma_2}\right) \quad \forall h \qquad (6.31)$$

To verify the above analysis, the WENO5 pressure and density solutions of the conservative two-fluid model (6.10) for a simple interface advection

$$v(x < x_0, t = 0) = v(x > x_0, t = 0) = 1.0 \qquad (6.32a)$$
$$p(x < x_0, t = 0) = p(x > x_0, t = 0) = 0.5 \qquad (6.32b)$$
$$\rho_1 \equiv \rho(x < x_0, t = 0) = 16.0 \neq \rho_2 \equiv \rho(x > x_0, t = 0) = 4.0 \qquad (6.32c)$$
$$\gamma_1 \equiv \gamma(x < x_0, t = 0) = 1.4 \neq \gamma_2 \equiv \gamma(x > x_0, t = 0) = 6.68 \qquad (6.32d)$$
$$\pi_1 \equiv \pi(x < x_0, t = 0) = 0.15 \neq \pi_2 \equiv \pi(x > x_0, t = 0) = 0 \qquad (6.32e)$$

for three resolutions of $N = 100, 200$, and 400 are shown in Fig. 6.1. The figure reveals that spurious oscillations in the pressure and density do not diminish even with a first-order rate.

Therefore, the pressure equilibrium across the interface for the simple advection problem is not maintained, and the conservative model is inconsistent.

6.1.5 Consistency of the Quasi-Conservative Model with the Simple Advection

To specialize the quasi-conservative model to the simple advection of an interface, we first write the system (6.9) in an equivalent form where the term in the advection equations are written as a conservative form plus source term as

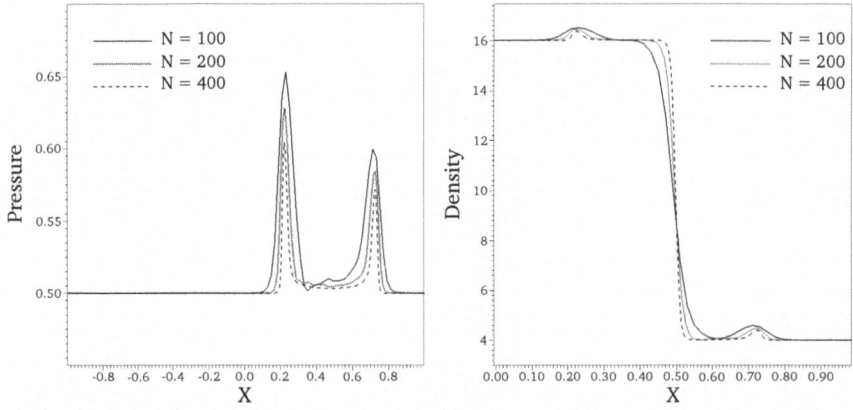

Fig. 6.1 The WENO5 solutions of the two-fluid advection equations given in Eq. (6.32) for three resolutions of $N = 100, 200$, and 400 using the conservative model (6.10), revealing spurious oscillations in the pressure and density that does not vanish even with a first-order rate. These results were first appeared in the author's earlier work [4]

$$\frac{\partial \rho}{\partial t} + \frac{\partial (\rho v)}{\partial x} = 0 \tag{6.33a}$$

$$\frac{\partial (\rho v)}{\partial t} + \frac{\partial (\rho v^2 + p)}{\partial x} = 0 \tag{6.33b}$$

$$\frac{\partial [\rho(e + \frac{v^2}{2})]}{\partial t} + \frac{\partial [(\rho e + \frac{\rho v^2}{2} + p)v]}{\partial x} = 0 \tag{6.33c}$$

$$\frac{\partial \Gamma}{\partial t} + \frac{\partial v \Gamma}{\partial x} - \Gamma \frac{\partial v}{\partial x} = 0 \tag{6.33d}$$

$$\frac{\partial \Pi}{\partial t} + \frac{\partial v \Pi}{\partial x} - \Pi \frac{\partial v}{\partial x} = 0 \tag{6.33e}$$

While the expressions for ρ' and v' are the same as the conservative model, the expressions for Γ' and Π' are different. Since velocity is uniform across the interface, the expression for Γ' and Π' derived as

$$\Gamma' = \Gamma - \tau v \frac{\delta \Gamma}{\delta x} - \tau \frac{\delta^2 \Gamma}{\delta x^2}, \tag{6.34a}$$

$$\Pi' = \Pi - \tau v \frac{\delta \Pi}{\delta x} - \tau \frac{\delta^2 \Pi}{\delta x^2} \tag{6.34b}$$

The pressure at the next time step then is

$$p' = \frac{E' - \frac{1}{2}\rho' v^2}{\Gamma'} - \frac{\Pi'}{\Gamma'} \tag{6.35a}$$

6.2 Consistent Numerical Method for the Two-Fluid System

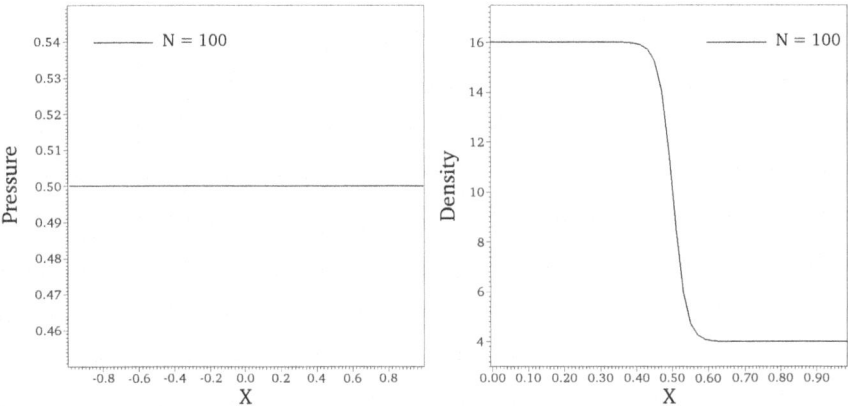

Fig. 6.2 The WENO5 solutions of the quasi-conservative model (6.9) two-fluid advection equations for the interface advection given in Eq. (6.32) for the resolutions of $N = 100$. The uniform pressure and non-oscillatory density capturing are evident. These results were first appeared in the author's earlier work [4]

$$= \frac{E - \tau p v \frac{\delta \Gamma}{\delta x} - \tau v \frac{\delta \Pi}{\delta x} - \frac{\tau v^3}{2} \frac{\delta \rho}{\delta x} - \tau p \frac{\delta^2 \Gamma}{\delta x^2} - \tau \frac{\delta^2 \Pi}{\delta x^2} - \frac{\tau v^2}{2} \frac{\delta^2 \rho}{\delta x^2} - \frac{1}{2} \rho' v^2}{\Gamma'} - \frac{\Pi'}{\Gamma'}$$
(6.35b)

$$= \frac{E + p(\Gamma' - \Gamma) - \Pi - \frac{1}{2} \rho v^2}{\Gamma'}$$
(6.35c)

$$= \frac{p\Gamma + \Pi + \frac{1}{2}\rho v^2 + p(\Gamma' - \Gamma) - \Pi - \frac{1}{2}\rho v^2}{\Gamma'} = p$$
(6.35d)

To verify the above analysis, the WENO5 solutions of the quasi-conservative model (6.9) for the simple interface advection problem (6.32) are shown in Fig. 6.2. The figure reveals oscillation-free capturing of the uniform pressure and the two-fluid interface.

Therefore, *the quasi-conservative model (6.33) is a consistent model and must be used for the compressible two-fluid systems.*

Note that from Eqs. (6.35), *it is clear that the particular functions Γ and Π given in (6.11) are essential for the consistency of the model, and any other function yields inconsistent model.*

6.2 Consistent Numerical Method for the Two-Fluid System

The conservative flux terms in the quasi-conservative two-fluid model (6.33) are discretized using the high-order WENO or TENO LF scheme developed in the previous chapter. Note that adding the two advection-type equations does not change the eigenvalues of the system. The eigenvectors however will be different and for the reconstruction on the characteristic space, these eigenvectors must be determined. We leave the derivation of the eigenvectors of matrices $[Q]$ and $[Q^{-1}]$ to an exercise at the end of the chapter.

The numerical scheme for the approximation of the source terms in Eqs. (6.33d) and (6.33e) required some care. We demand two requirements: (1) the source term to be approximated with the same high level of accuracy as other terms, and (2) the source term approximation must yield uniform interfacial parameters γ and Π in the interface-free regions since the purpose of the advection equations is solely the capturing of the interface. The requirements are satisfied by the following scheme

$$h \left(\Gamma \frac{\partial v}{\partial x} \right)_{x_i} \approx \Gamma_i \sum_{j=1}^{j=r+1} d_j \left[\left(\frac{v^+_{\text{WENO},i+j+1/2} + v^-_{\text{WENO},i+j+1/2}}{2} \right) - \left(\frac{v^+_{\text{WENO},i-j-1/2} + v^-_{\text{WENO},i-j-1/2}}{2} \right) \right] \quad (6.36)$$

where v^+_{WENO} and v^-_{WENO} correspond to the left- and right-biased WENO approximations that are consistent with the positive and negative characteristic speeds. A similar scheme is used for Π. The scheme in (6.36) is introduced in Ref. [1]. Note that in the finite volume formulation, it is difficult to devise a genuinely high-order accurate scheme for the source term and those introduced in Ref. [6] and further developed in Ref. [5], only results in a second-order accurate scheme for the source term.

6.3 Numerical Examples on High-Order Scheme for Two-Fluid System

Here, we present one- and two-dimensional numerical examples demonstrating the effectiveness of high-order schemes for two-fluid compressible systems. All computations are performed using the fourth-order Runge-Kutta method. The results are also presented in the author's article [1].

6.3.1 1D Strong Shock Helium-Air Interface Interaction

To assess the non-oscillatory and conservation properties of the proposed scheme, we consider the problem of a strong shock ($M = 8.96$) moving through light and high specific heat ratio gas (e.g., helium) and interacting with the interface of two

6.3 Numerical Examples on High-Order Scheme for Two-Fluid System

gases (e.g., helium-air interface) [10, 11]. The spatial domain is chosen to be $[-1, 1]$ and the initial conditions are

$$(\rho, v, p, \gamma, \Pi)(0, x) = \begin{cases} (0.384, 27.086, 100.176, 1.67, 0) & x \leq -0.8, \\ (0.1, 0, 1.0, 1.67, 0) & -0.8 \leq x \leq -0.2, \\ (1.0, 0, 1.0, 1.4, 0) & x > -0.2 \end{cases}$$
(6.37)

The computed density, pressure, and interface location (captured by transport of γ) with the total number of cells $N = 200$ and approximation order of $n = 5$ and the corresponding analytical solutions for the final time of $t = 0.07$ (shock arrives at the interface at $t = 0.0164$) are shown in Fig. 6.3. Non-oscillatory solutions are observed.

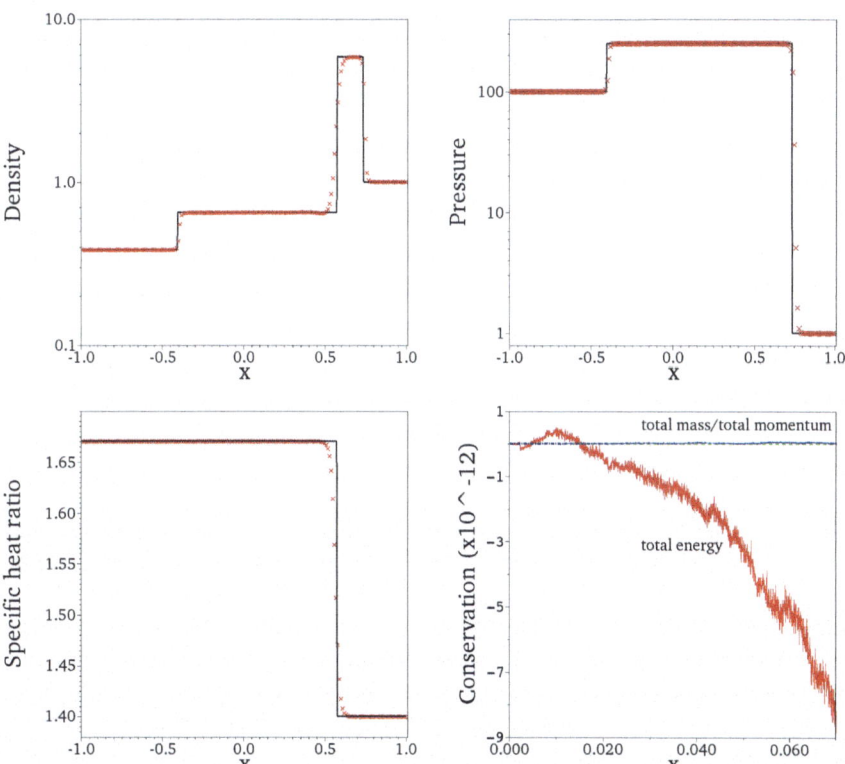

Fig. 6.3 $M = 8.96$ shock moving through helium interacting with helium-air interface with $N = 200$ cells and approximation order $n = 5$ at the final time of $T = 0.07$; top left panel, density; top right, pressure; bottom left, γ showing the interface; and bottom right, the conservation of total energy, momentum, and mass. Solid lines signify the analytical solutions. For conservation, the initial values and boundary fluxes are subtracted. These results were first appeared in the author's earlier work [1]

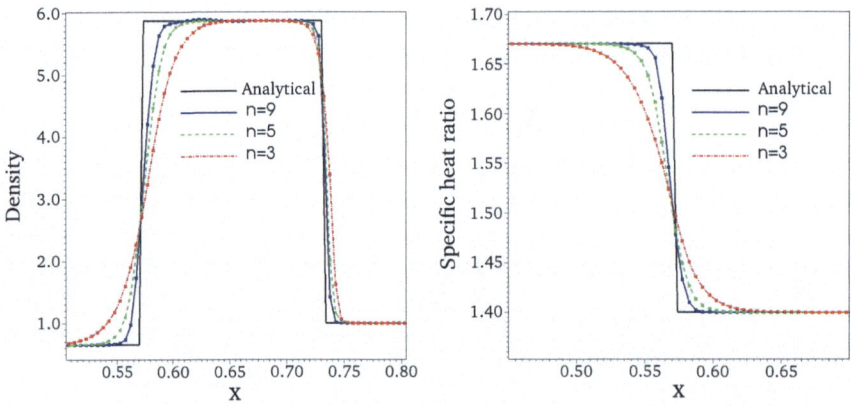

Fig. 6.4 A zoomed-in view of density and γ computed using various approximation orders n for $M = 8.96$ shock helium-air interface interaction with $N = 400$ cells at the final time of $T = 0.07$, clearly demonstrating the advantage of higher order schemes in sharper capturing of discontinuities. These results were first appeared in the author's earlier work [1]

Table 6.1 Density error and CPU time for various orders and number of cells in computing shock helium-air interaction problem at $t = 0.07$. These results were first appeared in the author's earlier work [1]

Order	$N = 200$		$N = 400$		$N = 800$	
	L_1 Error	CPU time (s)	L_1 error	CPU time (s)	L_1 error	CPU time (s)
$n = 3$	0.23	0.7	0.13	3	0.07	10
$n = 5$	0.15	1	0.07	5	0.03	21
$n = 9$	0.1	2	0.05	9	0.02	36

The effect of increasing order has also been investigated, resulting in sharper capturing of material interfaces and shocks as shown in Fig. 6.4. More qualitative comparisons of various orders and various numbers of grid cells are reported in Table 6.1. In this table, the L_1 errors in density and the corresponding CPU times for orders $n = 3, 5$ and 9 and $N = 200, 400$ and 800 are given. As expected, the data in the table shows that all approximation orders yield a first-order convergence, which is a consequence of applying a shock-capturing scheme to problems with discontinuous solutions. The data also reveals that higher order schemes ($n = 5$ and 9) outperform lower order schemes ($n = 3$) for achieving a higher level of accuracy per unit cost (CPU time). The computations for all orders were carried out with a CFL number of 0.5. Results and CPU times were unchanged for CFL numbers smaller than 0.5. The advantage of the higher order method will be even higher for multi-dimensional problems, where CPU time scales with N^{d+1}, where $d = 1, 2$ or 3 for one-, two- or three-dimensional calculations, respectively. For clarity, this convergence result is also plotted in Fig. 6.5.

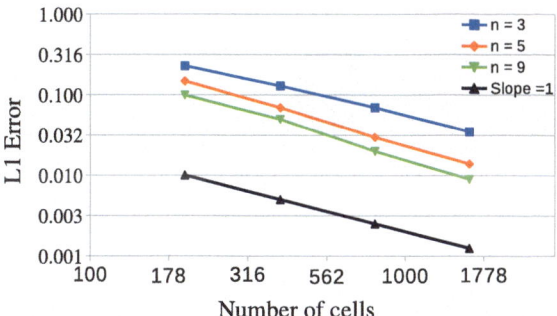

Fig. 6.5 L_1 error versus the number of grid points for strong shock helium-air interface interaction obtained using solutions based on approximation orders $n = 3, 5$ and 9, clearly showing the superiority of the high-order methods. For comparison, the linear line (slope = 1) is shown to confirm the first-order convergence of the method. These results were first appeared in the author's earlier work [1]

The excellent conservation properties of the scheme are also evident from density and pressure plots. For further investigation, the absolute value of total mass, momentum, and energy over the whole spatial domain and at any time step minus the corresponding initial values and the corresponding fluxes are computed. The results for $N = 200$ and the fifth-order method are shown in the bottom-right panel (Fig. 6.3). As seen, the conservation errors are of the order of double-precision errors (errors in mass and momentum are of the order of $1.0e - 15$, and that of energy is of the order of $1.0e - 12$). Similar conservation behavior for various finite difference orders and grid resolutions are also observed, verifying the excellent conservation properties of the proposed scheme.

6.3.2 Two-Dimensional Air Shock Cylindrical Helium Bubble Interaction

The two-dimensional test case involves computing the two-material Euler equations of gas dynamics with the material composition for a $M = 1.22$ shock wave traveling in a body of air and interacting with a cylindrical volume of helium, which is contaminated with 28% air by mass. The governing equations are the two-dimensional version of the two-fluid system (6.33).

This test problem features various linear and nonlinear dynamics, and thus, it is suitable for validation of numerical schemes for multi-component flows as performed by many [7, 12–14] and detailed experimental studies are also available [15]. The computational domain is chosen to coincide with other studies, and it is a rectangular domain of $[-0.0445\,m, 0.0445\,m] \times [-0.2225\,m, 0.2225\,m]$. The air shock is moving from left to right with its initial location of $x = -0.1m$; the helium bubble

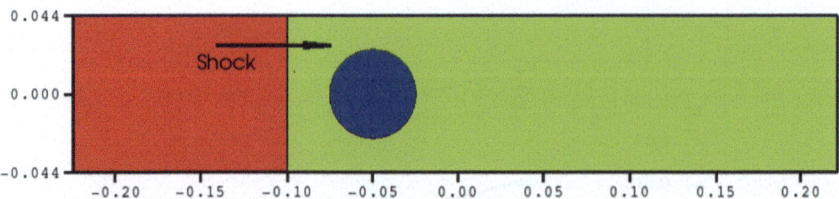

Fig. 6.6 Initial configuration of the air shock and the helium bubble for the shock-bubble interaction in a shock tube. This diagram was first appeared in the author's earlier work [1]

has a radius of 0.025 and is placed at $(-0.05m, 0m)$ (Fig. 6.6). The field variables are initialized as

$$(\rho, v_x, v_y, p, \gamma, \Pi) = \begin{cases} (1.024 \text{ kg/m}^3, 0, 0, 101325 \text{ Pa}, 1.4,) & \text{Pre-shock,} \\ (1.658 \text{ kg/m}^3, 114.49 \text{ m/s}, 0, 159060 \text{ Pa}, 1.4, 0) & \text{Post-shock,} \\ (0.22 \text{ kg/m}^3, 0, 0, 101325 \text{ Pa}, 1.64, 0) & \text{Helium bubble,} \end{cases}$$
(6.38)

The bubble density and specific heat are calculated using the corresponding properties of air and helium at the pressure of 1 am and temperature of 289 K with 28% air mass in the mixture. These initial configurations correspond to the initial conditions of the fifth-order finite volume computations by Coralic and Colonius, who used the mixture-theory model with single velocity, pressure, and temperature [7]. The interaction features an air jet impinging on the interface at the early stages, followed by the generation of a vortex ring and eventual Kelvin-Helmholtz instabilities at later times.

The computations were carried out on the whole domain, unlike many previous studies [7] that considered only half of the domain with symmetry condition assumptions along the center axes. To visualize the flow and accentuate its weak features, we compute numerical Schlieren images using the divergence of the density field. Following Quirk and Karni [16], for numerical Schlieren, the following shading function is used:

$$\phi = \exp(-k \frac{|\nabla \rho|}{|\nabla \rho|_{max}})$$
(6.39)

where $k = 600$ for helium and 120 for air. The function ϕ is computed for all grid points. The Schlieren images obtained using computations with our ninth-order accurate finite difference scheme with the grid spacing of $\Delta x = 139 \mu m$ shown in Fig. 6.7 for time snapshots of $t = 28, 52, 245, 673$, and $983 \mu s$. All images are very similar to the experimental shadowgraph images obtained by Haas and Sturtevant [15]. In addition, they are very similar to those of Coralic and Colonius [7], wherein they used a finite volume WENO scheme. Slight differences with the images of Coralic and Colonius are, however, observed for later times, $t = 678$, and $983 \mu s$. In particular, as seen from our results, the latter stages of dynamics, where Kelvin-Helmholtz instabilities develop, lack axial symmetry.

6.3 Numerical Examples on High-Order Scheme for Two-Fluid System

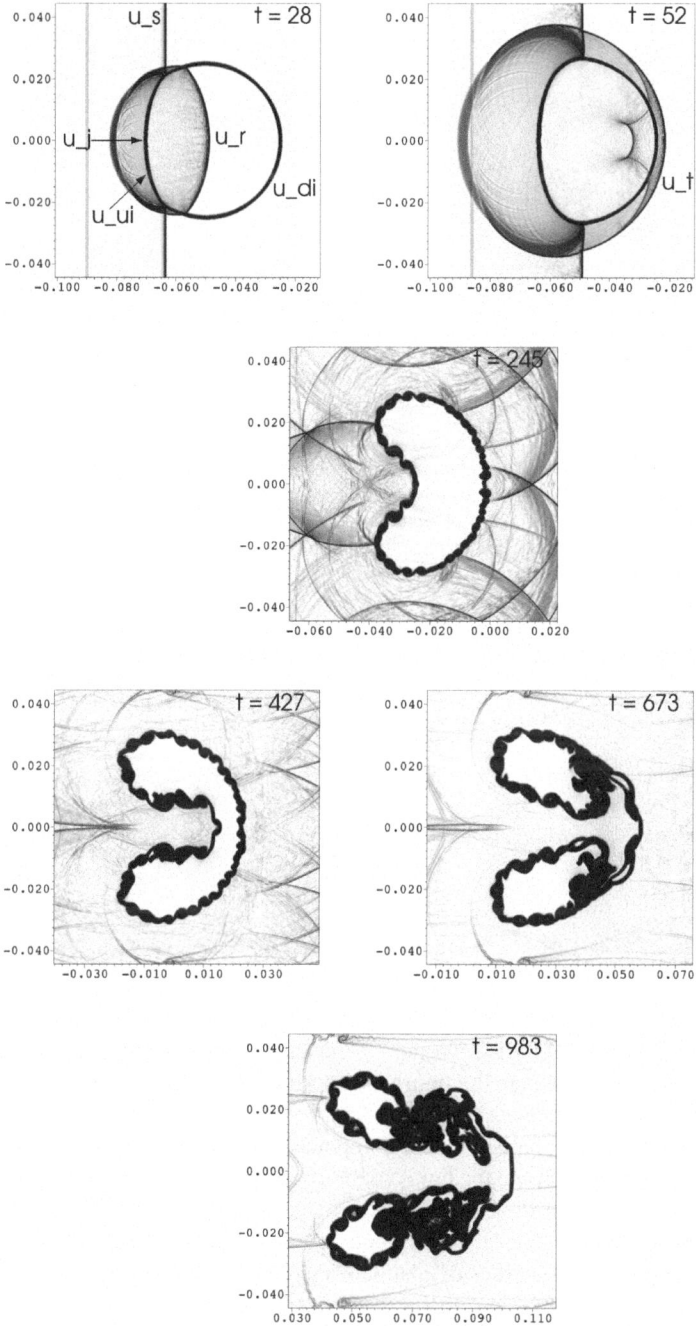

Fig. 6.7 Temporal snapshot of air shock/helium bubble interaction computed using approximation order $n = 9$ and depicted using the numerical Schlieren imaging technique of Quirk and Karni [16]. The time of each shot is given in µs on the top left corner of the panel. These results were first appeared in the author's earlier work [1]

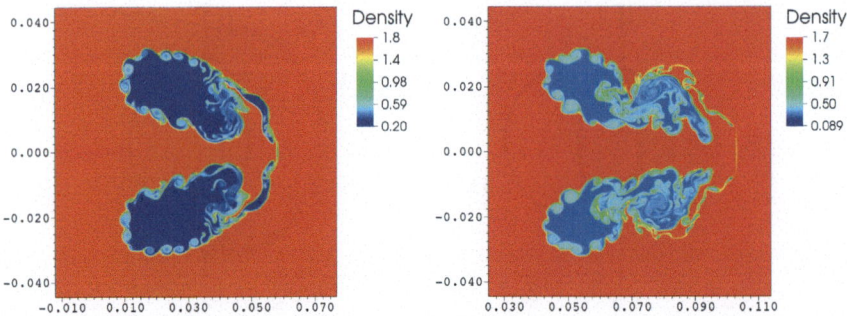

Fig. 6.8 Density pseudocolors for shock/helium bubble interaction at $t = 733\mu s$ and computed using $n = 9$ on the left and $t = 982\mu s$ on the right, illustrating the non-oscillatory computation of the interfacial instabilities. These results were first appeared in the author's earlier work [1]

Table 6.2 Velocity of various dynamical features for air shock helium bubble interaction computed using the present high-order finite difference method with the resolution of $\Delta x = 278 \mu s$. For comparison, the corresponding data obtained using the finite volume simulation of Coralic and Colonius [7] with significantly higher resolution of $\Delta x = 50 \, \mu s$ and experimental data of Haas and Sturtevant [15] are also presented. $u_s, u_r, u_t, u_{ui}, u_{di}$, and u_j are velocities of incident shock (averaged over $[-60, 0]\mu s$), the rarefacted shock (averaged over $[0, 52]\mu s$), the transmitted shock (averaged over $[52, 240]\mu s$), the upstream interface (averaged over $[10, 52]\mu s$), the downstream interface (averaged over $[140, 240]\mu s$), and the jet head (averaged over $[140, 240]\mu s$), respectively, and measured along the center line $x = 0$ in m/s. These results were first appeared in the author's earlier work [1]

Data	u_s	u_r	u_t	u_{ui}	u_{di}	u_j
Present FD simulation ($\Delta x = 278 \, \mu m$)	420	960	371	178	150	223
FV simulation ($\Delta x = 50 \, \mu m$, [7])	420	945	379	173	145	230
Experiment ([15])	410 ± 41	900 ± 90	393 ± 39	170 ± 17	145 ± 15	230 ± 23

To further emphasize the effectiveness of our scheme in non-oscillatory capturing of the interfacial instabilities, we depict the density pseudocolor at $t = 733$ and $983\mu s$ in Fig. 6.8. As seen in the figure, no spurious oscillations of any form exist in the computed density field.

The quantitative comparison of our computations with the experimental results of Haas and Sturtevent [15] and the finite volume computations of Coralic and Colonius [7] are also presented. In Table 6.2, the strength of various interaction features, namely the velocity of incident shock u_s, the velocity of rarefacted and transmitted shocks (u_r and u_t, respectively), velocities of upstream and downstream interfaces (u_{ui} and u_{di}, respectively), and the jet head velocity (u_j), all, are compared with the corresponding data from the experiment and the computations [7, 15]. These features are labeled in the first two snapshots of Fig. 6.7. In all cases, very close agreements within 7% mean values of experimental data and that of computations are obtained. This is significant in view of the fact that the grid spacing in our computation is approximately six-fold coarser than that of finite volume computations of Coralic and Colonius, implying significant savings in required CPU times and the overall efficiency of our scheme. This higher level of efficiency can be attributed to the higher order accuracy of our scheme (ninth-order finite difference) compared to the lower order accuracy of Coralic and Colonius (fifth-order finite volume).

6.4 Positivity-Preserving Property of Numerical Scheme for the Two-Fluid System

The above numerical examples do not require positivity-preservation property. However, there are many other cases involving compressible two-fluid systems that require positivity preservation, as will be shown by numerical examples later. In this section, we first discuss the positivity-preservation of the first-order and high-order schemes for the two-fluid system (6.33), before presenting some numerical examples.

6.4.1 Positivity Preservation of the First-Order Scheme

Considering that the eigenvalues of system (6.33) are v and $v \pm c$, the hyperbolicity preservation only requires positivity of square of sound speed. The positivity of density is required for consistency with the physical significance of density. Also, as discussed in Remark 9, the condition $\gamma > 1$ must be satisfied for consistency with the near-vacuum speed of sound. Using these physical consistency requirements and the expression for the speed of sound (6.5), the admissible set for positivity preservation are those satisfying

$$\rho > 0 \quad \rho(\gamma - 1) > 0 \quad (p + \pi) > 0, \tag{6.40}$$

or equivalently,

$$\rho > 0 \quad (\gamma - 1) > 0 \quad \frac{p + \pi}{\gamma - 1} > 0 \tag{6.41}$$

We may refer to $\frac{p+\pi}{\gamma-1}$ as the modified pressure. Note that the admissible set is convex. As the last positivity requirement reveals, the positivity of pressure is not required for the two-fluid systems involving a liquid. As discussed in Remark 10, this is consistent with the physics of liquid under tension where the pressure is negative.

The first-order LF scheme preserves the positivity of the density as discussed in the previous chapter. We next show that the condition $(\gamma - 1) > 1$ is also satisfied using the first-order scheme for the advection equation because the scheme is maximum-principle preserving.

6.4.1.1 Maximum-Principle Preservation of the First-Order Scheme for the Interfacial Functions

The first-order scheme for the advection equations in (6.33) is expressed as

$$\Gamma_i^{n+1} \equiv \mathcal{H}_\Gamma(\Gamma_i^n, \Gamma_{j-1}^n, \Gamma_{j+1}^n) = \Gamma_i^n - \lambda \left[(v_i^n \Gamma_i^n + \alpha \Gamma_i^n + v_{i+1}^n \Gamma_{i+1}^n - \alpha \Gamma_{i+1}^n) - \right.$$
$$\left. (v_{i-1}^n \Gamma_{i-1}^n + \alpha \Gamma_{i-1}^n + v_i^n \Gamma_i^n - \alpha \Gamma_i^n) \right] +$$
$$+ \lambda \Gamma_i^n (v_{i+1}^n - v_{i-1}^n) \tag{6.42}$$

An analogous scheme is adopted for the advection of Π. This scheme is trivially consistent, i.e.,

$$\mathcal{H}_\Gamma(a, a, a) = a \tag{6.43}$$

The monotonic behavior of $\mathcal{H}_\Gamma(\Gamma_i^n, \Gamma_{i-1}^n, \Gamma_{i+1}^n)$ is established under the condition

$$[1 - \lambda(2\alpha - v_{i-1} + v_{i+1})] \geq 0 \iff \Delta t \leq \frac{\Delta x}{\alpha + \max(v)} \tag{6.44}$$

which is within half of Δt that is required for the CFL condition. Therefore, analogous to the case of scalar conservation law discussed in Sect. 5.1.1, the monotonicity and consistency properties ensure the maximum-principle preserving property in this case.

6.4.1.2 Positivity of the Modified Pressure

As proved in Ref. [3], the positivity of the modified pressure is not ensured for all cases. Only if the material parameters satisfy a concavity condition, the admissible set is convex, the positivity of the modified pressure is ensured. The proof is rather tedious and here we only discuss the condition for the positivity based on Lemma 1 in Ref. [3].

Lemma 2 *Let us denote the independent conservative variable vector of the compressible two-material system by $\bar{u} = [\rho, \rho v, E, \Gamma, \Pi]^T$. The modified pressure $\frac{p+\pi}{\gamma-1} \equiv \mathcal{E}(\bar{u})$ is a concave function of $\bar{u}$, i.e., for any two conservative variable vectors $\bar{u}_1$ and $\bar{u}_2$ and any $\theta \in [0, 1]$,*

$$\mathcal{E}(\theta \bar{u}_1 + (1-\theta) \bar{u}_2) \geq \theta \mathcal{E}(\bar{u}_1) + (1-\theta) \mathcal{E}(\bar{u}_2) \tag{6.45}$$

if the following condition on the parameters of the two-material model is satisfied:

$$\pi_1 \geq \pi_2 \iff \gamma_1 \geq \gamma_2 \tag{6.46}$$

Remark 11 The condition (6.46) is satisfied when the two-fluid model consists of two ideal gases, two liquids with identical πs, and also many combinations of liquids and gases such as the most common case of liquid water and air. However, in general, it is not satisfied. Hence, the two-material model may not be concave for an arbitrarily chosen two-material configuration.

Remark 12 The concavity of the modified pressure, along with the concavity of density and order parameters, implies that the admissible set (6.41) is convex.

Proof The proof is given in Ref. [3] and left to one of the exercises.

6.4.2 Positivity Preservation of the High-Order Scheme

The positivity of the density is enforced following the flux-limiting algorithm presented for the maximum-principle preservation in the high-order scalar conservation law introduced in Sect. 5.2.2. The second condition of $(\gamma - 1) > 1$ is equivalent to the condition on the interfacial function $\Gamma > 0$; hence, a similar flux-limiting algorithm can be used for the positivity enforcement of the interfacial function Γ. The only difference is the discretized terms arising from the source term in the advection equation that must be added to the terms from the conservative fluxes to form the total effective fluxes. After obtaining the effective fluxes, the limiting algorithm presented in 5.2.2 can be applied without any modification.

The modified pressure is a concave function of the limiting parameters similar to the pressure in the single fluid case as long as the concavity condition on the material parameters is satisfied (Lemma 2). Therefore, the algorithm for enforcing the positivity of the pressure in the single-fluid case presented in Sect. 5.3.3 can be extended straightforwardly to the modified pressure in the two-fluid model. The details are similar and documented clearly in Ref. [4] for both one- and two-dimensional computations.

6.5 Numerical Examples on Positivity Preservation for Two-Fluid System

Here, we present the verification results for the positivity-preservation of the high-order scheme for the two-fluid system in one- and two-dimensional spaces. These numerical examples are also presented in our recent article [4]. We also present the impact of the positivity-preservation limiter on the overall CPU times for a practically relevant two-dimensional problem of shock-bubble interaction.

6.5.1 One-Dimensional Water and Air Shock, Interface Problem

We verify the positivity-preserving performance for a two-phase problem. The initial data consists of a region of quiescent air in equilibrium with water at atmospheric conditions. A Mach 2.5 shock is propagating through the water and impinges upon the interface. The water and air properties are chosen as

$$(\rho, v, p, \Gamma, \Pi) = \begin{cases} (1449 \frac{\text{kg}}{\text{m}^3}, 1281 \frac{\text{m}}{\text{s}}, 5.26 \times 10^9 \text{ Pa}, 4.40, 6.15 \times 10^8 \text{ Pa}) & \text{post-shock} \\ (998 \frac{\text{kg}}{\text{m}^3}, 0, 1.01 \times 10^5 \text{ Pa}, 4.40, 6.15 \times 10^8 \text{ Pa}) & \text{pre-shock} \\ (1.22 \frac{\text{kg}}{\text{m}^3}, 0, 1.01 \times 10^5 \text{ Pa}, 1.4, 0) & \text{quiescent air} \end{cases}$$
(6.47)

The post-shock conditions have been calculated using the Rankine-Hugoniot relations for the stiffened-gas equation of state derived in [10]. The computational domain is $[-1\text{m}, 1\text{m}]$, and the initial locations of the shock front and liquid/air interface are selected as $x = -0.1\text{m}$ and $x = 0$ respectively. The equation of state parameters for water (γ and Π) have been chosen to coincide with the values presented in [17] for the given Mach number regime.

We found that the positivity limiters were only required for the seventh-order and higher schemes; both the third and fifth-order schemes were able to provide solutions without the limiters. The eleventh and fifth-order solutions for density, pressure, and the two-material properties are presented in Figs. 6.9 and 6.10. The visible distinguishing features are the interface near $x = 0.5$ and the rarefaction wave propagating leftward into the water. This expansion wave is the result of the initial shock impinging upon the liquid/gas interface. Because of the scaling of the problem, the actual shock front is not discernible; although a zoomed-in view reveals the shock propagating within the gas ahead of the interface near $x = 0.8\text{m}$.

Figure 6.11 depicts the material properties at the water/air interface for several different solution orders. The higher order methods capture the interface more sharply; this is especially prominent for Π, in which the lower order solutions are notably less sharp on the right side of the discontinuity.

These results demonstrate the high-order positivity-preserving qualities of the scheme for flows involving a liquid/vapor interface. This is an important test, as it forms the basis for the 2D shock/bubble interaction problems involving gaseous bubbles immersed in liquid. Using our scheme, we solved this challenging problem in the seventh, ninth, and eleventh orders, which was impossible using the base scheme without limiters.

6.5 Numerical Examples on Positivity Preservation for Two-Fluid System

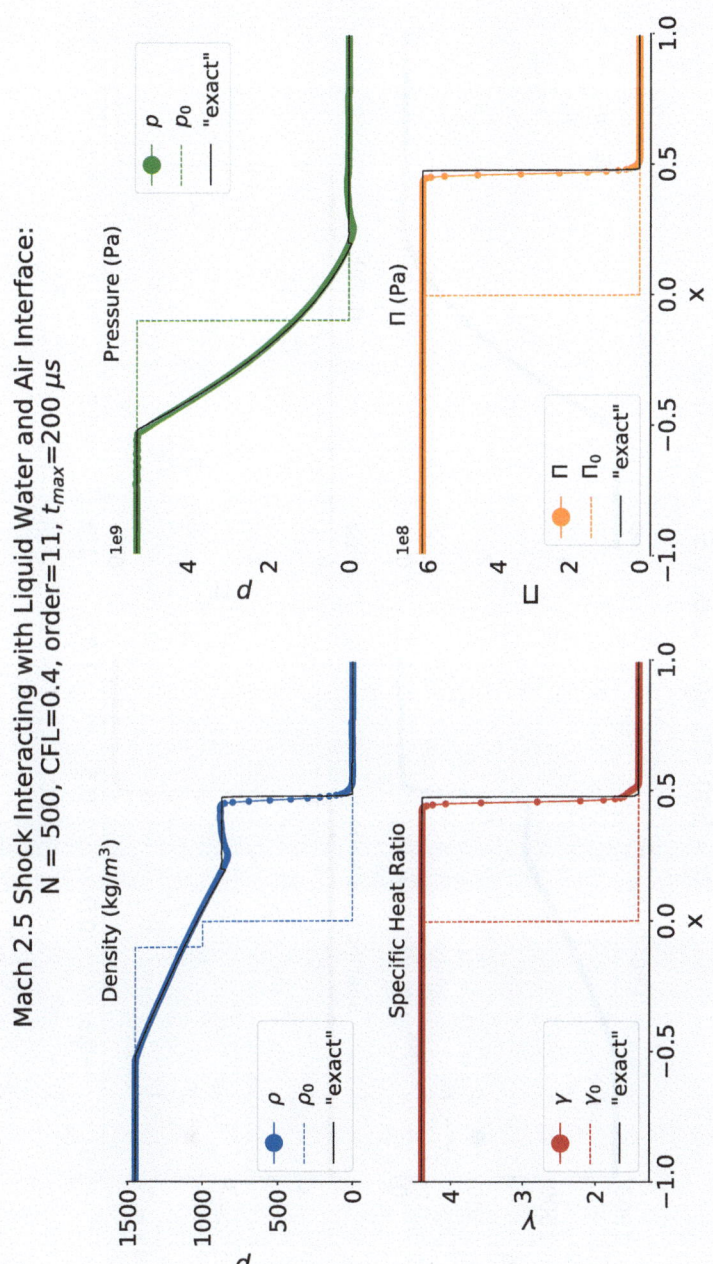

Fig. 6.9 Eleventh-order solution to the shock liquid/air interface problem. These results were first appeared in the author's earlier work [4]

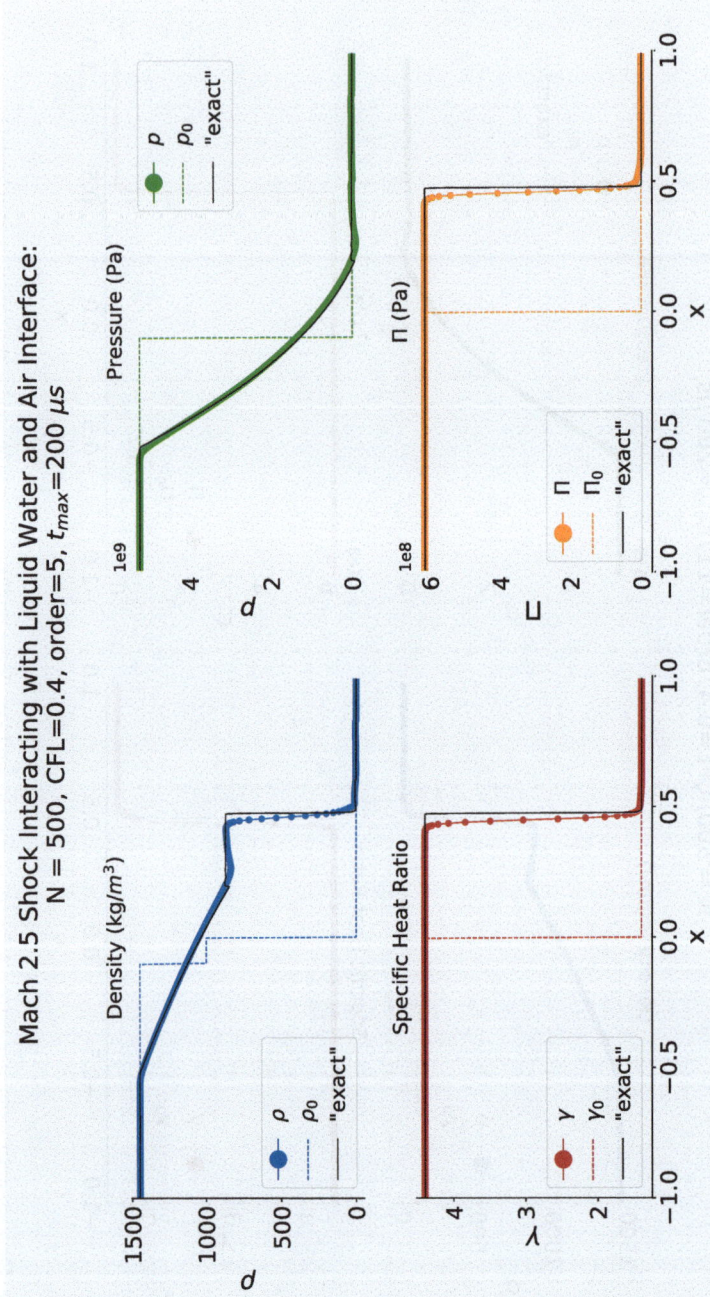

Fig. 6.10 Fifth-order solution to the shock liquid/air interface problem. These results were first appeared in the author's earlier work [4]

6.5 Numerical Examples on Positivity Preservation for Two-Fluid System

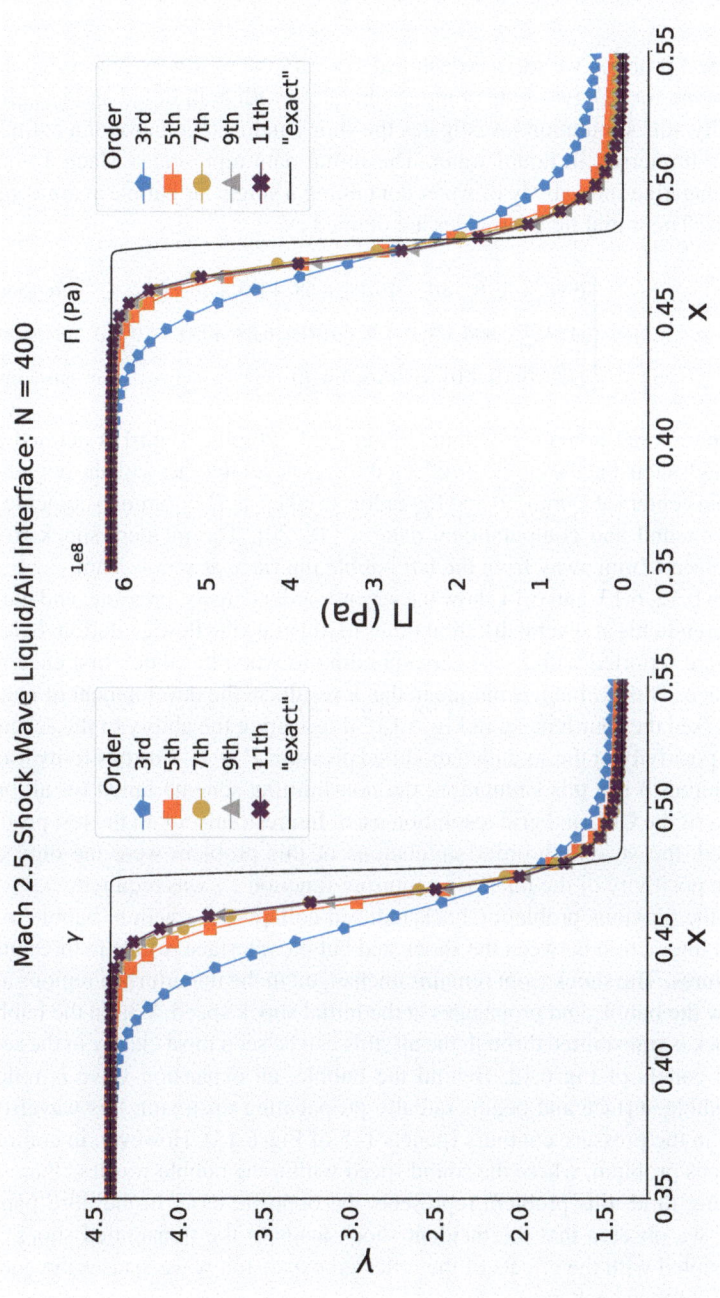

Fig. 6.11 Comparison of high-order solutions for specific heat ratio and Π for liquid/air interface problem. These results were first appeared in the author's earlier work [4]

6.5.2 Two-Dimensional High Mach Shock Air Bubble Interaction

The final test problem we solve is designed to verify the positivity-preserving nature of the scheme for 2D, two-fluid compressible flow with both liquid and gas phases. Specifically, this simulation investigates the shock-induced collapse of a cylindrical air bubble immersed in liquid water. The initial data represent a Mach 1.9 shock wave propagating into a body of water containing a single air bubble at atmospheric conditions. The initial field variables are defined as

$$(\rho, v_x, v_y, p, \Gamma, \Pi) = \begin{cases} (998 \frac{kg}{m^3}, 0, 0, 1.01 \times 10^5 \text{Pa}, 6.68, 4050 \times 10^5 \text{Pa}) & \text{pre-shock} \\ (1231 \frac{kg}{m^3}, 600.3 \frac{m}{s}, 0, 1.9 \times 10^9 \text{Pa}, 6.68, 4050 \times 10^5 \text{Pa}) & \text{post-shock} \\ (1.22 \frac{kg}{m^3}, 0, 1.01 \times 10^5 \text{Pa}, 1.4, 0) & \text{quiescent air} \end{cases}$$
(6.48)

The computational domain $[-20\text{mm}, 20\text{mm}] \times [-20\text{mm}, 20\text{mm}]$ is defined to be large enough such that the incident/reflected waves never interact with the boundaries. The bubble centers at (3mm, 0) and the radius is taken as $R = 6$mm to coincide with the experimental and computational data of [18–20]. The incident shock front is initially placed 1mm away from the left bubble interface at $x = -1$mm.

Figures 6.12, 6.13, and 6.14 show the seventh order density, pressure, and numerical Schlieren fields at several different times to illustrate the flow evolution. Note that all times are reported with $t = 0$ corresponding to when the shock first encounters the interface. This problem is unique in that it results in the development of negative pressures (see the fourth panel in Fig. 6.13), thus testing the ability of the scheme to maintain positivity of the so-called modified pressure $P + \Pi$. The positivity limiters were required to run this simulation; the non-limiting scheme failed for all orders regardless of the CFL and grid resolution used. Interestingly, of all the test problems we studied, the very high-order simulations of this problem were the only cases where the positivity of the interface capturing function Γ_1 was required.

As in the previous problem (shock-induced collapse of a helium bubble in air), the initial interaction between the shock and bubble interface results in three unique wave features. The shock front remains unchanged in the unperturbed regions above and below the bubble and propagates at the initial shock speed. Within the bubble, a weak shock is transmitted through the air; this can be seen most clearly in the second and third panels of Fig. 6.12. Behind the bubble, an expansion wave is reflected off the bubble surface and begins radially propagating upstream; this wavefront is observed in the pressure contours (panels 1–5 of Fig. 6.13). However, in contrast to the previous problem, where the sound speed within the bubble was less than in the surrounding fluid, this problem represents the opposite case. In the third panel of Fig. 6.14, we observe that the incident shock leads to the transmitted shock. This effect, coupled with the effects of the reflected expansion wave, causes the concave "bending" of the shock front.

6.5 Numerical Examples on Positivity Preservation for Two-Fluid System

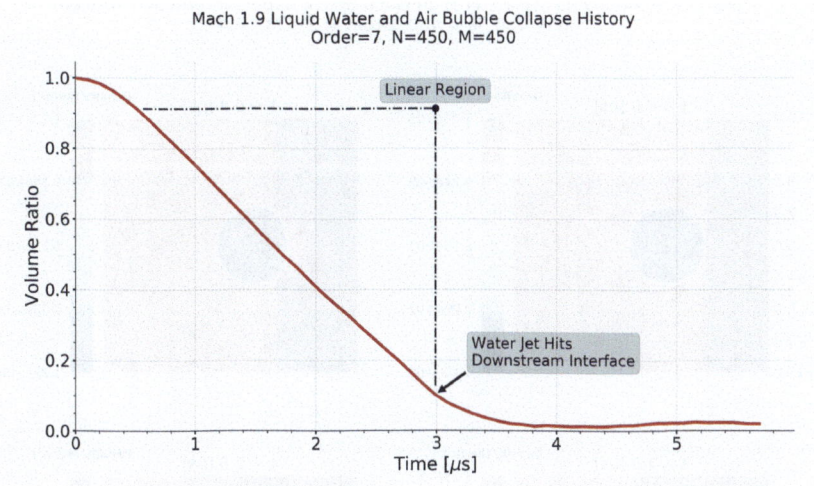

Fig. 6.12 Water/air shock-bubble interaction—seventh-order density solutions. These results were first appeared in the author's earlier work [4]

In the third panel of Fig. 6.12, we can see the early formation of a water jet developing along the $y = 0$ line. At this point in the collapse, the jet is manifested as a slight indentation at the center of the bubble's deformed left face. This jet rapidly cuts through the center of the air bubble in the fourth panel of Figs. 6.12 and 6.14 and in the fifth panel, the water jet has completely severed the air bubble in half. A shock wave resulting from this event is faintly seen propagating downstream in panel 5 of the numerical Schlieren contours.

In the literature, this event has been used to characterize the bubble collapse time. Specifically, [18] defines the collapse time, τ, as the duration between the initial shock/interface interaction and the bubble splitting. In their experiments using high-speed Schlieren photography, they found this time to be about 3.5 μs. The numerical studies of [19, 20] measured this collapse time at 3.1 μs and 3.0 μs, respectively, matching well with our result of 3.0μs. The evolution of the normalized bubble volume is presented in Fig. 6.15. At each time, the volume is calculated simply by summing the total number of cells bounded by the bubble. The key feature to highlight in this plot is the linearity of the bubble collapse. Both [18, 19] observed that for a cylindrical bubble, the volume ratio decreases linearly and that the collapse time, τ, marks the transition to a nonlinear regime. This trend is clearly observed in Fig. 6.15.

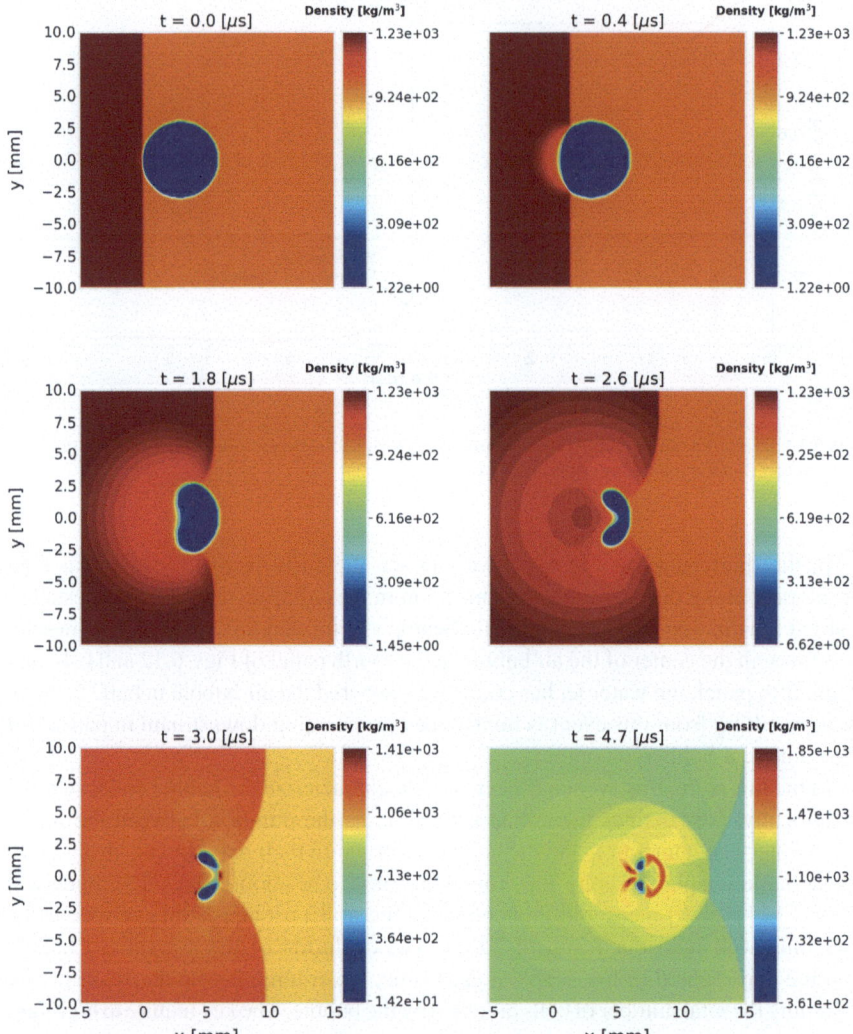

Fig. 6.13 Water/air shock-bubble interaction—seventh-order pressure solutions. These results were first appeared in the author's earlier work [4]

6.5 Numerical Examples on Positivity Preservation for Two-Fluid System

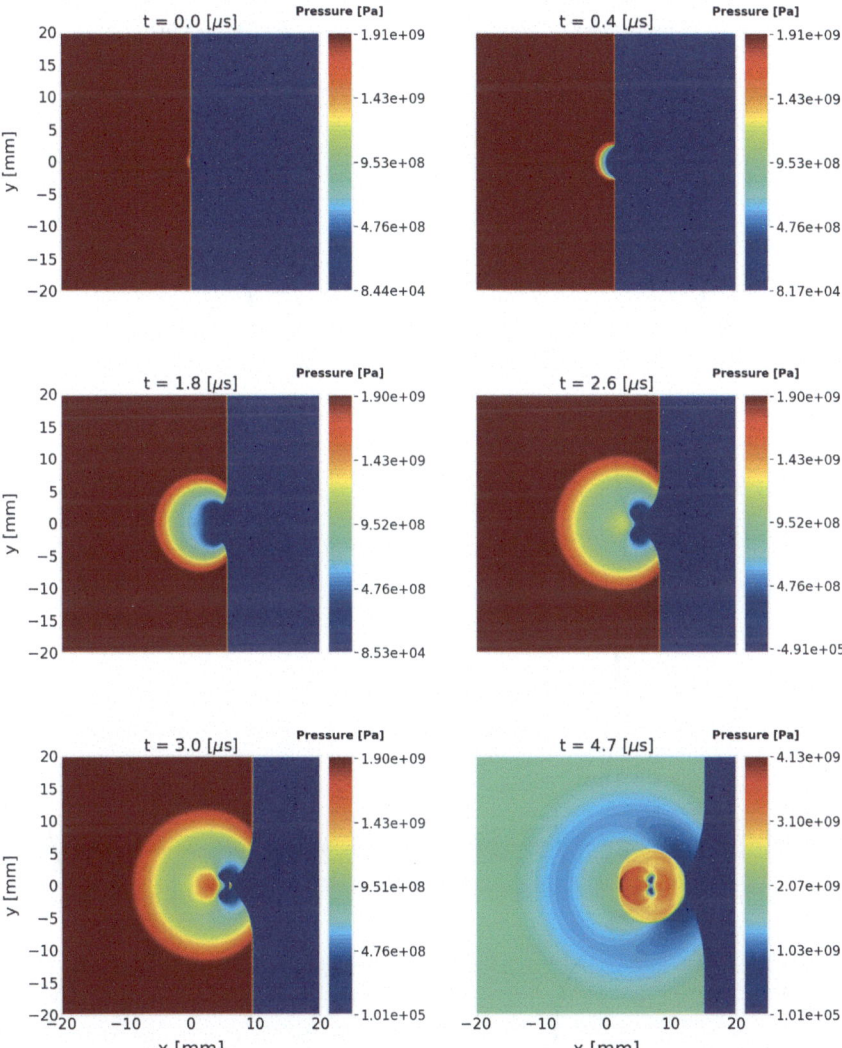

Fig. 6.14 Water/air shock-bubble interaction—numerical Schlieren. These results were first appeared in the author's earlier work [4]

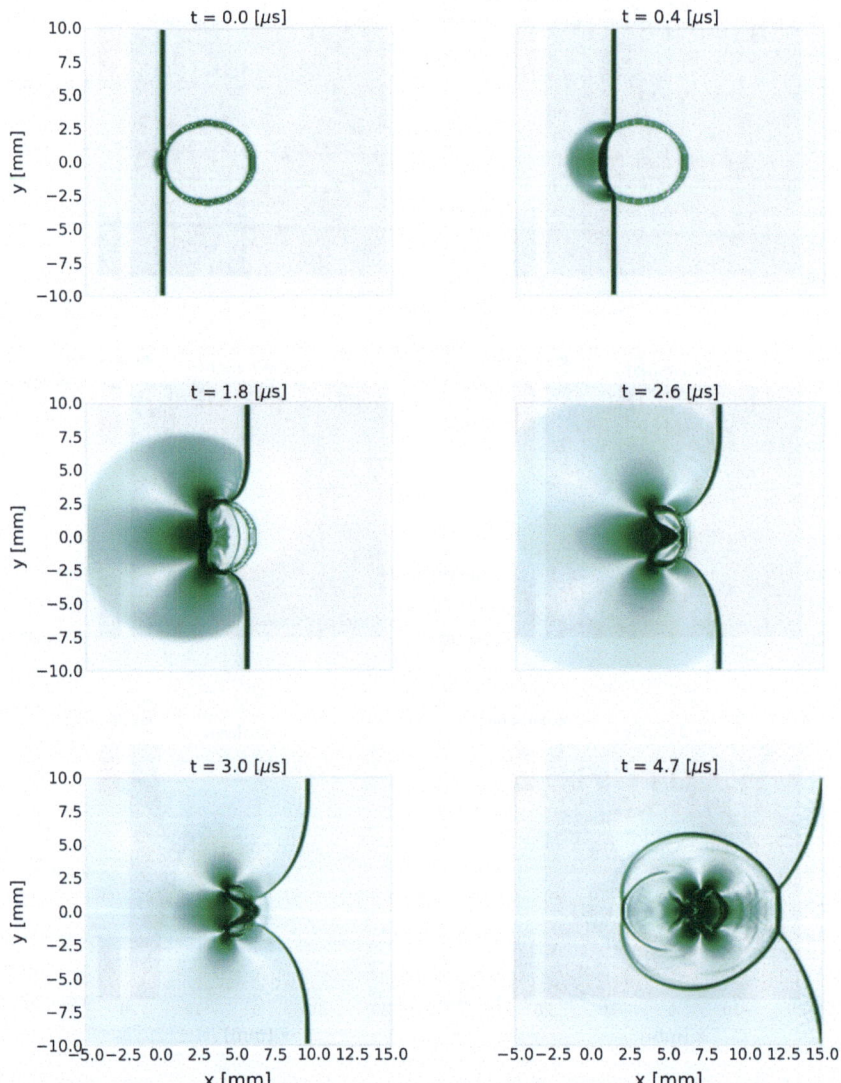

Fig. 6.15 Air bubble collapse history for water/air shock-bubble problem. These results were first appeared in the author's earlier work [4]

6.5.3 Impact on CPU Time

The flux-limiting process's impact on the scheme's overall CPU time is also investigated. Table 6.3 shows our scheme's average CPU time per time step with and without the positivity algorithm. Results are given for both the helium/air shock-bubble problem and the liquid shock air bubble interaction (Sect. 6.5.2). For the latter problem, negative density and modified pressure are obtained after the first step unless the limiters are used; thus, all reported CPU times are based on the first step. For all cases, the percentage difference was less than 10%

For the fifth-order simulation, only the first time step resulted in the fluxes being limited. This number increased to three during the seventh-order simulation. Both simulations ran for at least 800-time steps, so the number of steps that demanded positivity preservation represents a small percentage.

Although many factors affect the actual CPU time for a particular computation, measuring and comparing the CPU time with and without the limiter in one code and with one compiler is still useful. The takeaway message from the 10% extra cost reported above is that for any implementation, we expect a small increase in the computational cost when using the positivity preservation limiter.

6.6 Exercises

1. Derive the eigenvalues and eigenvectors of the two-fluid system (6.33) and the matrices of the left and right eigenvectors $[Q]$ and $[Q^{-1}]$.
2. Derive the characteristic variables for the two-fluid model (6.33).
3. Prove Lemma 2.
4. Show the inconsistency of the two-fluid model in determining the specific heat ratio in a diffused interface region.

Table 6.3 Comparison of CPU cost per time step for the scheme with limiters and without limiters for $(N, M) = (600, 150)$. These results were first appeared in the author's earlier work [4]

Order		Helium/Air		Water/Air	
		CPU Time [s]	% difference (%)	CPU Time [s]	% difference (%)
5	With limiters	6.04		6.06	
	Without limiters	5.52	9.35	5.53	9.71
7	With limiters	8.19		8.23	
	Without limiters	7.67	6.91	7.66	7.37

References

1. Shahbazi, K.: High-order finite difference scheme for compressible multi-component flow computations. Comput. Fluids **190**, 425–439 (2019)
2. Abgrall, R.: How to prevent pressure oscillations in multicomponent flow calculations: A quasi conservative approach. J. Comput. Phys. **125**(1), 150–160 (1996)
3. Shahbazi, K.: Positivity preservation of a first-order scheme for a quasi-conservative compressible two-material model. SIAM J. Sci. Comput. **43**(4), B1029–B1055 (2021)
4. Boe, D., Shahbazi, K.: A positivity preserving high-order finite difference method for compressible two-fluid flows. Numer. Methods Partial Differ. Equ. **39**(6), 4087–4125 (2023)
5. Shahbazi, K.: Robust second-order scheme for multi-phase flow computations. J. Comput. Phys. **339**, 163–178 (2017)
6. Johnsen, E., Colonius, T.: Implementation of WENO schemes in compressible multicomponent flow problems. J. Comput. Phys. **219**(2), 715–732 (2006)
7. Coralic, V., Colonius, T.: Finite-volume WENO scheme for viscous compressible multicomponent flows. J. Comput. Phys. **274**, 95–121 (2014)
8. Harlow, F., Amsded, A.: Fluid dynamics. Technical report, Los Alamos (1971)
9. Cocchi, J.P., Saurel, R., Loraud, J.C.: Treatment of interface problems with godunov-type schemes. Shock Waves **5**, 347–357 (1996)
10. Johnsen, E.: Numerical simulations of non-spherical bubble collapse with applications to shockwave lithotripsy. Ph.D. thesis, California Institute of Technology, Pasedina (2007)
11. Liu, T.G., Khoo, B.C., Yeo, K.S.: Ghost fluid method for strong shock impacting on material interface. J. Comput. Phys. **190**(2), 651–681 (2003)
12. Hejazialhosseini, B., Rossinelli, D., Bergdorf, M., Koumoutsakos, P.: High order finite volume methods on wavelet-adapted grids with local time-stepping on multicore architectures for the simulation of shock-bubble interactions. J. Comput. Phys. **229**(22), 8364–8383 (2010)
13. So, K.K., Hu, X.Y., Adams, N.A.: Anti-diffusion interface sharpening technique for two-phase compressible flow simulations. J. Comput. Phys. **231**(11), 4304–4323 (2012)
14. Terashima, H., Tryggvason, G.: A front-tracking/ghost-fluid method for fluid interfaces in compressible flows. J. Comput. Phys. **228**(11), 4012–4037 (2009)
15. Haas, J.-F., Sturtevant, B.: Interaction of weak shock waves with cylindrical and spherical gas inhomogeneities. J. Fluid Mech. **181**, 41–76 (1987)
16. Quirk, J.J., Karni, S.: On the dynamics of a shock–bubble interaction. J. Fluid Mech. **318**, 129–163 (1996)
17. Saurel, R., Abgrall, R.: A simple method for compressible multifluid flows. SIAM J. Sci. Comput. **21**(3), 1115–1145 (1999)
18. Bourne, N.K., Field, J.E.: Shock-induced collapse of single cavities in liquids. J. Fluid Mech. **244**, 225–240 (1992)
19. Ball, G.J., Howell, B.P., Leighton, T.G., Schofield, M.J.: Shock-induced collapse of a cylindrical air cavity in water: a free-Lagrange simulation. Shock Waves **10**(4), 265–276 (2000)
20. Wang, C., Shu, C.-W.: An interface treating technique for compressible multi-medium flow with Runge-Kutta discontinuous Galerkin method. J. Comput. Phys. **229**(23), 8823–8843 (2010)

Chapter 7
Schemes for Compressible Two-Fluid Navier-Stokes Equations

This chapter addresses how to devise high-order, positivity-preserving finite difference schemes for unsteady compressible Navier-Stokes equations. We consider the equation of states beyond ideal gas law, including the stiffened-gas equation of state used to represent non-ideal gases and liquids. After presenting the viscous stresses, viscous work terms, and heat conduction terms, we discuss how to compute them using a general Lax-Friedrichs type discretization that yields positivity-preserving solutions. This LF-type discretization of the viscous term is essential for positivity preservation in two and three dimensions as the simple second-order central finite difference viscous discretization can only be proved to be positivity preserving in one dimension and proofs beyond one dimension are not possible due to the mixed derivatives in the compressible viscous stresses [1]. The discretization is presented for both single-fluid and two-fluid cases.

7.1 Multi-dimensional Compressible Navier-Stokes Equations

The unsteady, compressible Navier-Stokes equations without external forces in the conservative form are written as

$$U_t + \nabla \cdot F^a = \nabla \cdot F^d \tag{7.1}$$

The conservative variables are $U = (\rho, \rho u, \rho v, \rho w, e_{\text{total}})^T = (\rho, \rho \vec{V}, e_{\text{total}})^T$, where ρ is the density, $\vec{V} = (u, v, w)^T$ is the velocity vector, and e_{total} is the total specific energy. The advective and diffusive fluxes are

$$F^a = \begin{pmatrix} \rho u \\ \rho u \otimes u + p \mathbb{I} \\ (e_{\text{total}} + p) u \end{pmatrix} \tag{7.2}$$

and

$$F^d = \begin{pmatrix} 0 \\ \tau \\ \vec{V} \cdot \tau - q \end{pmatrix} \quad (7.3)$$

In Eq. (7.2), p is the pressure, $\mathbb{I}$ is the identity matrix, and

$$\vec{V} \otimes \vec{V} = \begin{pmatrix} u^2 & uv & uw \\ vu & v^2 & vw \\ wu & wv & w^2 \end{pmatrix} \quad (7.4)$$

In Eq. (7.3), the viscous stress τ is

$$\tau = \begin{pmatrix} \tau_{11} & \tau_{12} & \tau_{13} \\ \tau_{21} & \tau_{22} & \tau_{23} \\ \tau_{31} & \tau_{32} & \tau_{33} \end{pmatrix} \quad (7.5)$$

where with the assumption of Newtonian fluid and Stokes hypothesis, the viscous shear stress is related to the dynamic shear viscosity coefficient and velocity gradient as

$$\tau_{11} = 2\mu \left(u_x - \frac{1}{3} \nabla \cdot \boldsymbol{u} \right) \quad (7.6a)$$

$$\tau_{22} = 2\mu \left(v_y - \frac{1}{3} \nabla \cdot \boldsymbol{u} \right) \quad (7.6b)$$

$$\tau_{33} = 2\mu \left(w_z - \frac{1}{3} \nabla \cdot \boldsymbol{u} \right) \quad (7.6c)$$

$$\tau_{12} = \mu(u_y + v_x) \quad (7.6d)$$

$$\tau_{13} = \mu(u_z + w_x) \quad (7.6e)$$

$$\tau_{23} = \mu(v_z + w_y) \quad (7.6f)$$

The form (7.6) ensures a reference-invariant representation of viscous stresses [2].

The heat flux q in Eq. (7.3) is defined using Fourier's law of heat conduction in the form

$$q = -\kappa \nabla T \quad (7.7)$$

where T is the temperature and κ is the thermal conductivity coefficient.

7.2 Property Relations

The system of NS equations (7.1) must be complemented with thermodynamically consistent equations of state. Specifically, property relations are required to relate pressure and temperature to the internal energy $e = e_{\text{total}} - (u^2 + v^2 + w^2)/2$ and the

7.2 Property Relations

density. Since the pressure, density, and temperature are easily measurable properties, equations of states are often given as a relationship between p, T, and ρ. Knowing $p = p(T, \rho)$, an expression for the internal energy can be obtained using Eq. (8.47) in the form

$$de = c_v dT - \frac{1}{\rho^2}\left[T\left(\frac{\partial p}{\partial T}\right)_\rho - p\right]d\rho \tag{7.8}$$

where c_v is the specific heat coefficients at constant volume. Now, using the given equation of state $p = p(T, \rho)$ eliminating temperature T from Eq. (7.8) yields an expression in the form $p = p(e, \rho)$ and similarly eliminating pressure p from (7.8) yields an expression in the form $T = T(e, \rho)$. Expressions $p = p(e, \rho)$ and $T = T(e, \rho)$ can then be used to compute pressure and temperature using the computed internal energy and density.

7.2.1 Example of Property Relations for the Stiffened-Gas Equation of State

Consider the stiffened-gas equation of state in the form

$$p = c_v(\gamma - 1)\rho T - \pi \tag{7.9}$$

where $\gamma = c_p/c_v$ for gases with c_p being the specific heat at constant pressure [3]. The parameter π models the inter-molecular bonds in non-ideal gases and liquids. Using the above procedure, we have

$$e = c_v T + \frac{\pi}{\rho} + e_0 \tag{7.10}$$

or

$$T = \frac{1}{c_v}\left(e - e_0 - \frac{\pi}{\rho}\right) \tag{7.11}$$

and

$$p = (\gamma - 1)\rho(e - e_0) - \gamma\pi \tag{7.12}$$

where e_0 is a reference energy arising from the integration. Hence, the three material parameters of c_v, γ, and e_0 must be known for the fluid. In addition, the dynamic viscosity μ and the thermal conductivity κ coefficients are also needed.

Remark 7.1 Note that instead of computing temperature from Eq. (7.11), one may alternatively compute temperature from Eq. (7.9) after computing pressure from Eq. (7.12). However, this is not a preferred approach in some two-fluid model computations where the equilibrium of interfacial temperature is desired [4], as further discussed in the two-fluid modeling of viscous compressible flows.

7.3 Diffusive Flux Discretization

We first provide a first-order LF-type scheme for the viscous flux discretization used in Ref. [1]. We then provide the high-order version of the discretization. The discretization follows closely that of the advection fluxes given in Chap. 2 with a difference in the definition of the stabilization parameter. This approach of using an LF-type discretization for the viscous term has similarity to the interior penalty method for the discontinuous Galerkin discretization of the diffusive terms [5, 6]. This approach allows the establishment of the positivity of the first-order scheme as proved in [1].

7.3.1 First-Order Scheme

Similar to the LF discretization of the advection flux, we first split the viscous flux F^d into left (positive) and right (negative) fluxes as

$$F^d = \frac{1}{2}(F^{d,L} + F^{d,R}) \tag{7.13}$$

with

$$F^{d,L} = F^d - \beta U \qquad F^{d,R} = F^d + \beta U \tag{7.14}$$

where β is a stabilization parameter to be defined appropriately. Note that opposite to the advection fluxes, the diffusive fluxes appear on the right-hand side of System (7.1); hence, the left and right fluxes are formed with the opposite sign of the stabilization parameter; compare the diffusive flux splitting (7.14) with the advection flux splitting (5.7). A first-order accurate mid-point approximations of the left and right fluxes are then obtained using the left and right values of fluxes as

$$F^d_{i+\frac{1}{2}} = \frac{1}{2}(F^{d,L}_{i+\frac{1}{2}} + F^{d,R}_{i+\frac{1}{2}}) \tag{7.15}$$

where

$$F^{d,L}_{i+\frac{1}{2}} = F^d_i - \beta U_i \tag{7.16}$$

$$F^{d,R}_{i+\frac{1}{2}} = F^d_{i+1} + \beta U_{i+1} \tag{7.17}$$

Then, the derivative of the diffusive flux is approximated using the mid-point approximate fluxes as

7.3 Diffusive Flux Discretization

$$\left.\frac{\partial \boldsymbol{F}^d}{\partial x}\right|_{x_i} \approx \frac{1}{h}\left[\boldsymbol{F}^d_{i+\frac{1}{2}} - \boldsymbol{F}^d_{i-\frac{1}{2}}\right] \tag{7.18}$$

$$= \frac{1}{2h}\left[\boldsymbol{F}^d_i - \beta \boldsymbol{U}_i + \boldsymbol{F}^d_{i+1} + \beta \boldsymbol{U}_{i+1} - (\boldsymbol{F}^d_{i-1} - \beta \boldsymbol{U}_{i-1}) - (\boldsymbol{F}^d_i + \beta \boldsymbol{U}_i)\right] \tag{7.19}$$

$$= \frac{1}{2h}\left[(\boldsymbol{F}^d_{i+1} + \beta \boldsymbol{U}_{i+1}) - (\boldsymbol{F}^d_{i-1} - \beta \boldsymbol{U}_{i-1}) - 2\beta \boldsymbol{U}_i\right] \tag{7.20}$$

Now, for the first-order forward Euler temporal discretization of the NS equations, we have

$$\boldsymbol{U}^{n+1}_i = \frac{1}{2}\boldsymbol{U}^n_i - \frac{\tau}{2h}\left[(\boldsymbol{F}^{a,n}_{i+1} - \alpha \boldsymbol{U}^n_{i+1}) - (\boldsymbol{F}^{a,n}_{i-1} + \alpha \boldsymbol{U}^n_{i-1}) + 2\alpha \boldsymbol{U}^n_i\right] \tag{7.21}$$

$$+ \frac{1}{2}\boldsymbol{U}^n_i + \frac{\tau}{2h}\left[(\boldsymbol{F}^{d,n}_{i+1} + \beta \boldsymbol{U}^n_{i+1}) - (\boldsymbol{F}^{d,n}_{i-1} - \beta \boldsymbol{U}^n_{i-1}) - 2\beta \boldsymbol{U}^n_i\right] \tag{7.22}$$

Or, rearranging yields

$$\boldsymbol{U}^{n+1}_i = \frac{1}{2}\left[1 - \frac{\tau\alpha}{h}\right]\boldsymbol{U}^n_i + \frac{\tau\alpha}{2h}\left[\boldsymbol{U}^n_{i+1} - \frac{\boldsymbol{F}^{a,n}_{i+1}}{\alpha}\right] + \frac{\tau\alpha}{2h}\left[\frac{\boldsymbol{F}^{a,n}_{i-1}}{\alpha} + \boldsymbol{U}^n_{i-1}\right] \tag{7.23}$$

$$+ \frac{1}{2}\left[1 - \frac{\tau\beta}{h}\right]\boldsymbol{U}^n_i + \frac{\tau\beta}{2h}\left[\boldsymbol{U}^n_{i+1} + \frac{\boldsymbol{F}^{d,n}_{i+1}}{\beta}\right] + \frac{\tau\beta}{2h}\left[\boldsymbol{U}^n_{i-1} - \frac{\boldsymbol{F}^{d,n}_{i-1}}{\beta}\right] \tag{7.24}$$

Similar to the discretization of pure advection, if α and β are chosen sufficiently large and under the CFL condition $\frac{\Delta t}{h}\max(\alpha, \beta) < 1$, the discretization is positivity preserving because $\boldsymbol{U}^{n+1}_i$ is a convex combination of six vectors that each yield a positive solution.

The stabilization parameter α choice is discussed in Chap. 2. It remains to define the stabilization parameter β and the computation of the diffusive fluxes as the diffusive fluxes depend on both conservative variable and derivative of velocity and temperature. Reference [1] introduced an expression for β for positivity preservation for ideal gases. We extend what is given in Ref. [1] to the stiffened-gas equation of state as follows:

$$\beta > \max\left\{\frac{1}{2\rho^2 e}\left[\sqrt{\rho^2(\boldsymbol{q}\cdot\boldsymbol{n})^2 + 2\rho^2 e(||\boldsymbol{\tau}\cdot\boldsymbol{n}||^2 + \pi^2)} + \rho|\boldsymbol{q}\cdot\boldsymbol{n}|\right]\right\} \tag{7.25}$$

The maximum is taken over the conservative variables and derivatives of the velocity and temperature of the points in the stencil of the LF scheme (i.e., $i-1, i$, and $i+1$ points), leading to a local version of the LF-type scheme. The global version of the scheme is obtained by taking the maximum over all the grid points.

The derivative of the velocity vector in the viscous stress and the derivative of the temperature in the heat conduction flux is obtained by a simple second-order derivative approximation. For instance, u_x is approximated as

$$u_x|_i \approx \frac{u_{i+1} - u_{i-1}}{2h} \tag{7.26}$$

Remark 7.2 In multiple space dimensions, the discretization of the diffusive fluxes is carried out direction by direction, as performed in the finite difference approach. Positivity preservation in each direction, along with Jensen's inequality, ensures the positivity preservation of the multi-dimensional scheme.

7.3.2 High-Order Scheme

We first write the LF splitting of the nonlinear viscous flux $\boldsymbol{F}^d(U)$ as

$$\boldsymbol{F}^d = \frac{1}{2}(\boldsymbol{F}^{d,L} + \boldsymbol{F}^{d,R}) \tag{7.27}$$

with

$$\boldsymbol{F}^{d,L} = \boldsymbol{F}^d - \beta U \qquad \boldsymbol{F}^{d,R} = \boldsymbol{F}^d + \beta U \tag{7.28}$$

Then, the high-order LF scheme has the following form

$$\left.\frac{\partial \boldsymbol{F}^d}{\partial x}\right|_{x_i} \approx \frac{1}{h}\sum_{j=1}^{r+1} d_j \left[\frac{1}{2}\left(\boldsymbol{F}^{d,L}_{i+j-\frac{1}{2}} + \boldsymbol{F}^{d,R}_{i+j-\frac{1}{2}}\right) - \frac{1}{2}\left(\boldsymbol{F}^{d,L}_{i-j+\frac{1}{2}} + \boldsymbol{F}^{d,R}_{i-j+\frac{1}{2}}\right) \right] \tag{7.29}$$

where coefficient d_j are given in Table 2.2. Here, $\boldsymbol{F}^{d,L}_{i+j-\frac{1}{2}}$ and $\boldsymbol{F}^{d,R}_{i-j+\frac{1}{2}}$ can be obtained from left and right WENO approximation of the mid-point values of the primitive variables and the derivative of the velocity for the viscous fluxes and the derivative of the temperature for the heat flux. The derivatives of the velocity and temperature are obtained from the high-order derivative approximation using the central WENO reconstruction of the velocity, density, and internal energy. The central WENO reconstruction can be computed by averaging the left and right WENO reconstructions of the velocity and temperature.

7.3.3 Example of High-Order Positivity Preserving Scheme for the Navier-Stokes Equations

To verify the positivity-preservation algorithm in the presence of both advective and diffusive fluxes, we solve the viscous interacting blast waves, and the results are discussed below. Specifically, we simulate the collision of a right-moving Mach 199 shock wave with a left-moving Mach 63 shock wave. This test is a modified version of the interacting blast wave problem discussed in [7, 8]. Specifically, we have moved

7.3 Diffusive Flux Discretization

the two initial blast waves closer together to reduce the effects of numerical diffusion and preserve the sharpness of the shock discontinuities at the collision, resulting in a more challenging test for the positivity limiters. Owing to the large pressure ratios across the shocks and near vacuum pressure values, this problem is designed to verify the preservation of pressure positivity. The domain is [0, 1], and the specific heat ratio is selected to correspond to air ($\gamma = 1.4$). The initial data are

$$(\rho, v, p) = \begin{cases} (1, 0, 1000) & 0 \leq x < 0.6 \\ (1, 0, \frac{1}{100}) & 0.6 \leq x < 0.7 \\ (1, 0, 100) & 0.7 \leq x \leq 1 \end{cases} \quad (7.30)$$

The specific heat values used are $c_p = 1.005$ and $c_v = 0.718$ corresponding to air. Figure 7.1 shows the computed density for the blast wave interaction problem by solving the Navier-Stokes equations. The results for two sets of values of diffusion coefficients are presented, illustrating the impact of the diffusive terms. For the first test case, viscosity and thermal conductivity are $\mu = 1.8(10)^{-5}$ and $k = 2.3(10)^{-5}$, and for the second, $\mu = 1.8(10)^{-4}$ and $k = 2.3(10)^{-4}$. For each case, the positivity limiters are required for all orders, and without positivity limiters, a negative square of sound speed has been generated, resulting in computational overflow.

All results shown in Fig. 7.1 were obtained with a CFL condition of 0.4. The results (errors and CPU times) did not vary when smaller values of CFL number were used. The computations for all orders were stable for the CFL condition of 0.5. In these test problems, the advection part determines the linear stability condition (because advection is dominant compared to the diffusion due to low diffusive coefficients), and the CFL of 0.5 is the maximum CFL under which the advective part of the NS equations is positivity preserving for all orders. For this test problem, we found that the fifth-order method outperforms other approximation orders for an efficiency comparison measured by the CPU times when the error first drops below 5%. However, the optimum approximation order is problem-dependent and depends on the desired accuracy level. Higher order methods tend to become more efficient at higher desired accuracy. For instance, for shock-entropy problems, the ninth-order accurate WENO method is more efficient than other lower approximation orders, as shown in Ref. [9]. Also, in Sect. 5.3.1, we showed that for the propagation of an ultrasound wave, the fifth- and ninth-order schemes give comparable accuracy at comparable CPU times, and both outperform the third-order scheme. Similar efficiency results are also obtained in Sect. 6.3.1, where various approximation orders are used to compute the 1D strong shock helium-air interface interaction problem.

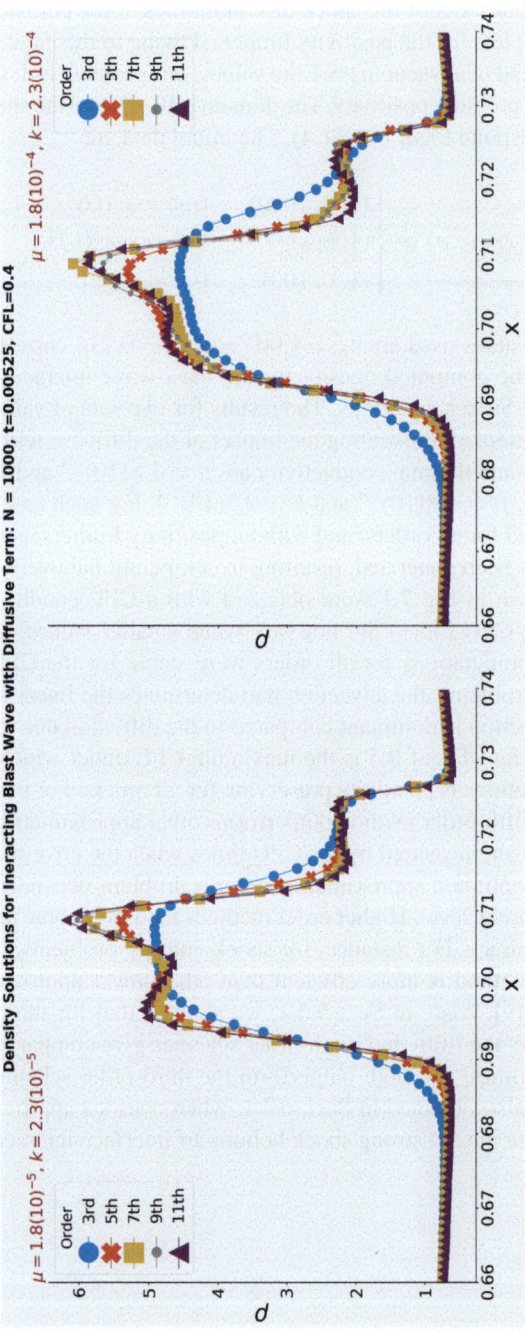

Fig. 7.1 High-order positivity preservation solution of compressible Navier-Stokes equations for the blast wave problems with a spatial grid of $N = 1000$, and various orders 3, 5, 7, 9, and 11 for two set of values of viscosity and thermal conductivity coefficients, μ, and κ

7.4 Two-Fluid Compressible Viscous Model

For the two-fluid version of compressible viscous model, in addition to the mass, momentum, and energy equations given in (7.1), five extra equations evolving material parameters arising from modeling the compressible media with the stiffened-gas equation of state (Eq. (7.10)) are required [4]. Specifically, maintaining an interfacial pressure equilibrium requires solving advection-type equations for $\Gamma = 1/(\gamma - 1)$ and $\Pi = \gamma \pi/(\gamma - 1)$ similar to the friction-less (Euler model) and a conservative-type equation for ρe_0. On the other hand, for maintaining interfacial temperature equilibrium, an advection-type equation for π is required, and two conservative-type equations for ρc_v and ρe_0. Furthermore, the pressure must be calculated using Eq. (7.12) with the computed parameters e_0, Γ and Π, while temperature must be computed using Eq. (7.11) with the computed parameters c_v, e_0, and π.

Specifically, for modeling two-fluid cases, the additional equations beyond the single-fluid model NS equation are

$$\Gamma_t + \vec{V} \cdot \nabla \Gamma = 0 \tag{7.31a}$$

$$\Pi_t + \vec{V} \cdot \nabla \Pi = 0 \tag{7.31b}$$

$$\pi_t + \vec{V} \cdot \nabla \pi = 0 \tag{7.31c}$$

$$(\rho c_v) + \nabla \cdot (\rho c_v \vec{V}) = 0 \tag{7.31d}$$

$$(\rho e_0) + \nabla \cdot (\rho e_0 \vec{V}) = 0 \tag{7.31e}$$

Note that pressure is determined using Eq. (7.12) with Γ, Π, and e_0 obtained from Eqs. (7.31a), (7.31b), and (7.31e), respectively, while the temperature is determined using Eq. (7.11) with π, c_v and e_0 obtained from Eqs. (7.31c), (7.31d), and (7.31e), respectively. Reference [4] provides an analysis similar to what is presented in Sect. 6.1.5 for the Euler model. The analysis demonstrates that this model yields accurate computations of the pressure and temperature in material interfaces, particularly in capturing interfacial temperature and pressure equilibrium.

7.4.1 Positivity of the Two-Fluid Viscous Model

The positivity proof of the first-order LF scheme for ρ, Π and Γ presented for the Euler system in Sect. 6.4 applies directly to the positivity of ρ, c_v, e_0 and Γ, Π and π in the two-fluid viscous model. The positivity of the square of the speed of sound also follows from the concavity condition on the material parameters presented in Lemma 2 (Chap. 6).

Finally, the positivity of temperature is proved as follows. Using the definition of pressure (7.12),

$$\frac{p + \pi}{\gamma - 1} > 0 \implies \rho(e - e_0) - \pi > 0 \tag{7.32}$$

The definition of temperature (7.11) and the positivity of ρ and c_v yield

$$T = \frac{1}{c_v}[\rho(e - e_0) - \pi] > 0 \tag{7.33}$$

7.4.2 Computation of Thermal Conductivity and Viscosity Coefficient

In the diffuse interfacial region, heat conduction and viscous fluxes scale with the volume fraction of each fluid. Since the volume fractions are not explicitly captured in the two-fluid model presented here, we need to evolve either the thermal conductivity coefficient or dynamics viscosity coefficient using an advection-type equation. For instance, using the thermal conductivity coefficient κ, we solve

$$\kappa_t + \vec{V} \cdot \nabla \kappa = 0 \tag{7.34}$$

Then using

$$\kappa = z\kappa_1 + (1 - z)\kappa_2 \tag{7.35}$$

the volume fraction of fluid 1, z, is obtained as

$$z = \frac{\kappa - \kappa_2}{\kappa_1 - \kappa_2} \tag{7.36}$$

Clearly, due to the monotonicity property of the advection discretization presented in Sect. 6.4, $0 < \kappa < 1$. Then, using z, the viscosity coefficient in the interfacial region is computed as

$$\mu = z\mu_1 + (1 - z)\mu_2 \tag{7.37}$$

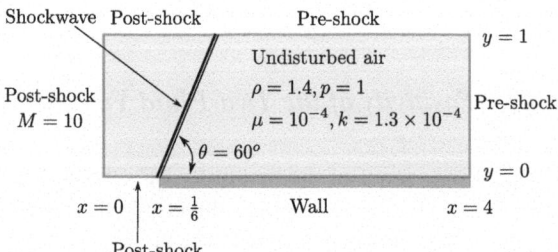

Fig. 7.2 Schematics of the double Mach reflection of a strong shock

7.5 Exercises

1. Let us describe the initial condition for a viscous double Mach reflection of a strong shock featuring an oblique shockwave boundary layer interaction. The schematic of this problem is given in Fig. 7.2. The non-viscous version of this problem was introduced in [10]. The objective is to use this initial condition and solve the compressible Navier-Stokes equations using a first-order and high-order LF-type schemes.

   Specifically, an oblique shock forms at an angle $\theta = 60°$ with a no-slip wall. The undisturbed density and pressure upstream of the shock are ρ_1 and p_1, and the Mach number of the shock in the shock reference frame is $M_1 = 10$. The density and pressure post-shock, ρ_2, and p_2 are determined using the normal shock relations:

   $$\frac{p_2}{p_1} = \frac{2\gamma}{\gamma+1} M_{1n}^2 - \frac{\gamma-1}{\gamma+1} \tag{7.38}$$

   $$\frac{\rho_2}{\rho_1} = \frac{(\gamma+1)M_{1n}^2}{(\gamma-1)M_{1n}^2+2} \tag{7.39}$$

   where

   $$M_{1n} = M_1 \sin\theta \tag{7.40}$$

   The ratio of normal velocities in the moving shock reference frame is

   $$\frac{V_{2n}}{V_{1n}} = \frac{(\gamma-1)M_{1n}^2+2}{(\gamma+1)M_{1n}^2} \tag{7.41}$$

   We need to determine the post-shock velocity in the laboratory (stationary) reference frame. Since the gas in front of the shock is stationary in the laboratory reference frame, the post-shock velocity $\vec{V}_{\text{post-shock}}$ in the laboratory frame is

   $$\vec{V}_{\text{post-shock}} = \vec{V}_2 - \vec{V}_1 = (V_{1n}\vec{n} + V_{1t}\vec{t}) - (V_{2n}\vec{n} + V_{2t}\vec{t}) \tag{7.42}$$

   and since $V_{t2}\vec{t} = V_{t1}\vec{t}$, then the post-shock velocity is

   $$\vec{V}_{\text{post-shock}} = (V_{2n} - V_{1n})\vec{n} = (V_{2n} - V_{1n})\cos\theta\,\vec{i} + (V_{2n} - V_{1n})\sin\theta\,\vec{j} \tag{7.43}$$

   The shock velocity in the laboratory (stationary) reference frame is

   $$V_s = \sqrt{\frac{\gamma p}{\rho}} M_1 \tag{7.44}$$

   The normal components of the velocity is determined as

$$V_{1n} = -V_s \sin \theta \tag{7.45}$$

The negative sign is because the left x direction is considered positive (Fig. 7.2). (**a**) Write a computer code based on the first-order LF-type scheme for the two-dimensional compressible Navier-Stokes equations with ideal gas law, constant viscosity, and thermal conductivity coefficients. (**b**) Write a computer code based on a high-order LF-type scheme for the two-dimensional compressible Navier-Stokes equations with ideal gas law, constant viscosity, and thermal conductivity coefficients. (**c**) Verify that the first-order scheme with a proper choice of stabilization parameters is positivity preserving for this problem. (**d**) Determine if the high-order scheme requires a positivity-preserving limiter for this problem.

References

1. Zhang, X.: On positivity-preserving high order discontinuous galerkin schemes for compressible navier-stokes equations. J. Comput. Phys. **328**, 301–343 (2017)
2. Acheson, D.J.: Elementary Fluid Dynamics. Oxford Applied Mathematics and Computing Science Series. Oxford University Press (1990)
3. Harlow, F., Amsded, A.: Fluid dynamics. Technical report, Los Alamos (1971)
4. Shahaboddin Alahyari Beig and Eric Johnsen: Maintaining interface equilibrium conditions in compressible multiphase flows using interface capturing. J. Comput. Phys. **302**, 548–566 (2015)
5. Shahbazi, K.: An explicit expression for the penalty parameter of the interior penalty method. J. Comput. Phys. **205**(2), 401–407 (2005)
6. Arnold, D.N.: An interior penalty finite element method with discontinuous elements. SIAM J. Numer. Anal. **19**(4), 742–760 (1982)
7. Woodward, P., Colella, P.: The numerical simulation of two-dimensional fluid flow with strong shocks. J. Comput. Phys. **54**(1), 115–173 (1984)
8. Gerolymos, G.A., Sénéchal, D., Vallet, I.: Very-high-order weno schemes. J. Comput. Phys. **228**, 8481–8524 (2009)
9. Shahbazi, K., Albin, N., Bruno, O.P., Hesthaven, J.S.: Multi-domain fourier-continuation/weno hybrid solver for conservation laws. J. Comput. Phys. **230**(24), 8779–8796 (2011)
10. Woodward, P., Colella, P.: The numerical simulation of two-dimensional fluid flow with strong shocks. J. Comput. Phys. **54**(1), 115–173 (1984)

Chapter 8
Thermodynamic Property Relations for Two-Phase Modeling

The two-fluid model introduced in Chap. 6 is limited to stiffened-gas equations of states. Although simple, the stiffened-gas equation of state has a limited range of validity [1]. Models supporting more general equations of states are thus needed. Moreover, the two-fluid model in Chap. 6 does not support capturing phase change. To construct and understand models supporting higher flexibility in the choice of equations of states and the type of interactions among phases, we need to discuss the fundamental thermodynamics property relations, equilibrium conditions, and derivation of thermodynamics properties from experimentally obtained general equations of states. The computation of such complex models also relies on knowing equations for properties such as internal energy, entropy, chemical (or Gibbs) potential, and sound speed for a general equation of state.

This chapter reviews the fundamental thermodynamic property relations. It explains how to derive expressions for the internal energy, enthalpy, and entropy of a substance, knowing a general (non-ideal) equation of state. It also discusses the total equilibrium conditions of mechanical, thermal, and phase equilibriums. A phase equilibrium is achieved when the chemical (Gibbs) potential between the two phases is equal. Euler's thermodynamics equation illustrates how to find the chemical (Gibbs) potential from the expressions of the internal energy and entropy for a substance with a general equation of a state. Furthermore, approaches for finding the speed of sounds from a general equation of state are also discussed. As we have seen in the previous chapters, the speed of sound not only characterizes the waves in the compressible transport medium, but it is also essential in designing stable numerical schemes based on Lax-Friedrichs approaches.

The first four sections of this chapter are based on the classic book of Callen [2].

8.1 Fundamental Relations

Our discussion on fundamental property relations relies on three postulates (assumptions):

- The entropy of a system is a continuous, differentiable, and monotonically increasing function of the system's internal energy, mass, and volume.
- The entropy of a system is additive over its constituent subsystems.
- If/when the total energy is kept constant, a process proceeds toward an equilibrium state that poses maximum total entropy.

For a system of a pure substance with a single phase, using the first postulate, we may write the relation for the entropy as

$$S = S(U, V, M) \tag{8.1}$$

where S, U, V, and M are the entropy, internal energy, volume, and the system's mass, respectively. They are all extensive properties. The first postulate allows us to find internal energy uniquely based on the other properties in the form

$$U = U(S, V, M) \tag{8.2}$$

The infinitesimal energy change is written as

$$dU = \left(\frac{\partial U}{\partial S}\right)_{V,M} dS + \left(\frac{\partial U}{\partial V}\right)_{S,M} dV + \left(\frac{\partial U}{\partial M}\right)_{S,V} dM \tag{8.3}$$

where the subscripts denote properties that are kept fixed during the partial differentiation. Or, if we define

$$T = \left(\frac{\partial U}{\partial S}\right)_{V,M} \tag{8.4a}$$

$$P = -\left(\frac{\partial U}{\partial V}\right)_{S,M} \tag{8.4b}$$

$$\mu = \left(\frac{\partial U}{\partial M}\right)_{S,V} \tag{8.4c}$$

we obtain the Gibbs property relation in the form

$$dU = TdS - PdV + \mu dM \tag{8.5}$$

Here, T is temperature, P is the pressure, and μ is called electrochemical or chemical potential. Relations (8.1) and (8.2) are called fundamental property relations since knowing an explicit expression for the fundamental relation allows us to derive all intensive properties of T, P, and μ. Of course, as will be discussed in the next section, these definitions of pressure, temperature, and chemical potential given in (8.4) are consistent with our intuitive understanding of these properties and experimental observations.

8.2 Conditions of Equilibrium

Next, we will show that the principle of maximum entropy and the additive properties of entropy and energy are equivalent to thermal, mechanical (or pressure), and chemical potential equilibrium.

8.2 Conditions of Equilibrium

In this section, we derive the conditions of equilibrium. Intuitively, the temperatures of systems in a thermal contact are equal at the equilibrium. However, let us find a complete picture, i.e., what are all the consequences of using fundamental property relations and thermodynamics postulates applied to an equilibrium state? Moreover, we will show that the thermodynamics postulates, along with definitions of temperature, pressure, and chemical potential, Eq. (8.4), yield results consistent with observable reality, including uniform temperature, pressure, and chemical potential for an equilibrium state.

Consider two systems at equilibrium separated with a movable, heat conducting, and porous wall (Fig. 8.1), otherwise isolated from their surroundings. Due to the molecular random motions, the two systems can interact with each other via infinitesimal heat transfer (are in thermal contact), infinitesimal mass transfer (are in diffusive contact), and infinitesimal volume change, but otherwise, they are isolated from their surroundings. What are the conditions at the equilibrium? The fundamental relation for each system is

$$U = U(S, V, M) \tag{8.6}$$

The infinitesimal changes in energy $dU_1 = -dU_2$, volume $dV_1 = -dV_2$, and mass $dM_1 = -dM_2$ are independent of each other, as changes in U can result from changes in entropy S. The fundamental relation for each system yields

$$dU_1 = \left(\frac{\partial U_1}{\partial S_1}\right)_{V_1, M_1} dS_1 + \left(\frac{\partial U_1}{\partial V_1}\right)_{S_1, M_1} dV_1 + \left(\frac{\partial U_1}{\partial M_1}\right)_{S_1, V_1} dM_1 \tag{8.7}$$

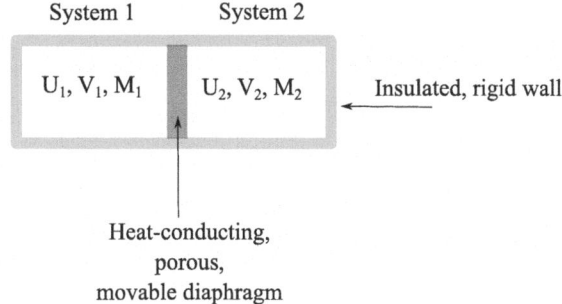

Fig. 8.1 A schematic of two systems at equilibrium separated by a heat conducting, porous, and movable diaphragm

$$dU_2 = \left(\frac{\partial U_2}{\partial S_2}\right)_{V_2, M_2} dS_2 + \left(\frac{\partial U_2}{\partial V_2}\right)_{S_2, M_2} dV_2 + \left(\frac{\partial U_2}{\partial M_2}\right)_{S_2, V_2} dM_2 \qquad (8.8)$$

Rearranging and using temperature, pressure, and chemical definitions from (8.4) yield

$$dS_1 = \frac{dU_1}{T_1} + \frac{P_1 \, dV_1}{T_1} - \frac{\mu_1 \, dM_1}{T_1} \qquad (8.9)$$

$$dS_2 = \frac{dU_2}{T_2} + \frac{P_2 \, dV_2}{T_2} - \frac{\mu_2 \, dM_2}{T_2} \qquad (8.10)$$

The total entropy is the sum of the entropy of each system:

$$dS_{\text{total}} = dS_1 + dS_2 = \frac{dU_1}{T_1} + \frac{dU_2}{T_2} + \frac{P_1 \, dV_1}{T_1} + \frac{P_2 \, dV_2}{T_2} - \frac{\mu_1 \, dM_1}{T_1} - \frac{\mu_2 \, dM_2}{T_2} \qquad (8.11)$$

Since $dU_1 = -dU_2$, $dV_1 = -dV_2$, and $dM_1 = -dM_2$, we have

$$dS_{\text{total}} = \left(\frac{1}{T_1} - \frac{1}{T_2}\right) dU_1 + \left(\frac{P_1}{T_1} - \frac{P_2}{T_2}\right) dV_1 + \left(\frac{\mu_2}{T_2} - \frac{\mu_1}{T_1}\right) dM_1 \qquad (8.12)$$

The entropy-maximum postulate then yields

$$dS_{\text{total}} = 0 \implies \left(\frac{1}{T_1} - \frac{1}{T_2}\right) dU_1 + \left(\frac{P_1}{T_1} - \frac{P_2}{T_2}\right) dV_1 + \left(\frac{\mu_2}{T_2} - \frac{\mu_1}{T_1}\right) dM_1 = 0 \qquad (8.13)$$

This result is valid for any arbitrary independent values of dU_1, dV_1, and dM_1. Hence, it must be valid for the special case of $dV_1 = dM_1 = 0$, which results in

$$\left(\frac{1}{T_1} - \frac{1}{T_2}\right) dU_1 = 0 \qquad (8.14)$$

The arbitrariness of dU_1 then yields

$$\frac{1}{T_1} - \frac{1}{T_2} = 0 \qquad (8.15)$$

or

$$T_1 = T_2 \qquad (8.16)$$

confirming that the temperature of two systems must be the same at equilibrium, which can be verified with laboratory experiments. Substituting the temperature equality into Eq. (8.13) then yields

$$\left(\frac{P_1}{T_1} - \frac{P_2}{T_2}\right) dV_1 + \left(\frac{\mu_2}{T_2} - \frac{\mu_1}{T_1}\right) dM_1 = 0 \qquad (8.17)$$

which must be valid for any values of dV_1 and dM_1. For the special case of $dM_1 = 0$, Eq. (8.17) results in

$$\left(\frac{P_1}{T_1} - \frac{P_2}{T_2}\right) dV_1 = 0 \tag{8.18}$$

The arbitrariness of dV_1 and the temperature equality then yields the pressure equality

$$P_1 = P_2 \tag{8.19}$$

Alternatively, specializing Eq. (8.17) to $dV_1 = 0$ and an arbitrary value of dM_1 finally yields

$$\mu_1 = \mu_2 \tag{8.20}$$

Therefore, if we accept the entropy extremum property and the definitions of temperature, pressure, and chemical potential, Eq. (8.4), we obtain the thermal, mechanical, and chemical equilibrium and vice versa, i.e.,

$$dS_{\text{total}} = 0 \implies (T_1 = T_2), (P_1 = P_2), (\mu_1 = \mu_2) \tag{8.21}$$

something that can be confirmed using laboratory experiments. Thermodynamics postulates are thus consistent with observable reality.

8.3 Euler's Thermodynamics Equation

Is the chemical potential related to other thermodynamic properties? To find out, we need the second postulate of thermodynamics. The second postulate states that the entropy of a composite system is additive over the constituent subsystems. The additive property applied to spatially separate subsystems leads to the following: The entropy of a simple system is a homogeneous first-order function of the extensive parameters, or the internal energy of a simple system is a homogeneous first-order function of the extensive parameters. If a system's extensive parameters are multiplied by a constant λ, the entropy or the internal energy is multiplied by the same constant. Or

$$U(\lambda S, \lambda V, \lambda M) = \lambda\, U(S, V, M) \tag{8.22}$$

Taking the differential with respect to λ yields

$$\frac{\partial U(\lambda S, \lambda V, \lambda M)}{\partial(\lambda S)} \frac{\partial \lambda S}{\partial \lambda} + \frac{\partial U(\lambda S, \lambda V, \lambda M)}{\partial(\lambda V)} \frac{\partial \lambda V}{\partial \lambda} + \frac{\partial U(\lambda S, \lambda V, \lambda M)}{\partial(\lambda M)} \frac{\partial \lambda M}{\partial \lambda} = U(S, V, M) \tag{8.23}$$

$$U = TS - PV + \mu M \tag{8.24}$$

This equation is called Euler's form of the fundamental relation. Dividing by M yields Euler's form in intensive parameters as

$$u = Ts - Pv + \mu \qquad (8.25)$$

Or

$$\mu = u + Pv - Ts = h - Ts \qquad (8.26)$$

$(h - Ts)$ is denoted by g and is called the Gibbs potential or the Gibbs function. Hence, the Gibbs potential coincides with the chemical potential for a single species system.

Choosing λ to be equal to $\frac{1}{M}$ in Eq. (8.22) yields

$$U(\frac{S}{M}, \frac{V}{M}, \frac{M}{M}) = \frac{U(S, V, M)}{M} \qquad (8.27)$$

Or

$$u = u(s, v) \qquad (8.28)$$

is the fundamental relation in intensive parameters, u, s, and v.

8.4 Alternative Forms of Fundamental Relations

The fundamental relation $u = u(s, v)$ reveals that to determine the internal energy of a system, complete knowledge of entropy and specific volume is required. These two properties are, however, difficult to measure in laboratory experiments. The pressure and temperature of a system are easier to measure, and thus, it is useful to represent the fundamental relations as a function of the pressure and temperature, which are the partial derivatives of the internal energy with respect to the volume and entropy, respectively. This problem reduces to a transformation of the functional relation $f = f(x)$ to $\psi = \psi(\frac{df}{dx})$ without any loss in the knowledge of the curve $f = f(x)$. This transformation is called the Legendre transformation, and it is $\psi = f - (\frac{df}{dx})x$, which represents the y-values of the family of tangent lines enveloping the curve (Fig. 8.2). To see why ψ is only a function $\frac{df}{dx} = f'(x)$, we have

$$f' = f'(x) \implies x = x(f') \implies f = f(x) = f(x(f')) = F(f') \qquad (8.29)$$

Hence,

$$\psi = f - \left(\frac{df}{dx}\right)x = F(f') - f'x(f') = \psi(f') = \psi\left(\frac{df}{dx}\right) \qquad (8.30)$$

This transformation requires that the function f is continuous and differentiable.

We now apply Legendre transformation to $u = u(s, v)$. Recall that enthalpy is a combination property

$$h = u + Pv \qquad (8.31)$$

8.4 Alternative Forms of Fundamental Relations

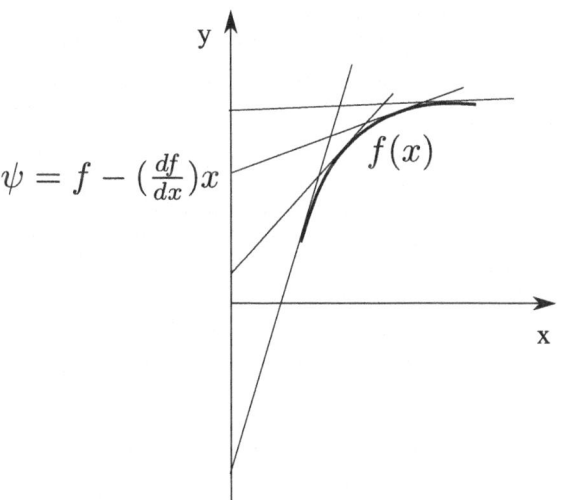

Fig. 8.2 An illustration of Legendre transformation of function $f = f(x)$ onto $\psi = f - (\frac{df}{dx})x$, being the y values where the tangent to the curve $f = f(x)$ intersects the y axes

h is a Legendre transformation of the fundamental relation $u = u(s, v)$,

$$h = u - \left(\frac{\partial u}{\partial v}\right)_s v = u + Pv \tag{8.32}$$

Moreover, it contains all the knowledge of the fundamental relation and is just another representation of the fundamental relation. Enthalpy is obtained by replacing v in the fundamental relation with a partial derivative of u with respect to v, i.e., $h = h(s, P)$. The actual relation is obtained after eliminating u and v from the following three equations

$$\left.\begin{array}{l} h = u + Pv \\ u = u(s, v) \\ P = -\left(\frac{\partial u}{\partial v}\right)_s = P(s, v) \\ \text{Eliminating } u \text{ and } v \end{array}\right\} \implies h = h(s, P) \tag{8.33}$$

In other words, $h = u - \left(\frac{\partial u}{\partial v}\right)_s v = u + Pv$ is a family of tangent surfaces enveloping the curved surface $u = u(s, v)$, and, thus, fully represent $u = u(s, v)$. Assuming s is fixed, $h = h(p)$ is a family of tangent curves enveloping the curve $u = u(v)$; thus, it fully represents the curve.

Two other properties are a combination of properties that are useful in characterizing certain processes. These are Helmholtz function (or potential) a and the Gibbs function (or potential) g, defined as

$$a \equiv u - Ts = a(v, T) \tag{8.34}$$

$$g \equiv h - Ts = u + Pv - Ts = g(T, P) \tag{8.35}$$

Helmholtz and Gibbs functions are also Legendre transformation of the fundamental relations $u = u(s, v)$. In the Helmholtz function, the entropy is replaced by the temperature, which is the partial derivative of the internal energy with respect to entropy. In the Gibbs function, both the entropy and specific volume are replaced by pressure and temperature. Like the enthalpy function, the Helmholtz and Gibbs functions contain all the thermodynamics information about the system and, thus, are alternative representations of the fundamental relations.

The Helmholtz potential is the most useful thermodynamics potential when the volume and temperature are constant, while the Gibbs potential is most useful when temperature and pressure are constant. As the previous section shows, the Gibbs potential is useful in characterizing the phase change and, generally, any form of mass (particle) transfer. The Gibbs potential will be used for determining the phase equilibrium in the two-phase flow modeling in Chap. 9.

8.5 Examples of Fundamental Equations

The ideal gas is often identified by the following two equations of state

$$u = c_v T \tag{8.36a}$$

$$Pv = RT \tag{8.36b}$$

We are interested in determining an explicit expression for the square of sound speed that is defined as

$$c^2 = \left(\frac{\partial p}{\partial \rho}\right)_s = -v^2 \left(\frac{\partial p}{\partial v}\right)_s \tag{8.37}$$

Using the definition of pressure (8.4), we can express the speed of sound as the second derivative of the internal energy as

$$c^2 = \left(\frac{\partial p}{\partial \rho}\right)_s = -v^2 \left(\frac{\partial p}{\partial v}\right)_s = v^2 \left(\frac{\partial^2 u}{\partial v^2}\right)_s \tag{8.38}$$

There is no direct way to derive an expression for the speed of sound using the two equations of state (8.36). We need first to use these two equations to obtain the fundamental relation in the form $u = u(s, v)$, which contains all the information we need about the ideal gas system.

Let us start with the differential form of the fundamental equation

$$du = Tds - Pdv \tag{8.39}$$

Replacing $T = \frac{u}{c_v}$ and $P = \frac{Ru}{c_v v}$ into the fundamental relation yields

$$du = \frac{u}{c_v}ds - \frac{Ru}{c_v v}dv \implies \frac{c_v du}{u} = ds - R\frac{dv}{v} \implies c_v \ln(u) = s - R\ln(v) + \text{const.} \tag{8.40}$$

The square of the speed of sound can now be derived as

$$c^2 = -v^2 \left(\frac{\partial P}{\partial v}\right)_s = v^2 \left(\frac{\partial^2 u}{\partial v^2}\right)_s = \gamma RT = \frac{\gamma P}{\rho} \tag{8.41}$$

We now consider the equation of states for liquid water. The simplest equation of state for liquid water is the so-called stiffened-gas equation of state in the form

$$u - q = c_v T + \frac{\pi}{\rho} \tag{8.42a}$$

$$P = (\gamma - 1)\rho u - \gamma \pi \tag{8.42b}$$

where γ, π and c_v are material parameters determined from fitting to experimental data and q is a reference energy. The parameter π represents the impact of inter-molecular bonds in liquids on the pressure. For a given material, parameters γ and π are obtained by fitting experimental data and theoretical expression of shock speed [3]. The stiffened-gas equation of state is accurate for high pressures. A more sophisticated equation of states involving more parameters is more suitable for low pressures. We may derive the fundamental equation and the speed of sound for liquid in a similar fashion as the case of an ideal gas. In the next section, we derive alternative definitions of speed of sound that can be used to obtain expressions of speed of sound directly from equations of state.

8.6 General Relations for du, dh, and ds

We have seen relations for an ideal gas's internal energy, enthalpy, and entropy change. For general substance, it is useful to represent the change with respect to change in some directly measurable properties. The candidates are (T, v) or (P, v) or (T, P).

Let us first derive an expression for the du. We take $u = u(T, v)$. Taking the full differential yields

$$du = \left(\frac{\partial u}{\partial T}\right)_v dT + \left(\frac{\partial u}{\partial v}\right)_T dv \tag{8.43}$$

We have $\left(\frac{\partial u}{\partial T}\right)_v = c_v$. Thus,

$$du = c_v dT + \left(\frac{\partial u}{\partial v}\right)_T dv \tag{8.44}$$

The specific heat at constant volume is a measurable property that is often measured and documented. Though measurable, the partial derivative in the second term, $(\partial u/\partial v)_T$, is not commonly reported, as achieving a constant temperature process requires having the system in contact with a large thermal reservoir. It is more desirable to write the second term using the derivative of pressure P, with respect to T at constant v, which is much easier to determine in a laboratory experiment. To this end, we first take the full differential of $s = s(T, v)$

$$ds = \left(\frac{\partial s}{\partial T}\right)_v dT + \left(\frac{\partial s}{\partial v}\right)_T dv \qquad (8.45)$$

Then, we replace ds from the above Eq. into $du = Tds - Pdv$ to obtain

$$du = \left(\frac{\partial s}{\partial T}\right)_v dT + \left[T\left(\frac{\partial s}{\partial v}\right)_T - P\right] dv \qquad (8.46)$$

Comparing with Eq. (8.44) yields $\left(\frac{\partial u}{\partial v}\right)_T = \left[T\left(\frac{\partial s}{\partial v}\right)_T - P\right]$. Now using the third Maxwell relation, $\left(\frac{\partial s}{\partial v}\right)_T = \left(\frac{\partial P}{\partial T}\right)_v$, we obtain

$$du = c_v dT + \left[T\left(\frac{\partial P}{\partial T}\right)_v - P\right] dv \qquad (8.47)$$

This equation is based on all measurable properties. Integration from state 1 to 2 yields the change in the internal energy

$$u_2 - u_1 = \int_{T_1}^{T_2} c_v dT + \int_{v_1}^{v_2} \left[T\left(\frac{\partial P}{\partial T}\right)_v - P\right] dv \qquad (8.48)$$

Using a similar approach but starting with $h = h(T, P)$ yields the following equation for dh

$$dh = c_p dT + \left[v - T\left(\frac{\partial v}{\partial T}\right)_P\right] dP \qquad (8.49)$$

and integrating yields

$$h_2 - h_1 = \int_{T_1}^{T_2} c_p dT + \int_{P_1}^{P_2} \left[v - T\left(\frac{\partial v}{\partial T}\right)_P\right] dP \qquad (8.50)$$

Also, for the entropy, we have

$$ds = \frac{c_v}{T} dT + \left(\frac{\partial P}{\partial T}\right)_v dv \qquad (8.51)$$

or

$$ds = \frac{c_p}{T} dT - \left(\frac{\partial v}{\partial T}\right)_P dP \qquad (8.52)$$

Similarly, integrating the last two equations yields

$$s_2 - s_1 = \int_{T_1}^{T_2} \frac{c_v}{T} dT + \int_{v_1}^{v_2} \left(\frac{\partial P}{\partial T}\right)_v dv \tag{8.53}$$

and

$$s_2 - s_1 = \int_{T_1}^{T_2} \frac{c_P}{T} dT - \int_{P_1}^{P_2} \left(\frac{\partial v}{\partial T}\right)_P dP \tag{8.54}$$

Eq. (8.54) is obtained by assuming $s = s(T, P)$. For a saturation state, this relation is not valid as $P_{\text{sat}} = P_{\text{sat}}(T)$, and dependency of s with specific volume is not represented by $s_{\text{sat}} = s(T, P_{\text{sat}}(T))$. Therefore, the valid functional relation for a saturated state is $s = s(T, v)$; hence, Eq. (8.53) is a valid equation for an entropy change in a saturated state.

Similarly, Eq. (8.50) is obtained by assuming $h = h(T, P)$. For a saturation state, this relation is not valid as $P_{\text{sat}} = P_{\text{sat}}(T)$, and dependency of h with specific volume is not represented by $h_{\text{sat}} = h(T, P_{\text{sat}}(T))$. Therefore, for a saturated state, we may use the Gibbs relation to determine an enthalpy change using the change in entropy as

$$dh = T\,ds + v\,dP^{\cancel{0}} \implies \int_f^g dh = \int_f^g T\,ds \implies h_{fg} = T_{\text{sat}} s_{fg} \tag{8.55}$$

Alternatively, at a saturated state, knowing the change in the internal energy using Eq. (8.48), the change in the enthalpy is determined using the definition of enthalpy as

$$h_2 - h_1 = u_2 - u_1 + P_{\text{sat}}(v_2 - v_1) \tag{8.56}$$

Also note that all terms are directly measurable in the equations for the internal energy change, enthalpy, and entropy; thus, these relations are very useful in finding internal energy, enthalpy, and entropy from experimental data.

8.7 Speed of Sound as a Function of T and P

Eliminating dT term from the two equations (8.51) and (8.52) yields

$$\left(\frac{1}{c_v} - \frac{1}{c_P}\right) ds = \frac{1}{c_v}\left(\frac{\partial P}{\partial T}\right)_v + \frac{1}{c_P}\left(\frac{\partial v}{\partial T}\right)_P \tag{8.57}$$

Setting $ds = 0$ as the sound speed involves a constant entropy process, yields

$$c^2 = -v^2 \left(\frac{\partial P}{\partial v}\right)_s = v^2 \frac{c_P}{c_v} \frac{\left(\frac{\partial P}{\partial T}\right)_v}{\left(\frac{\partial v}{\partial T}\right)_P} \tag{8.58}$$

Using the specific heat ratio $\gamma = c_P/c_v$ yields

$$c^2 = -v^2 \left(\frac{\partial P}{\partial v}\right)_s = \gamma v^2 \frac{\left(\frac{\partial P}{\partial T}\right)_v}{\left(\frac{\partial v}{\partial T}\right)_P} = -\gamma v^2 \left(\frac{\partial P}{\partial v}\right)_T \qquad (8.59)$$

Note that the last equality in Eq. (8.59) is achieved using the Cyclic relation.

8.8 Example: Internal Energy for Stiffened-Gas EoS

Consider the stiffened-gas EoS used as a simple model for liquids and solids [1] in the form

$$P = c_v(\gamma - 1)\rho T - \pi \qquad (8.60)$$

where c_v is the specific heat at constant volume, and $\gamma = c_p/c_v$ is the specific heat ratio and π represent the inter-molecular bond. All are material constants determined from fitting to experimental data. The stiffened-gas equation of state reduces to the ideal gas law for $\pi = 0$. For liquid water, Ref. [1] provides the following parameter values for liquid and vapor water:

$$c_{v,l} = 1816 \, \text{Jkg}^{-1}\text{K}^{-1}, \quad \gamma_l = 2.35, \quad \pi_l = 10^9 \, \text{Pa} \qquad (8.61)$$

$$c_{v,g} = 1043 \, \text{Jkg}^{-1}\text{K}^{-1}, \quad \gamma_g = 1.43, \quad \pi_g = 0 \, \text{Pa} \qquad (8.62)$$

Note that γ of the liquid water differs from the specific heat ratio of water at low pressures because the stiffened-gas equation of state is a simple model that accurately represents the behavior of the liquid only in a limited regime of high pressures [1] and not in the whole temperature-pressure domain.

We want to determine internal energy, enthalpy, and entropy expressions. Equation (8.47) reveals that to determine internal energy, we need to determine

$$\left(\frac{\partial P}{\partial T}\right)_v = c_v(\gamma - 1)\rho \qquad (8.63)$$

Now, substituting this into Eq. (8.47) yields

$$du = c_v dT + \left[T\left(\frac{\partial P}{\partial T}\right)_v - P\right] dv = c_v dT + [Tc_v(\gamma - 1)\rho - P] dv \qquad (8.64)$$

Using the stiffened-gas equation of state in the second term yields

$$du = c_v dT + [P + \pi - P] dv = c_v dT + \pi dv \qquad (8.65)$$

Integration and using q as a constant for the reference energy yields

$$u = c_v T + \frac{\pi}{\rho} + q \qquad (8.66)$$

This equation is similar to the expression for the internal energy of ideal gases but with an extra term due to the nonzero π in liquids and solids.

8.9 Exercises

1. Using the stiffened-gas equation of state given as (8.60), determine expressions of the speed of sound, entropy, and enthalpy.
2. An equation of state of a substance is given as $T = T(P, v)$. Derive an expression for the square of sound speed. Hint: Follow a similar procedure as when the equation of state is given as $u = u(P, v)$.
3. Consider a Nobel-Abel gas with the equation of state in the form

$$P = \frac{RT}{v - b} - \Pi \tag{8.67}$$

where b and Π are constants and R is the gas-specific constant. The Nodel-Abel equation of state contains the effect of agitation (RT term), the effect of repulsion ($(v - b)^{-1}$ term), and the attractive forces ($-\pi$ term). The repulsive forces are present in dense gases and liquids, and the attractive forces are significant in liquids. Also, assume the specific heats are constant. (a) Drive an equation for the internal energy and the entropy of this gas. (b) Drive an equation for the chemical (Gibbs) potential of this gas. (c) Drive an equation for the sound speed of this gas.
4. Consider a fluid with the van der Waals equation of state in the form

$$P = \frac{RT}{v - b} - \frac{d}{v^2} \tag{8.68}$$

where b and d are constants and R is the gas-specific constant. Assume the specific heats are also constant. The van der Waals equation of state is similar to the Nobel-Abel, but with an improved attractive term. (a) Drive an equation for the internal energy and the entropy of this gas. (b) Drive an equation for the chemical (Gibbs) potential of this gas. (c) Drive an equation for the sound speed of this gas.

References

1. Le Métayer, O., Massoni, J., Saurel, R.: Modelling evaporation fronts with reactive Riemann solvers. J. Comput. Phys. **205**(2), 567–610 (2005)
2. Callen, H.B.: Thermodynamics and an Introduction To Thermostatistics, 2nd edn. Wiley, New York, NY (1985)
3. Cocchi, J.P., Saurel, R., Loraud, J.C.: Treatment of interface problems with Godunov-type schemes. Shock Waves **5**, 347–357 (1996)

Chapter 9
Mixture Theory Modeling of Compressible Two-Phase Systems

In Chaps. 6 and 7, the discussed two-fluid models were limited to the stiffened-gas equation of states. The stiffened-gas equation of state, although simple, it does not have sufficient accuracy [1]. Furthermore, the discussed model does not support phase change. The mixture-theory modeling allows the use of a more general equation of states and mass transfer among different phases, and thus, results in higher fidelity of the computations. We discuss in this chapter a simple model based on mixture theory [2, 3] as an introduction to this rich and currently developing subject. Saurel and Pantano present a review of the diffuse-interface capturing of compressible two-phase flows [4].

A two-fluid or two-phase medium consists of two substances whose dynamics are characterized by the interaction between different components. The interactions may consist of a purely mechanical nature such as forces arising from pressure and velocity differences, or a nonmechanical nature that arises from chemical reactions, heat transfer, or mass transfer among different components. To capture the potential interactions of any type over the whole spatial domain, the mathematical models require the coexistence of the two components at all points over the domain or having a mixture of two components. On a discrete level, the coexistence of two components in each subsystem (on each cell on the discrete level), though with different quantities, is required. The mass fraction or mole fraction can characterize the relative quantity of each component. The mass fraction approach is consistent with the equations of mass, momentum, and energy for a single-component system, and thus, it is often used.

Without any external disturbances, the constituents of a two-component system at a macroscopic level are at equilibrium. The total equilibrium is interpreted as mechanical (equality of the velocity and pressure), thermal, and chemical equilibrium. Events from the system's boundary disturb the local equilibrium among the various components, and since the response rate is different for each component, the interactions among components will follow. Therefore, the mathematical model of the two-fluid or two-phase system must accommodate the differences in the intensive properties, and on the other hand, allow equilibrium processes. For instance, the model must consist of mass, momentum, and energy equations for each component for the three unknowns of velocity, density, and temperature (the pressure is obtained from the equation of state, while the Gibbs (chemical) potential can be obtained using Euler's form of the fundamental law equation, $g = h - Ts$). Also, for the general case, the mass fraction for each constituent is required to determine the system's state fully. Therefore, for a two-component system, for instance, two mass, two momentum, and two energy equations, along with a single advection-like equation for the mass of one of the components, are required, adding up to seven equations.

For ease, efficiency, and robustness of numerical computations, simpler models based on the equilibrium of some of the intensive properties are more desirable. The simplest mixture model considers the constituents in mechanical and thermal equilibrium but allows chemical potential non-equilibrium. These conditions yield a four-equation model for the two-component system consisting of two mass equations, one total momentum, and one total energy equation. Including two mass equations allows the modeling of mass transfer (or phase change or chemical reaction) between the two phases or two components.

In this chapter, we first derive the seven-equation model before describing the simplified four equations models for a mixture of two gases or a mixture of liquid and gas.

9.1 Mixture Theory Model for Two-Fluid System

We start with mass conservation to derive the conservation laws for each mixture component. We denote the mass fraction of Fluid 1 and 2 by y_1 and y_2, and the density of the mixture by ρ. Then, the differential mass of component 1 is expressed as

$$dM_1 = \frac{dM_1}{dM}\frac{dM}{dV} = y_1 \rho \tag{9.1}$$

where M is the total mass of the mixture occupying the volume V. Integration over the control volume then yields the total mass of component 1 as

$$M_1 = \int_V y_1 \rho dV \tag{9.2}$$

The Reynolds Transport Equation then yields the conservation of mass of component 1 for the control volume in the form

$$\frac{d}{dt}\int_{CV} y_1 \rho dV + \int_{CS} y_1 \rho \vec{V}_{\text{rel}} \cdot \vec{n} dA = 0 \tag{9.3}$$

9.1 Mixture Theory Model for Two-Fluid System

From this equation, the differential form of the mass equation in one dimension is obtained as

$$\frac{\partial(y_1\rho)}{\partial t} + \frac{\partial(y_1\rho V_1)}{\partial x} = 0 \tag{9.4}$$

Similarly, the conservation mass for Fluid 2 can be derived. Moreover, the conservation of momentum and energy for each component can also be derived, giving rise to the conservation law statement for each of the components in the mixture of the two-fluid in the form

$$\frac{\partial(y_1\rho)}{\partial t} + \frac{\partial(y_1\rho V_1)}{\partial x} = 0 \tag{9.5a}$$

$$\frac{\partial(y_2\rho)}{\partial t} + \frac{\partial(y_2\rho V_2)}{\partial x} = 0 \tag{9.5b}$$

$$\frac{\partial(y_1\rho V_1)}{\partial t} + \frac{\partial(y_1\rho V_1^2 + P_1)}{\partial x} = 0 \tag{9.5c}$$

$$\frac{\partial(y_2\rho V_2)}{\partial t} + \frac{\partial(y_2\rho V_2^2 + P_2)}{\partial x} = 0 \tag{9.5d}$$

$$\frac{\partial(y_1\rho E_1)}{\partial t} + \frac{\partial(y_1\rho E_1 + P_1)V_1}{\partial x} = 0 \tag{9.5e}$$

$$\frac{\partial(y_2\rho E_2)}{\partial t} + \frac{\partial(y_2\rho E_2 + P_2)V_2}{\partial x} = 0 \tag{9.5f}$$

In System (9.5), $E_i = e_i + \frac{V_i^2}{2}$. This system has six equations (two mass, two momentum, and two energy equations). To determine the number of unknowns, we need to relate the density of the mixture ρ to that of each fluid. In the case of a mixture of two gases, it is sensible to follow the Dalton law and assume that both gases occupy the entire volume, i.e., $V_1 = V_2 = V$. In the case of a gas-liquid mixture, it is sensible to follow the Amagat law and assume that each fluid occupies a fraction of the volume, i.e., $V_1 + V_2 = V$. A distinct expression for the density follows depending on whether the Dalton law or Amagat law is used. For the Dalton law, we have

$$\rho = \frac{M}{V} = \frac{M}{M_1}\frac{M_1}{V} = \frac{M}{M_1}\frac{M_1}{V} = \frac{\rho_1}{y_1} \tag{9.6}$$

or

$$\rho_1 = y_1\rho \qquad \rho_2 = y_2\rho \tag{9.7}$$

and

$$\rho_1 + \rho_2 = y_1\rho + y_2\rho = (y_1 + y_2)\rho = \rho \tag{9.8}$$

On the other hand, for the Amagat law, we have

$$y_1\rho = \frac{M_1}{M}\frac{M}{V} = \frac{M_1}{V_1}\frac{V_1}{V} = \rho_1\alpha_1 \tag{9.9}$$

where $\alpha_1 = V_1/V$ is the volume fraction of Fluid 1. Hence, for the Amagat law, we have

$$y_1\rho = \alpha_1\rho_1 \qquad y_2\rho = \rho_2\alpha_2 \qquad \rho = \alpha_1\rho_1 + \alpha_2\rho_2 \tag{9.10}$$

From (9.7) and (9.10), it is revealed that regardless of the law of mixture, knowing the density of the mixture and the mass fraction (or volume fraction) yields the density of each fluid.

This, in turn, reveals that the two-fluid system (9.5) encompasses seven unknowns namely, ρ_1, ρ_2, V_1, V_2, P_1, P_2, and y_1 or (α_1). Note that the internal energies are not unknown; for each fluid, the internal energy is related to that fluid's density and pressure via an equation of state, i.e., $e_i = e_i(\rho_i, p_i)$ for $i = 1$ and 2. However, the two-fluid system (9.5) consists of only six equations. One extra equation is thus required for the well-posedness. An advection equation for the mass fraction or volume fraction is used to complement the system and renders the system well posed.

9.1.1 Reduction Under the Instantaneous Equilibrium

For instantaneous or a very fast relaxation of mechanical and thermal differences between the two fluids, the mechanical and thermal equilibrium must be satisfied:

$$V_1 = V_2 = V, \qquad P_1 = P_2 = P, \qquad T_1 = T_2 = T \tag{9.11}$$

The justification for the instantaneous relaxation is discussed in Sect. 9.2.

Enforcing these equilibrium conditions allows reducing the two-fluid system (9.5) to the following system

$$\frac{\partial(y_1\rho)}{\partial t} + \frac{\partial(y_1\rho V)}{\partial x} = 0 \tag{9.12a}$$

$$\frac{\partial(\rho)}{\partial t} + \frac{\partial(\rho V)}{\partial x} = 0 \tag{9.12b}$$

$$\frac{\partial(\rho V)}{\partial t} + \frac{\partial(\rho V^2 + P)}{\partial x} = 0 \tag{9.12c}$$

$$\frac{\partial(\rho E)}{\partial t} + \frac{\partial(\rho E + P)V}{\partial x} = 0 \tag{9.12d}$$

where $E = e + V^2/2$. The second, third, and fourth equations are obtained by adding the two mass, momentum, and energy equations in the system (9.5) with the use of $\rho = y_1\rho + y_2\rho$. Also, consistent with the additive property of the energy, we have denoted the mixture energy in (9.12d) as

$$E = y_1 E_1 + y_2 E_2 \tag{9.13}$$

9.1 Mixture Theory Model for Two-Fluid System

Therefore, the reduced system consists of the mixture's mass, momentum, energy equations, and the mass equation for one of the fluids. The system consists of four unknowns and four equations. Also, note that this system is valid for the Dalton or Amagat law.

9.1.2 System of Two Ideal Gases

Let us consider a mixture of two ideal gases with R_1 and R_2 as their gas-specific constants and consider the case of instantaneous thermal and mechanical relaxation. The two-fluid system (9.12), which is valid for instantaneous relaxation, is complemented with

$$\rho_1 + \rho_2 = \rho \implies \frac{P}{R_1 T} + \frac{P}{R_2 T} = \frac{P}{RT} \implies \frac{R_1 R_2}{R_1 + R_2} = R \tag{9.14}$$

The last statement reveals that enforcing the instantaneous relaxation for two ideal gases requires the mixture's gas-specific constant to be set to $R = (R_1 R_2)/(R_1 + R_2)$.

Therefore, the following system must be solved for the mixture of two ideal gases under the instantaneous relaxation:

$$\frac{\partial(y_1 \rho)}{\partial t} + \frac{\partial(y_1 \rho V)}{\partial x} = 0, \tag{9.15a}$$

$$\frac{\partial(\rho)}{\partial t} + \frac{\partial(\rho V)}{\partial x} = 0 \tag{9.15b}$$

$$\frac{\partial(\rho V)}{\partial t} + \frac{\partial(\rho V^2 + P)}{\partial x} = 0 \tag{9.15c}$$

$$\frac{\partial(\rho E)}{\partial t} + \frac{\partial(\rho E + P)V}{\partial x} = 0 \tag{9.15d}$$

$$P = \rho R T, \quad e = c_v T \tag{9.15e}$$

where

$$c_v = y_1 c_{v,1} + y_2 c_{v,2}, \quad R = (R_1 R_2)/(R_1 + R_2) \tag{9.16}$$

Note that the system consists of six equations (including four conservation laws and two property relations) and six unknowns ρ, y_1, V, P, T, and e.

Remark 9.1 The system (9.15) has a well-defined characteristic speed, namely V, $V + c$, and $V - c$, where the c is the speed of sound in the mixture and given by

$$c^2 = \frac{\gamma P}{\rho} \tag{9.17}$$

where $\gamma = c_p/c_v$ is the specific heat ratio of the mixture with c_v of the mixture is computed using (9.16). c_p is computed similarly as $c_p = y_1 c_{p,1} + y_2 c_{p,2}$.

Remark 9.2 The system (9.15) cannot capture the sharp interfaces. As discussed in Section 6.1.4, the conservative model is inconsistent with the simple advection of a sharp interface, and only the two-fluid system (6.9) is consistent with the advection of an interface. However, the model (9.15) yields a nonoscillatory solution as long as the interface is sufficiently diffused and resolved with the grid resolution.

9.1.3 System of Gas and Liquid

For a gas-liquid mixture, the instantaneous mechanical and thermal relaxations, along with the Amagat law of additive volumes, yield the following system for the gas-liquid mixture:

$$\frac{\partial (y_1 \rho)}{\partial t} + \frac{\partial (y_1 \rho V)}{\partial x} = 0 \tag{9.18a}$$

$$\frac{\partial (\rho)}{\partial t} + \frac{\partial (\rho V)}{\partial x} = 0 \tag{9.18b}$$

$$\frac{\partial (\rho V)}{\partial t} + \frac{\partial (\rho V^2 + P)}{\partial x} = 0 \tag{9.18c}$$

$$\frac{\partial (\rho E)}{\partial t} + \frac{\partial (\rho E + P)V}{\partial x} = 0 \tag{9.18d}$$

$$v = y_1 v_1 + y_2 v_2, \quad e = y_1 e_1 + y_2 e_2 \tag{9.18e}$$

where $v = 1/\rho$ and $v_i = 1/\rho_i$ signify the specific volume of the mixture and Fluid i. The first identity in (9.18e) is obtained as

$$1 = \alpha_1 + \alpha_2 \implies 1 = \frac{V_1}{V} + \frac{V_2}{V} = \frac{M_1 v_1}{V} + \frac{M_2 v_2}{V} \tag{9.19}$$

Multiplication with V and division by M then yields

$$\frac{V}{M} = \frac{M_1 v_1}{M} + \frac{M_2 v_2}{M} \implies v = y_1 v_1 + y_2 v_2 \tag{9.20}$$

Remark 9.3 Using (9.10), the system (9.18) can be equivalently expressed based on the mass conservation equation of each component in the form of

$$\frac{\partial (\alpha_1 \rho_1)}{\partial t} + \frac{\partial (\alpha_1 \rho_1 V)}{\partial x} = 0 \tag{9.21a}$$

$$\frac{\partial (\alpha_2 \rho_2)}{\partial t} + \frac{\partial (\alpha_2 \rho_2 V)}{\partial x} = 0 \tag{9.21b}$$

9.1 Mixture Theory Model for Two-Fluid System

$$\frac{\partial(\rho V)}{\partial t} + \frac{\partial(\rho V^2 + P)}{\partial x} = 0 \quad (9.21c)$$

$$\frac{\partial(\rho E)}{\partial t} + \frac{\partial(\rho E + P)V}{\partial x} = 0 \quad (9.21d)$$

$$v = y_1 v_1 + y_2 v_2, \quad e = y_1 e_1 + y_2 e_2 \quad (9.21e)$$

where $1/v = \rho = \alpha_1 \rho_1 + \alpha_2 \rho_2$ and $y_1 = \alpha_1 \rho_1/\rho$ based on (9.9). Note that the specific volume and energy of each component can be expressed as a function of the equilibrium temperature and pressure in the form

$$v_1 = v_1(T, P), \quad v_2 = v_2(T, P), \quad e_1 = e_1(T, P) \quad e_2 = e_2(T, P), \quad (9.22)$$

where the explicit forms of $v_1(T, P)$, $v_2(T, P)$, $e_1(T, P)$, and $e_2(T, P)$ are obtained from the equation of states. The system (9.21) consists of six equations (including four conservation laws) and also involves six unknowns $\alpha_1 \rho_1$, $\alpha_2 \rho_2$, V, P, T, and e.

Remark 9.4 Reference [2] discusses the speed of sound for the model (9.21). Finding the eigenvalues of the quasi-linear form of the four-equation model yields a speed of sound that its upper bound is Wood's speed of sound [5] in the form

$$\frac{1}{\rho c^2} = \frac{\alpha_1}{\rho_1 c_1^2} + \frac{\alpha_2}{\rho_2 c_2^2} \quad (9.23)$$

For example, $\alpha = 0.5$, $\rho_1 = 1000$ kgm^{-1}, $\rho_2 = 1$ kgm^{-1}, $c_1 = 1500$ ms^{-1}, and $c_2 = 300$ ms^{-1} yield a mixture speed of sound $c \approx 20$ ms^{-1}, which is significantly smaller than the sound speed in each of pure component of the mixture.

Remark 9.5 The system (9.21) also yields an oscillatory solution for the advection of a sharp interface. A diffuse-interface resolved within the grid points is required for oscillation-free solutions. Saurel et al. [6] discuss that diffuse interfaces physically exist, for instance, in the evaporating front; hence, the model (9.21) can be an appropriate model for such flows.

9.1.4 Modeling Mass Transfer or Phase Change

The system (9.21) leads to the transport of phases or components without any phase changes. From the equilibrium thermodynamics, we deduce that at a specific location, if there is only liquid (vapor), vaporization (condensation) occurs if the temperature is above (below) the saturation temperature at the given pressure, or if the pressure is lower (higher) than the saturation pressure at the given temperature. The phase change yields the equality of the Gibbs (chemical) potential. Hence, Modeling the mass transfer or phase change requires an extra condition for the equality of the Gibbs potentials ($g_1 = g_2$) to be added to the equilibrium conditions:

$$T = T_1 = T_2 \qquad (9.24a)$$
$$P = P_1 = P_2 \qquad (9.24b)$$
$$e = y_1 e_1(T, P) + y_2 e_2(T, P) \qquad (9.24c)$$
$$\alpha_1 + \alpha_2 = 1 \quad \text{or} \quad v = y_1 v_1(T, P) + y_2 v_2(T, P) \qquad (9.24d)$$
$$g_1 = g_2 \quad \text{or} \quad h_1(T, P) - T s_1(T, P) = h_2(T, P) - T s_2(T, P) \qquad (9.24e)$$

Solving system (9.24) yields (y_1, P, T). The last equation in the system (9.24) is the saturation curve obtained based on the adopted equation of state. Exercise 3 involves the derivation of the saturation curve for a specific equation of state.

9.1.5 Modeling Heat Conduction in Gas-Liquid Mixture with Amagat Law

Heat conduction plays an important role in boiling flows. Furthermore, in the presence of heat conduction at the interface of a pure liquid and pure gas, the single temperature model is consistent, while without heat conduction, the single temperature model, (9.11), is not valid. Therefore, we need to describe how to include heat conduction in gas-liquid mixtures.

As explained earlier, in a liquid-gas mixture governed by the Amagat law, the volume of each phase dV_i is a fraction of the total volume dV, i.e.,

$$dV_i = \alpha_i dV \qquad (9.25)$$

and the surface of area of each phase's volume is a fraction of the surface area of the mixture's volume, i.e.,

$$dA = dA_1 + dA_2 \qquad (9.26)$$

The net rate of heat conduction flux exiting the surface of a mixture volume can be written as the sum of the heat fluxes from each phase as

$$\text{total heat exiting} = \sum_{i=1}^{2} \int_{A_i} \vec{q}_i \cdot \vec{n}_i dA_i \qquad (9.27)$$

The divergence theorem then transfers the surface integrals to the volume integrals as

$$\text{total heat exiting} = \sum_{i=1}^{2} \int_{V_i} \nabla \cdot (\vec{q}_i dV_i) \qquad (9.28)$$

Substituting for dV_i from (9.25) then yields

$$\text{total heat exiting} = \int_{V_{mix}} \nabla \cdot (\alpha_1 \boldsymbol{q}_1 + \alpha_2 \boldsymbol{q}_2) dV \qquad (9.29)$$

Finally, using Fourier's law heat condition, we have

$$\text{total heat exiting} = \int_{V_{mix}} \nabla \cdot [(\alpha_1 \kappa_1 + \alpha_2 \kappa_2) \nabla T] dV \qquad (9.30)$$

We now define the effective thermal conductivity of a gas-liquid mixture governed by the Amagat law as

$$\kappa_{\text{mixture}} = \alpha_1 \kappa_1 + \alpha_2 \kappa_2 \qquad (9.31)$$

and the effective heat flux in the mixture as

$$\boldsymbol{q}_{\text{mixture}} = \kappa_{\text{mixture}} \nabla T \qquad (9.32)$$

Therefore, the one-dimensional four-equation model with heat conduction is written as

$$\frac{\partial \rho}{\partial t} + \frac{\partial (\rho V)}{\partial x} = 0 \qquad (9.33a)$$

$$\frac{\partial (y_1 \rho)}{\partial t} + \frac{\partial (y_1 \rho V)}{\partial x} = 0 \qquad (9.33b)$$

$$\frac{\partial (\rho V)}{\partial t} + \frac{\partial (\rho V^2 + P)}{\partial x} = 0 \qquad (9.33c)$$

$$\frac{\partial (\rho E)}{\partial t} + \frac{\partial [(\rho E + P)V + q_{\text{mixture}}]}{\partial x} = 0 \qquad (9.33d)$$

Remark 9.6 The effective viscosity coefficient for a mixture of gas-liquid governed by the Amagat law is similarly defined as

$$\mu_{\text{mixture}} = \alpha_1 \mu_1 + \alpha_2 \mu_2. \qquad (9.34)$$

9.2 Justification for Single Temperature, Pressure, and Velocity Model

The instantaneous equilibrium of pressure, velocity, and temperature requires that the mixture consists of very fine liquid and gas particles. Very fine particles ensure that relaxation times are significantly smaller than the macroscopic time scales for a given application. Experimental observations of cavitating and flashing flows near macroscopic interfaces support these assumptions [7]. Let us find an estimate for the fineness of the liquid droplets for the instantaneous thermal relaxation to be valid.

Considering conduction as the sole heat transfer mechanism, solving the equation of heat conduction for a spherical droplet cooling down into the surrounding vapor yields the following effective heat flux from the sphere to the vapor

$$q = \frac{2\kappa_v}{d} T \qquad (9.35)$$

where T is excess temperature between the droplet and vapor, κ_v the thermal conductivity of the vapor, and d diameter of the droplet. From the energy balance for the droplet, we have

$$\rho_d C_d V \frac{dT}{dt} = -\frac{2 A \kappa_v}{d} T \qquad (9.36)$$

where ρ_d and C_d are the droplet's density and specific heat, respectively. Substituting the droplet volume $V = \frac{\pi d^3}{6}$ and its surface area of $A = \pi d^2$, we obtain

$$\frac{dT}{dt} = -\frac{12\kappa_v}{\pi \rho_d C_d d^2} T \qquad (9.37)$$

The solution is exponentially decaying in time

$$T(t) = T_0 e^{-\alpha t} \qquad (9.38)$$

where the decay rate is

$$\alpha_{\text{decay}} = \frac{12\kappa_v}{\pi \rho_d C_d d^2} \qquad (9.39)$$

In a time as large as threefold the inverse of the decay rate, the initial excess temperature drops approximately ten-fold. Therefore, the time scale of thermal equilibrium is

$$\tau_{\text{th}} \approx \frac{\rho_d C_d d^2}{10 \kappa_v} \qquad (9.40)$$

Taking $\kappa_v \approx 0.04$ Wm^{-1}K^{-1}, and $\rho_d = 1000$ kg m^{-3} and $C_d = 4000$ Jkg^{-1}K^{-1} yields

$$\tau_{\text{th}} \approx 10^7 d^2 \qquad (9.41)$$

Hence, a micron-size droplet yields thermal equilibrium in 10 μs. Note the heat flux in Eq. (9.35), which is only the heat conduction from an individual droplet in a pool of vapor, underestimates the heat flux. Therefore, the thermal equilibrium time for the mixture of droplet and vapor might be significantly higher.

Comparing this with pressure relaxation time, which scales as

$$\tau_p = \mathcal{O}\left(\frac{d}{c}\right) \tag{9.42}$$

where the speed of sound in the mixture $c \approx 10 - 100$ ms^{-1} yields relaxation times below microseconds for pressure equilibrium in a droplet-vapor mixture with micron-size droplets. Hence, pressure equilibrium might be faster than the thermal equilibrium for mixtures with micron-size droplets or larger droplets. A similar analysis leads to a comparable time scale for the velocity relaxation time.

Therefore, since we do not directly compute the mixture dynamics because the direct computations are limited to sizes larger than the computational grid spacing, thermal and pressure equilibrium enforcement can be interpreted as a sub-grid scale modeling of the sub-grid mixture dynamics. Therefore, the validity of such a model falls in a regime where the separation of macroscopic temporal scales from the sub-grid mixture scales occurs.

9.3 Two-Phase Model with Finite Phase Change Rate

The system (9.18) is valid for the instantaneous phase change (or infinite rate of phase change relaxation). For a finite phase change relaxation rate, the mass conservation of Fluid 1, (9.18a), is modified as follows [6]:

$$\frac{\partial(y_1\rho)}{\partial t} + \frac{\partial(y_1\rho V)}{\partial x} = \underbrace{\rho K(g_2 - g_1)}_{\text{phase change term}} \tag{9.43}$$

where K is the phase change relaxation rate. Therefore, the two-phase model with heat conduction and finite relaxation rate has the following form

$$\frac{\partial(y_1\rho)}{\partial t} + \frac{\partial(y_1\rho V)}{\partial x} = \rho K(g_2 - g_1) \tag{9.44a}$$

$$\frac{\partial(\rho)}{\partial t} + \frac{\partial(\rho V)}{\partial x} = 0 \tag{9.44b}$$

$$\frac{\partial(\rho V)}{\partial t} + \frac{\partial(\rho V^2 + P)}{\partial x} = 0 \tag{9.44c}$$

$$\frac{\partial(\rho E)}{\partial t} + \frac{\partial[(\rho E + P)V + q_{\text{mixture}}]}{\partial x} = 0 \tag{9.44d}$$

$$v = y_1 v_1 + y_2 v_2, \quad e = y_1 e_1 + y_2 e_2 \tag{9.44e}$$

This model allows the quantification of the relaxation rate impact on the flow dynamics.

9.3.1 Numerical Scheme for Two-Phase Flow Model

The numerical solution of the two-phase model with heat transfer and finite phase change rate, system (9.44), follows the method of lines consisting of the spatial discretization, followed by the temporal integration. The discretizations of the advective and diffusive fluxes are carried out using the high-order spatial discretization described in Chaps. 5 and 7. The resultant system of ODEs can be stiff because of multiple time scales corresponding to the advection, heat conduction, and phase change rate. The multi-implicit spectral deferred correction (MISDC) scheme presented in Subsection 4.2.4 effectively solves the ODE system. If the phase change possesses the fastest rate, followed by the heat conduction and advection, the phase change term and the heat conduction are treated implicitly, while the advection is treated explicitly. Following the methodology of MISDC, the phase change term is integrated over the finest quadrature points; the diffusive term is over the second finest quadrature points, and the advective terms are over the coarsest quadrature points. The resultant scheme is high-order in both space and time.

9.4 Exercises

1. Find the eigenvalues and eigenvectors of the four-equation model. Also, determine an expression for speed of sound.
2. Consider a two-phase flow model with the gas phase being modeled as an ideal gas and the liquid phase being modeled as a stiffened gas. Find an expression for the saturation curve, i.e., $P_{\text{sat}} = P(T_{\text{sat}})$
3. Consider the Nobel-Abel-Stiffened-Gas equation of state in the form

$$P(\rho, e) = \frac{(\gamma - 1)\rho e}{1 - \rho b} - \gamma \Pi \quad (9.45)$$

   where b, γ, and Π are constants. Also, consider $e = C_v T$ as the second EoS. (**a**) Derive an expression for the speed of sound. (**b**) Derive an expression for the saturation curve.
4. Provide a complete algorithm for solving the four-equation model with phase change.

References

1. Le Métayer, O., Saurel, R.: The Noble-Abel Stiffened-Gas equation of state. Phys. Fluids. **28**(4), 046102, 04 (2016)
2. Le Martelot, S., Saurel, R., Nkonga, B.: Towards the direct numerical simulation of nucleate boiling flows. Int. J. Multiph. Flow **66**, 62–78 (2014)

3. Saurel, R., Boivin, P., Le Métayer, O.: A general formulation for cavitating, boiling and evaporating flows. Comput. Fluids **128**, 53–64 (2016)
4. Saurel, R., Pantano, C.: Diffuse-interface capturing methods for compressible two-phase flows. Annu. Rev. Fluid Mech. **50**(1), 105–130 (2018)
5. Wood, A.B.: A Textbook of Sound. G. Bell and Sons Ltd, London (1930)
6. Saurel, R., Boivin, P., Le Métayer, O.: A general formulation for cavitating, boiling and evaporating flows. Comput. Fluids **128**, 53–64 (2016)
7. Simoes-Moreira, J.R., Shepherd, J.E.: Evaporation waves in superheated dodecane. J. Fluid Mech. **382**, 63–86 (1999)

Appendix
Basic Schemes for Model Problems

This appendix reviews essential concepts in numerical approximations of model problems. For ordinary differential equations, we discuss low-order explicit and implicit schemes and the suitability of each depending on the nature of the problem and highlight the issue of multi-scale problems and stiffness. For spatial derivatives, we discuss low-order schemes for advection and diffusion problems. We analyze the stability of the low-order schemes and explain the dissipative and dispersive nature of approximations.

A.1 Temporal Schemes

Disctretizing the spatial derivatives in the time-dependent PDEs governing the compressible flows yields systems of ODEs in time that need to be solved. Therefore, it is natural to begin the discussion of the numerical methods for the compressible flow equations by numerical algorithms for a system of ODEs. We first discuss the exact solution of the ODEs before any discussions of numerical approximations. For numerical approximations, we present the design of the algorithm, followed by its convergence analysis. The convergence analysis consists of two stages: the discussion of local truncation error (the consistency analysis) and whether a small local error leads to a small global error (the stability analysis). Consistency and stability are two ingredients required for establishing the convergence of a numerical solution to the exact solution [1, 2]. A scheme is stable if small local errors yield small global errors compared to the ODE's exact solution.

The presentation starts by constructing low-order schemes and their convergence properties, including accuracy and stability analysis for scalar and vector ODEs. The concept of stiffness is introduced, and the role of implicit schemes for overcoming the difficulty of solving stiff problems is discussed.

The discussion in the appendix is based on the classical results, which can be found in several books; for instance, see Ref. [1].

A.1.1 First- and Second-Order Schemes for ODEs

Let us consider the general formulation of an initial value problem for a system of ODEs in the form

$$w'(t) = f(t, w(t)) \quad t > 0 \quad w(0) = w_0 \tag{A.1}$$

with $w(0)$ and f being m-dimensional vectors.

The exact integration (A.1) from time step t_n to t_{n+1} with the step size of Δt is

$$w(t_{n+1}) = w(t_n) + \int_{t_n}^{t_{n+1}} f(\tau, w(\tau))d\tau \tag{A.2}$$

The integral can be approximated in several ways. Using the parameter $0 \le \theta \le 1$, we use the following approach to approximate the integral

$$w(t_{n+1}) = w(t_n) + \Delta t(1-\theta)f(t_n, w(t_n)) + \Delta t\theta f(t_{n+1}, w(t_{n+1})) + \rho_n \tag{A.3}$$

where we have used a fraction of f at t_n and a fraction of f at t_{n+1}. The approximation of the integral yields error ρ_n called the local truncation error. Dropping the local truncation error ter in Eq. (A.3), we obtain an approximate solution w_{n+1} to the exact solution $w(t_{n+1})$ in the form

$$w_{n+1} = w_n + \Delta t(1-\theta)f(t_n, w_n) + \Delta t\theta f(t_{n+1}, w_{n+1}) \tag{A.4}$$

The scheme (A.4) is called θ-method. We determine an expression for the local truncation error ρ_n by replacing $f(t_i, w(t_i))$ with $w'(t_i)$ to obtain

$$w(t_{n+1}) = w(t_n) + \Delta t(1-\theta)w'(t_n) + \Delta t\theta w'(t_{n+1}) + \rho_n \tag{A.5}$$

Then developing the Taylor series of $w(t_{n+1})$ and $w'(t_{n+1})$ about t_n yields

$$\sum_{k=0} \frac{\Delta t^k}{k!} w^{(k)}(t_n) = w(t_n) + \Delta t(1-\theta)w'(t_n) + \Delta t\theta \sum_{k=0} \frac{\Delta t^k}{k!} w^{(k+1)}(t_n) + \rho_n \tag{A.6}$$

where $w^{(k)}$ signifies the k-th derivative of w. Finally, simplification leads to

$$\rho_n = (\theta - \frac{1}{2})w''(t_n)\Delta t^2 + \frac{3\theta - 2}{6}w'''(t_n)\Delta t^3 + \cdots \tag{A.7}$$

The value of parameter θ determines the local truncation error; $\theta = 0$ yields the forward (Euler) scheme, $\theta = 1$ the backward (Euler) scheme, and $\theta = 1/2$ is the trapezoidal scheme. The word "forward" signifies that the sole values at t_n approximate the integral forward to t_{n+1}. Conversely, The word "backward" is used to signify that the sole values at t_{n+1} are used to approximate the integral backward to t_n. Provided the function $f(t, w(t))$ being smooth, the local truncation error of the forward and backward Euler scheme is $\mathcal{O}(\Delta t^2)$. Hence, they are of consistency order one. On the other hand, the local truncation error of the trapezoidal rule is $\mathcal{O}(\Delta t^3)$, and thus, it is of consistency order of two. (In general, if the truncation error of a scheme is $\mathcal{O}(\Delta t^{r+1})$, the scheme possesses a consistency order of r.)

Determination of w_{n+1} using the forward Euler scheme requires the evaluation of $f(t, w(t))$ at time t_n, which is known. As a result, the forward Euler scheme is called an explicit scheme. On the other hand, in the backward Euler scheme, the determination of w_{n+1} requires the evaluation of $f(t, w(t))$ at time t_{n+1} which is unknown. As a result, the backward Euler scheme is called an implicit scheme. Values of $0 < \theta \leq 1$ all yield an implicit scheme. In particular, the trapezoidal rule with $\theta = 1/2$ is also an implicit scheme.

The truncation error can be significant for any finite Δt however small if the function is not smooth, i.e., its derivative is very high or infinite. In Chap. 2, We discuss the approximation of non-smooth functions such as shock waves in detail. For now, we understand that the magnitudes of the solution's derivatives influence the size of the error.

After establishing the consistency of the θ-method, we need to analyze the stability of the scheme, i.e., in order to ensure that the scheme's solution converges to the exact solution when integration is performed to a final time, we need to determine how local truncation error grows over time. We first analyze the stability of the basic schemes for a single ODE, followed by a discussion of the stability of the scheme for a system of ODEs.

A.1.2 Stability of the Basic Schemes for a Single ODE

Consider the simple test equation

$$w' = \lambda w, \quad w = w(t) \in \mathbb{C} \tag{A.8}$$

which is obtained from (A.1) by replacing $f(t, w(t))$ with $\lambda w(t)$. We assume λ is a complex constant because using a Fourier mode $(\alpha_k(t)e^{i2\pi kx})$ to solve the advection-diffusion schemes yields an ODE with a complex parameter λ. Specifically, substitution of the single Fourier mode into the advection-diffusion equation $(v_t + av_x = dv_{xx})$ with the advection speed a and diffusion coefficient d yields the following ODE for the coefficient of the Fourier mode, α_k:

$$\alpha'_k = [-(2\pi k)^2 d - i2\pi ka]\alpha_k \tag{A.9}$$

which is of the form (A.8) with $\lambda = -(2\pi k)^2 d - i2\pi ka$

Exact solution

The exact solution of the ODE (A.8) is

$$w(t) = e^{\lambda t} w(0) \qquad (A.10)$$

The solution grows or decays exponentially for positive or negative $\Re(\lambda)$, respectively. For $\Re(\lambda) = 0$, the magnitude of the solution remains unchanged.

If the initial solution is disturbed, i.e., instead of $w(0)$, the initial condition is $\tilde{w}(0)$ with error $\epsilon(0) = w(0) - \tilde{w}(0)$, the disturbed solution at a later time, $\tilde{w}(t)$ also has the same expression:

$$\tilde{w}(t) = e^{\lambda t} \tilde{w}(0) \qquad (A.11)$$

Subtracting the last equation from Eq. (A.10) yields a similar expression for the error, $\epsilon(t) = w(t) - \tilde{w}(t)$:

$$\epsilon(t) = e^{\lambda t} \epsilon(0) \qquad (A.12)$$

Similar to the exact solution, the error magnitude grows, decays, or remains constant for a positive, a negative, or the zero value of $\Re(\lambda)$, respectively.

Stability

Applying the scheme (A.4) to (A.8) yields

$$w_{n+1} = w_n + \Delta t \lambda (1 - \theta) w_n + \Delta t \lambda \theta w_{n+1} \qquad (A.13)$$

Setting $z = \Delta t \lambda$ and simplifying yield

$$w_{n+1} = \left[\frac{1 + z(1 - \theta)}{1 - z\theta} \right] w_n \qquad (A.14)$$

Similarly, Eq. (A.3) yields the following for the exact solution

$$w(t_{n+1}) = \left[\frac{1 + z(1 - \theta)}{1 - z\theta} \right] w(t_n) + \left(\frac{1}{1 - z\theta} \right) \rho_n \qquad (A.15)$$

Denoting the errors at time step n by $\epsilon_n = w(t_n) - w_n$ and subtracting (A.13) from (A.15), we obtain

$$\epsilon_{n+1} = \left[\frac{1 + z(1 - \theta)}{1 - z\theta} \right] \epsilon_n + \left(\frac{1}{1 - z\theta} \right) \rho_n \qquad (A.16)$$

The function $R(z) = \frac{1 + z(1-\theta)}{1 - z\theta}$ determines the growth or decay of the error from one-time step to another, and thus, it is called the stability function of the scheme.

Recursively replacing the expression for ϵ_i into (A.16) then yields

$$\epsilon_n = \left(\frac{1+z(1-\theta)}{1-z\theta}\right)^n \epsilon_0 + \frac{1}{1-z\theta}\left[\left(\frac{1+z(1-\theta)}{1-z\theta}\right)^{n-1}\rho_1 + \cdots + \rho_{n-1}\right] \quad (A.17)$$

Taking the absolute values of both sides and using the triangular inequality yields

$$|\epsilon_n| \leq \left|\left(\frac{1+z(1-\theta)}{1-z\theta}\right)^n\right| |\epsilon_0| + \left|\frac{1}{1-z\theta}\left[\left(\frac{1+z(1-\theta)}{1-z\theta}\right)^{n-1}\rho_1 + \cdots + \rho_n\right]\right| \quad (A.18)$$

The second term on the right-hand side consists of n terms of each $\mathcal{O}(\Delta t^{r+1})$ where $r = 1$ for all values of θ except for $\theta = 1/2$. For $\theta = 1/2, r = 2$, corresponding to the trapezoidal rule. In the worst case that all terms add up, the second term is $\mathcal{O}(\Delta t^r)$, provided the stability function $R(z)$ is bounded. The boundedness or growth of the stability function is discussed below.

We consider two cases of $\Re(z) \leq 0$ and $\Re(z) > 0$. For $\Re(z) > 0$, the exact solution of (A.8) and its errors grow in time without a bound. For $\Re(z) \leq 0$, $z \in \mathbb{C}^-$, or z in the left-half plane in the complex plane, the exact solution and its error decay; hence, we need to analyze the stability of the θ-method. Specifically, we require that the modulus of the stability function be smaller than unity:

$$|R(z)| = \left|\frac{1+(1-\theta)z}{1-\theta z}\right| \leq 1, \quad z \in \mathbb{C}^- \quad (A.19)$$

Since the stability function $R(z)$ does not have any poles in the left-half complex plane (it is analytic or has a convergent Taylor series), then using the maximum modulus principle [3], the maximum of the modulus of the stability function is achieved on the boundary of the left-half complex plane[1], i.e.,

---

[1] Without getting into much complex analysis, we can intuitively explain the maximum modulus principle as follows. Consider the function $f(z)$ is analytic (has a convergent Taylor series). Then, we always have

$$f(z+h) = f(z) + hf^{(1)}(z) + \frac{h^2}{2}f^{(2)}(z) + \cdots \quad (A.20)$$

The main difference between a complex function and the real function that is relevant to the maximum modulus principle is that a complex variable can expand in any direction in the complex plane. In contrast, a real variable can only expand to the left or right. Now, if $f^{(1)}(z) \neq 0$, we have sufficiently small h such that

$$|f(z+h)| \approx |f(z) + hf^{(1)}(z)| = |f(z)| + |h||f^{(1)}(z)| > |f(z)| \quad (A.21)$$

If $f^{(1)}(z) = 0$, we can again find an h such that $h^2 f^{(2)}(z)$ is in the direction of $f(z)$ which yields

$$|f(z+h)| \approx |f(z) + \frac{h^2}{2}f^{(2)}(z)| = |f(z)| + |\frac{h^2}{2}f^{(2)}(z)| > |f(z)| \quad (A.22)$$

$$\max_{z \in \mathbb{C}^-} |R(z)| = \max_{y \in \mathbb{R}} |R(iy)| \tag{A.23}$$

For $z = iy$, we have

$$|R(iy)| = \left| \frac{1 + i(1-\theta)y}{1 - i\theta y} \right| \tag{A.24}$$

Stability demands

$$|R(iy)| = \left| \frac{1 + i(1-\theta)y}{1 - i\theta y} \right| \le 1 \tag{A.25}$$

Since the square of the real part of the numerator and denominator are the same, for the modulus to be smaller than one, the square of the imaginary part of the numerator must be smaller than that of the denominator, i.e.,

$$[(1-\theta)y]^2 \le (\theta y)^2 \implies \theta \ge \frac{1}{2} \tag{A.26}$$

Hence, all θ-methods with $\theta \ge 1/2$ yield a stability function with a modulus smaller than or equal to unity; hence, they are all unconditionally stable. The unconditionally stable methods are also called A-stable. The term A-stability was coined by Dahlquist [4]. According to Wanner [5], Dahlquist gave the following reason for this choice: "I didn't like all these 'strong', 'perfect', 'absolute', 'generalized', 'super', 'hyper', 'complete' and so on in mathematical definitions, I wanted something neutral; Moreover, having been impressed by David Young's 'property A', I chose the term 'A-stable'".

The trapezoidal rule has a stability function equal to one, i.e., the numerical solution magnitude remains constant (does not grow or decay in time). Hence, for $\Re(z) = 0$, the trapezoidal rule is consistent with the exact solution. In contrast, other unconditional schemes, such as the backward Euler scheme, yield decaying solution inconsistent with the exact solution.

The explicit Euler scheme is only conditionally stable, with a stability set being the interior of a unit circle centered at $x = -1$. Since the real part of the eigenvalue is assumed to be negative, we may express $\lambda = -\lambda_R + i\lambda_I$ with $\lambda_R > 0$, yielding the following expression for the stability of the explicit Euler scheme

$$|\lambda \Delta t + 1| < 1 \implies |-\lambda_R \Delta t + i\lambda_I \Delta t + 1| \le 1 \tag{A.27}$$

Taking the modulus of the left-hand side then yields

$$(-\lambda_R \Delta t + 1)^2 + (\lambda_I \Delta t)^2 \le 1 \implies (-\lambda_R \Delta t)^2 - 2\lambda_R \Delta t + 1 + (\lambda_I \Delta t)^2 \le 1 \tag{A.28}$$

Analogous argument is repeated if $f^{(2)}(z) = 0$. Therefore, the modulus of $f(z)$ grows as z goes toward the domain boundary over which $f(z)$ is defined. When $f(z)$ is a real function, the relation (A.22) does not hold because h^2 is always positive. For $f^{(2)}(z) < 0$, for any h (being positive or negative), the function magnitude shrinks.

Appendix: Basic Schemes for Model Problems

Simplification then yields

$$\Delta t[(\lambda_R^2 + \lambda_I^2)\Delta t - 2\lambda_R] \leq 0 \tag{A.29}$$

or

$$\Delta t \leq \frac{2}{\lambda_R + \frac{\lambda_I^2}{\lambda_R}} \tag{A.30}$$

For purely imaginary eigenvalue $\lambda = i\lambda_I$, the explicit Euler scheme is *unconditionally unstable*, i.e., no values of Δt renders the modulus of the stability function smaller than unity:

$$\theta = 0, \ z = iy \implies R(z) = 1 + z = 1 + iy \implies |R(z)| = (1+y^2)^{1/2} > 1 \ \forall y \in \mathbb{R} \tag{A.31}$$

This is relevant as systems with purely imaginary eigenvalues appear in the central difference discretization of dissipation-free advection-type PDEs.

Therefore, the bound (A.18) shows that the numerical error is bounded by the exact error plus a term proportional to Δt^r. The smallness of Δt guarantees the closeness of the numerical solution to the exact solution, or in other words, guarantees the scheme's accuracy (or convergence). Also notable is when the initial error is zero or zero within the machine precision and the scheme is stable. The global error is at least proportional to Δt^r. In general, for a stable scheme of local (or consistency) order of accuracy of $r + 1$, we deduce

$$|\epsilon_n| \leq c\Delta t^r + \mathcal{O}(\text{machine precision}) \tag{A.32}$$

where c is a constant proportional to the $(r + 1)$-derivative of the solution and does not depend on a time step size. The bound (A.32) clearly shows the anticipated error reduction rate as a function of decreasing time step size until the error drops to the level of machine precision accuracy. When the error becomes comparable to the machine precision accuracy, the error is polluted with the machine precision error, and a further reduction in the time step size does not result in the error reduction.

A.1.3 Stability of Basic Schemes for System of ODEs

We now consider a system of ODEs in the form

$$w'(t) = Aw(t) \quad t > 0, \quad w(0) = w_0 \tag{A.33}$$

where $w(t)$, $w'(t)$, and $w(0)$ are m-dimensional vectors and A is an $m \times m$ matrix. System (A.33) can arise from the spatial discretization of linear PDEs or the linearization of the nonlinear PDEs.

Exact solution

When the matrix A has the eigendecomposition $A = Q\Lambda Q^{-1}$, the system (A.33) can be rewritten as m uncoupled ODEs for characteristic variable $w_c(t) = Q^{-1}w(t)$ in the form

$$w_c'(t) = \Lambda w_c(t) \quad t > 0 \quad w_c(0) = w_{c,0} \tag{A.34}$$

or

$$w_{c,i}'(t) = \lambda_i w_{c,i}(t) \quad \forall i = 1, \cdots, m \quad t > 0 \quad w_{c,i}(0) = w_{c,0,i} \tag{A.35}$$

Similar to the scalar ODE, the solution is

$$w_{c,i}(t) = e^{\lambda_i t} w_{c,i}(0) \tag{A.36}$$

The solution grows or decays exponentially for positive or negative $\Re(\lambda)$, respectively. For $\Re(\lambda) = 0$, the magnitude of the solution remains unchanged.

If the initial solution is disturbed, i.e., instead of $w(0)$, the initial condition is $\tilde{w}(0)$ with error $\epsilon(0) = w(0) - \tilde{w}(0)$, the disturbed characteristic solution at a later time, $\tilde{w}_c(t)$ also has the same expression as

$$\tilde{w}_{c,i}(t) = e^{\lambda_i t} \tilde{w}_{c,i}(0) \tag{A.37}$$

Subtracting the last equation from Eq. (A.36) yields a similar expression for the error, $\epsilon_c(t) = w_c(t) - \tilde{w}_c(t)$

$$\epsilon_{c,i}(t) = e^{\lambda_i t} \epsilon_{c,i}(0) \tag{A.38}$$

Similar to the exact solution, the error magnitude grows, decays, or remains constant for a positive, a negative, or the zero value of $\Re(\lambda_i)$, respectively. The difference with the scalar case is the existence of m eigenvalues, and hence, potentially m rates of decay. In other words, there are multiple time scales; the more negative the real part of an eigenvalue, the higher the solution's or error's decay rate. In the case where disparity between the magnitude of the eigenvalues is large, the part of the solution corresponding to the eigenvalue with a large negative real part has negligibly small contribution to the solution and the error, and other eigenvalues dominate the solution.

Stability

We now turn to the θ-scheme for the system (A.33) which has the form

$$w_{n+1} = (I - \theta \Delta t A)^{-1}[I + (1-\theta)\Delta t A]w_n \tag{A.39}$$

Appendix: Basic Schemes for Model Problems

denoting $Z = \Delta t A$, we rewrite the scheme as

$$w_{n+1} = (I - \theta Z)^{-1}[I + (1 - \theta)Z]w_n \tag{A.40}$$

Similar to the scalar case, the error is expressed as

$$\epsilon_n = \left\{(I - \theta Z)^{-1}[I + (1 - \theta)Z]\right\}^n \epsilon_0 + (I - \theta Z)^{-1} \sum_{j=0}^{n-1} [I + (1 - \theta)Z]^j \rho_{n-j-1} \tag{A.41}$$

Using the eigendecomposition $Z = Q\Lambda Q^{-1}$ and Neumann series expansion [2] for $(I - \theta Z)^{-1} = \sum_{n=0}^{\infty} (\theta Z)^n$ when Z is bounded or $(I - \theta Z)^{-1} = -Z^{-1}(I - \theta^{-1}Z^{-1})^{-1} = \sum_{n=0}^{\infty} (\theta Z)^{-n}$ when Z is not bounded, we have

$$\epsilon_{c,n,i} = \left\{(I - \theta \lambda_i)^{-1}[I + (1 - \theta)\lambda_i]\right\}^n \epsilon_{c,0,i} + \sum_{j=0}^{n-1} \left[(I - \theta \lambda_i)^{-1}[I + (1 - \theta)Z]\right]^j \rho_{c,n-j-1,i} \tag{A.46}$$

The derivation is left to Exercise 2 at the end of the chapter. Similar to the scalar case, for $\theta \geq 1/2$, the θ-method is unconditionally stable, and for $\theta < 1/2$, the scheme is only conditionally stable. In particular, the forward Euler scheme with $\theta = 0$, has the stability condition as

$$\Delta t \leq \min_i \left(\frac{2}{\lambda_{i,R} + \frac{\lambda_{i,I}^2}{\lambda_{i,R}}} \right) \tag{A.47}$$

When the eigenvalues are widely spread out, i.e., the largest in absolute values of the real part of the eigenvalues is much larger than the smallest one, the problem is called stiff. The stiffness results in a requirement of very small time step size

---

[2] Consider the finite sum

$$\sum_{i=0}^{n} Z^i \tag{A.42}$$

for a bounded matrix Z. Then,

$$(I - Z)(\sum_{i=0}^{n} Z^i) = (I - Z^{n+1}) \tag{A.43}$$

Taking the limit $n \to \infty$ and noting that Z is bounded yield

$$\lim_{n \to \infty} (I - Z) \sum_{i=0}^{n} Z^i = \lim_{n \to \infty} (I - Z^{n+1}) = I \tag{A.44}$$

Therefore, for a bounded matrix Z

$$(I - Z)^{-1} = \sum_{i=0}^{\infty} Z^i \tag{A.45}$$

The series is called the Neumann series [6].

to satisfy the stability condition (A.47) corresponding to the eigenvalue with the largest in the absolute value of the real part (λ^*). The eigenvalue λ^* has a much smaller impact on the solution; however, it dictates a very small time step size and effectively determines the cost of computations.

When matrix A varies at every time step, for instance, when A arises from the local linearization of a nonlinear system, the eigenvalues vary from one-time step to another, and Δt must be calculated every time step based on current eigenvalues.

A.1.4 Stiffness

In the linear model problem (A.8), $|\lambda|$ represents the rate of change of solution or the time scale of the problem, and for the explicit or forward Euler scheme, λ dictates the smallness of time step for the stable solution as revealed by the inequality (A.47). For a system of differential equations, if the solution has negative eigenvalues of different sizes, the part of the solution associated with the largest absolute value eigenvalue decays the fastest. In contrast, the parts associated with smaller eigenvalues decay more slowly. Dahlquist [7] gives an example in the form

$$w'_1 = w_2 \qquad (A.48a)$$
$$w'_2 = -1000w_1 - 1001w_2 \qquad (A.48b)$$

The exact solution of the system is given as

$$w_1 = \underbrace{A\exp(-t)}_{\text{slow part}} + \underbrace{B\exp(-1000t)}_{\text{fast part}} \qquad (A.49)$$

which can be easily verified by direct substitution into the system. The solution consists of two distinct temporal scales: the slowly varying or decaying part (the first term) and the fast varying or decaying part (the second term). Examining the matrix's eigenvalues also reveals two distinct negative eigenvalues: $\lambda \approx -1000$ and $\lambda \approx -1$. The use of an explicit Euler scheme requires a very small time step proportional to the inverse of the first eigenvalue $|\lambda|$ as is evident from Eq. (A.47). Suppose one is concerned with the slow-evolving part of the solution, which is often the case as the fast solution feature dies out very quickly. In that case, the stability condition of the explicit Euler scheme is prohibitively restrictive and computationally inefficient. Problems such as System (A.48) that have distinct fast and slow decaying solutions (multi-scale features) are called stiff problems [7].

The stiffness can also appear in the numerical solution of a single differential equation with significantly different time scales in the solution. For the single differential equation, the multi-scale solution can appear as a result of a driving force. For instance, the following forced differential equation

$$w' = \lambda(\sin t - w), \qquad w(0) = 0 \tag{A.50}$$

with the exact solution

$$w(t) = \frac{\sin t - \frac{1}{\lambda}\cos t + \frac{1}{\lambda}\exp(-\lambda t)}{1 + \frac{1}{\lambda}} \tag{A.51}$$

clearly consists of two vastly different scales as long as $|\lambda|$ is very large [7]. For large values of λ, for instance $\lambda \geq 100$, the exponential term is less than 10^{-6} for $t \geq 0.1$. One might expect that for $t \geq 0.1$, the time step size $\Delta t = 0.1$ should yield a reasonable approximation. However, from the stability analysis of the explicit Euler scheme, it is obvious that a significantly smaller time step is required for a stable solution.

Therefore, the explicit schemes are not suitable for the stiff problems. Intuitively, an explicit scheme computes the solution at time t_{n+1} solely using the solution at time t_n, which, due to the large variations of the solution in the stiff problems, can have significant errors and results in instability. On the other hand, we expect the implicit schemes that at least partly evaluate the derivatives using the solution at t_{n+1} to be suitable for stiff problems. The analysis in (A.1.3) confirms this and reveals that θ-methods with $\theta \geq 1/2$ are unconditionally stable, regardless of how large is $|\Re(\lambda)|$.

A.1.5 Stiff problems and L-stability

A model of a linear stiff problem can be considered as the ODE (A.8) with $\lambda \to -\infty$ [4]. For the stiff problem with $\lambda \to -\infty$, it is assumed that Δt is not correspondingly small such that $z = \lambda \Delta t \to -\infty$. For the case of $z = \lambda \Delta t \to -\infty$, the θ-method exhibits two distinctive limiting behaviors depending on θ values:

$$\lim_{z \to -\infty} |R(z)| \to a, \quad 0 < a \leq 1, \text{ for } \frac{1}{2} \leq \theta < 1 \tag{A.52}$$

$$\lim_{z \to -\infty} |R(z)| \to 0, \qquad \text{for } \theta = 1 \tag{A.53}$$

Only the case of $\theta = 1$ yields solutions consistent with the exact solution in this limiting case of $z = \lambda \Delta t \to -\infty$, which is due to $\lim_{z \to -\infty} |R(z)| \to 0$. In contrast, other values of θ do not yield solutions consistent with the exact solution in this limiting case. Any A-stable scheme with $\lim_{z \to -\infty} |R(z)| \to 0$ is called L-stable [4]. L-stability is an essential requirement in the approximation of stiff problems, and the implicit (backward) Euler scheme is an example of L-stable scheme. The trapezoidal rule is only *A*-stable and not *L*-stable; hence, it can yield significant errors for the case of $z = \lambda \Delta t \to -\infty$.

A.2 Spatial Schemes

As we move from numerical solutions of ODEs to time-dependent PDEs, we must adopt some strategies for discretization or numerical approximation in space and time. One simple approach is to discretize the PDE in space, i.e., approximate the spatial derivatives using the solution at a discrete set of m points in space. This results in a system of ODEs of m equations, which can be integrated in time using a temporal numerical scheme, which can be any of those described in the previous section. This approach of discretization in space, followed by discretization in time, is called the method of lines. We focus on the method of lines here and discuss simple discretization schemes for the advection and diffusion equations. The advection and diffusion equations are sufficiently simple that they allow rigorous analysis of the stability and accuracy of several basic numerical schemes. The outline of the steps for the method of lines applied to a PDE in the form

$$\frac{\partial u}{\partial t} + \frac{\partial f(u, \partial u/\partial x)}{\partial x} = 0 \tag{A.54}$$

is as follows:

- First, enforce the PDEs on a discrete set of spatial points or spatial grid:

$$\left.\frac{\partial u}{\partial t}\right|_{t,x_j} + \left.\frac{\partial f(u, \partial u/\partial x)}{\partial x}\right|_{t,x_j} = 0 \quad j = 0, \cdots m \tag{A.55}$$

This leads to $(m + 1)$ ODEs.
- Second, approximate the spatial derivative using the discrete values of u_i as

$$\left.\frac{\partial f(u, \partial u/\partial x)}{\partial x}\right|_{t,x_j} \approx \left.\frac{Df[u_{j-l}, \cdots, u_{j+l'}]}{Dx}\right|_{t,x_i} \tag{A.56}$$

where D/Dx signifies the discrete spatial derivative computed using solution values at $(l + l' + 1)$ grid points.
- Third, after substituting the discrete spatial derivative operator into (A.55), analyze the characteristics of the coupled system of ODEs in the form

$$\left.\frac{\partial u}{\partial t}\right|_{t,x_j} + \left.\frac{Df[u_{j-l}, \cdots, u_{j+l'}]}{Dx}\right|_{t,x_j} = 0 \quad j = 0, \cdots, m \tag{A.57}$$

In other words, determine the eigenvalues of the matrix associated with the discrete spatial derivative operator. The nature of eigenvalues dictates the stability property

of spatial discretization. When $f(u, \partial u/\partial x)$ is a nonlinear function, the Jacobian of the linearized form dictates the linear stability properties of the spatial scheme.
- Finally, if the spatial discretization is stable, enforce the resultant system of ODEs on a discrete set of points in time or on a temporal grid as

$$\left.\frac{\partial u}{\partial t}\right|_{t_n, x_j} + \left.\frac{Df[u_{j-l}, \cdots, u_{j+l'}]}{Dx}\right|_{t_n, x_j} = 0 \quad j = 0, \cdots m, \ n = 1, \cdots, N \quad (A.58)$$

Then, discretize in time using a temporal scheme.

We next introduce basic low-order spatial discretization schemes and discuss their stability and accuracy for advection and diffusion equations. The discussion is based on the classical results, which can be found, for instance, in Ref. [1].

A.2.1 Advection equation

Consider the advection equation

$$\frac{\partial u}{\partial t} + a\frac{\partial u}{\partial x} = 0 \quad (A.59)$$

subjected to appropriate initial and boundary conditions. Here, a is a constant advection velocity, $x \in [0, 1]$, $t \in [0, T]$, and $u = u(t, x)$.

For an approximation of the spatial derivative, we start with the Taylor series of $u(t, x)$ in x and about x_i:

$$u(t, x) = u(t, x_j) + u_x(t, x_j)(x - x_j) + \frac{1}{2}u_{xx}(t, x_j)(x - x_j)^2 + \cdots \quad (A.60)$$

Evaluating this series at x_{j+1} yields

$$u(t, x_{j+1}) = u(t, x_j) + u_x(t, x_j)(x_{j+1} - x_j) + \frac{1}{2}u_{xx}(t, x_j)(x_{j+1} - x_j)^2 + \cdots \quad (A.61)$$

An expression for the first spatial derivative $u_x(t, x_j)$ then follows

$$u_x(t, x_j) = \frac{u(t, x_{j+1}) - u(t, x_j)}{x_{j+1} - x_j} - \frac{1}{2}u_{xx}(t, x_j)(x_{j+1} - x_j) + \cdots \quad (A.62)$$

Using a uniform sub-division of the x interval, or a uniform grid with spacing $h = (x_{j+1} - x_j)$, yields

$$u_x(t, x_j) = \frac{u(t, x_{j+1}) - u(t, x_j)}{h} - \frac{1}{2}u_{xx}(t, x_j)h + \cdots \quad (A.63)$$

Now, the first term on the right-hand side gives an approximation of the derivative

$$w_x(t, x_j) = \frac{u(t, x_{j+1}) - u(t, x_j)}{h} \tag{A.64}$$

with

$$w_x(t, x_j) = u_x(t, x_j) + O(h) \tag{A.65}$$

This last equation shows the local truncation error and also the consistency of the approximation, i.e., as the grid spacing goes to zero, the approximation approaches the exact solution:

$$h \to 0 \implies w_x(t, x_j) \to u_x(t, x_j) \tag{A.66}$$

Alternatively, one may express the Taylor series about x_{i-1} and evaluate the expression at x_i, yielding

$$u_x(t, x_j) = \frac{u(t, x_j) - u(t, x_{j-1})}{h} + \frac{1}{2} u_{xx}(t, x_j) h + \cdots \tag{A.67}$$

and

$$w_x(t, x_j) = \frac{u(t, x_j) - u(t, x_{j-1})}{h} \tag{A.68}$$

with the same truncation error.

The approximations (A.64) and (A.68) are referred to as the right and left approximations, respectively. For $a > 0$ ($a < 0$), (A.64) and (A.68) are alternatively referred to as downwind (upwind) and upwind (downwind) approximations, respectively. A central approximation can also be obtained by expressing the Taylor series about x_i and evaluating it twice, one at x_{i-1} and one at x_{i+1}, and summing both expressions. The central approximation is

$$w_x(t, x_j) = \frac{u(t, x_{j+1}) - u(t, x_{j-1})}{2h} \tag{A.69}$$

with

$$w_x(t, x_j) = u_x(t, x_j) + O(h^2) \tag{A.70}$$

Substituting the approximation of the spatial derivative in the advection equation gives

$$w'(t, x_j) = -a w_x(t, x_j) \tag{A.71}$$

where $w(t, x_j) \approx u(t, x_j)$ and $w(0, x) = u(0, t)$. Since this equation is valid for any i, we have a system of ODEs with a shorthand notation as

$$w'(t) = A w(t) \tag{A.72}$$

Appendix: Basic Schemes for Model Problems

where A is a matrix and $w(t)$ is a vector of all discrete values of the approximated solution, provided that we apply a proper boundary condition. As will be seen later, the stability analysis of different methods becomes easy using periodic boundary conditions.

Using periodic boundary condition, for the left approximation, the matrix A

$$A = \frac{a}{h} \begin{pmatrix} -1 & & & & 1 \\ 1 & -1 & & & \\ & \ddots & \ddots & & \\ & & 1 & -1 & \\ & & & 1 & -1 \end{pmatrix} \quad (A.73)$$

and the equations in the matrix form are

$$\begin{pmatrix} w'_0(t) \\ w'_1(t) \\ \vdots \\ w'_{m-2}(t) \\ w'_{m-1}(t) \end{pmatrix} = \frac{a}{h} \begin{pmatrix} -1 & & & & 1 \\ 1 & -1 & & & \\ & \ddots & \ddots & & \\ & & 1 & -1 & \\ & & & 1 & -1 \end{pmatrix} \begin{pmatrix} w_0(t) \\ w_1(t) \\ \vdots \\ w_{m-2}(t) \\ w_{m-1}(t) \end{pmatrix} \quad (A.74)$$

The equations for the right approximation are

$$\begin{pmatrix} w'_0(t) \\ w'_1(t) \\ \vdots \\ w'_{m-2}(t) \\ w'_{m-1}(t) \end{pmatrix} = \frac{a}{h} \begin{pmatrix} 1 & -1 & & & \\ & 1 & -1 & & \\ & & \ddots & \ddots & \\ & & & 1 & -1 \\ -1 & & & & 1 \end{pmatrix} \begin{pmatrix} w_0(t) \\ w_1(t) \\ \vdots \\ w_{m-2}(t) \\ w_{m-1}(t) \end{pmatrix} \quad (A.75)$$

and the equations for the central approximations are

$$\begin{pmatrix} w'_0(t) \\ w'_1(t) \\ \vdots \\ w'_{m-2}(t) \\ w'_{m-1}(t) \end{pmatrix} = \frac{a}{2h} \begin{pmatrix} 0 & -1 & & & 1 \\ 1 & 0 & -1 & & \\ & \ddots & \ddots & \ddots & \\ & & 1 & 0 & -1 \\ -1 & & & 1 & 0 \end{pmatrix} \begin{pmatrix} w_0(t) \\ w_1(t) \\ \vdots \\ w_{m-2}(t) \\ w_{m-1}(t) \end{pmatrix} \quad (A.76)$$

In the above equations, $w_i(t) = w(t, x_i)$, where

$$x_i = ih \qquad \forall\, i = 0, \cdots, m-1 \quad (A.77)$$

with

$$h = \frac{1}{m} \quad (A.78)$$

Stability of spatial discretization of advection equation

The solution to the simple advection equation subject to a single Fourier mode is the same Fourier mode but with a shift of $(-at)$. In other words, the magnitude of the solution does not change or does not grow or diminish over time; it only propagates with speed a. The solution of the semi-discretized system arising from a spatial discretization of the advection equation must mimic the behavior of the exact solution. We here focus on determining the eigenvalues of the various numerical schemes introduced in the previous section with the aim of determining their impacts on the stability of the numerical solution. We need first to determine the eigenvalues of the exact first spatial differentiation. Plugging a solution ansatz in the form

$$u(t, x) = \gamma(t)e^{i\,2\pi kx} \tag{A.79}$$

into the advection equation yields

$$\gamma' e^{i\,2\pi kx} + a(i2\pi k)\gamma e^{i\,2\pi kx} = 0 \tag{A.80}$$

After simplification, we obtain the following ODE

$$\gamma' = -i2\pi ka\,\gamma \tag{A.81}$$

which reveals the exact eigenvalue to be purely imaginary and in the form

$$\lambda_{\text{exact}} = -i2\pi ka \tag{A.82}$$

Since the eigenvalue is purely imaginary, the initial Fourier mode neither grows nor shrinks. The spatial scheme should have similar characteristics. In particular, a stable spatial numerical scheme yields a system (matrix) with eigenvalues that are not positive since a positive eigenvalue corresponds to exponential growths, inconsistent with the exact solution behavior, and thus unstable.

The matrices of the semi-discretized equations are in the form of a circulant matrix. That is,

$$A = \begin{pmatrix} c_0 & c_1 & \cdot & \cdot & c_{m-1} \\ c_{m-1} & c_0 & c_1 & \cdot & c_{m-2} \\ \cdot & c_{m-1} & c_0 & \cdot & \cdot \\ \cdot & \cdot & \cdot & \cdot & c_1 \\ c_1 & c_2 & \cdot & c_{m-1} & c_0 \end{pmatrix} \tag{A.83}$$

The eigenvalues of a circulant matrix are

$$\lambda_k = \sum_{j=0}^{m-1} c_j e^{2\pi ikx_j} \quad k = 1, \cdots, m \tag{A.84}$$

Appendix: Basic Schemes for Model Problems

where c_js are the entries of the circulant matrix, and the eigenvectors are the Fourier modes

$$\phi_k = (\psi_k(x_0), \psi_k(x_1), \cdots, \psi_k(x_{m-1})) \tag{A.85}$$

where

$$\psi_k(x_j) = e^{2\pi i k x_j} \tag{A.86}$$

Furthermore, the solution to a single Fourier mode as an initial condition is

$$w_k(t) = e^{At}\phi_k = e^{\lambda_k t}\phi_k \tag{A.87}$$

which can be shown with the repeated use of $A\phi_k = \lambda_k \phi_k$ as follows:

$$e^{At}\phi_k = \left[I + tA + \frac{1}{2}(tA)^2 + \frac{1}{3!}(tA)^3 + \cdots\right]\phi_k \tag{A.88}$$

$$= \phi_k + tA\phi_k + \frac{1}{2}(tA)^2\phi_k + \frac{1}{3!}(tA)^3\phi_k + \cdots \tag{A.89}$$

$$= \phi_k + t\lambda_k\phi_k + \frac{1}{2}(t\lambda_k)^2\phi_k + \frac{1}{3!}(t\lambda_k)^3\phi_k + \cdots \tag{A.90}$$

$$= \left[I + t\lambda_k + \frac{1}{2}(t\lambda_k)^2 + \frac{1}{3!}(t\lambda_k)^3 + \cdots\right]\phi_k \tag{A.91}$$

$$= e^{\lambda_k t}\phi_k \tag{A.92}$$

The eigenvalues for the left approximation (A.74) are

$$\lambda_k = \frac{a}{h}\left[-e^{2\pi i k x_0} + e^{2\pi i k x_{m-1}}\right] = \frac{a}{h}[\cos(2\pi k h) - 1] - \frac{ia}{h}\sin(2\pi k h) \tag{A.93}$$

Here, we used

$$e^{2\pi i k x_0} = e^{2\pi i k(0)h} = e^0 = 1 \tag{A.94}$$

and

$$e^{2\pi i k x_{m-1}} = e^{2\pi i k(m-1)h} = \cos(2\pi k(m-1)h) + i\sin(2\pi k(m-1)h) \tag{A.95}$$
$$= \cos(2\pi k m h - 2\pi k h) + i\sin(2\pi m h - 2\pi k h) \tag{A.96}$$
$$= \cos(2\pi k(-1)h) + i\sin(2\pi k(-1)h) \tag{A.97}$$
$$= \cos(2\pi k h) - i\sin(2\pi k h) \tag{A.98}$$

Now denoting $\lambda_k = y + iz$, where y and z being the real and imaginary parts of λ_k, we have

$$e^{\lambda_k t} = e^{(y+iz)t} = e^{yt}e^{izt} \tag{A.99}$$

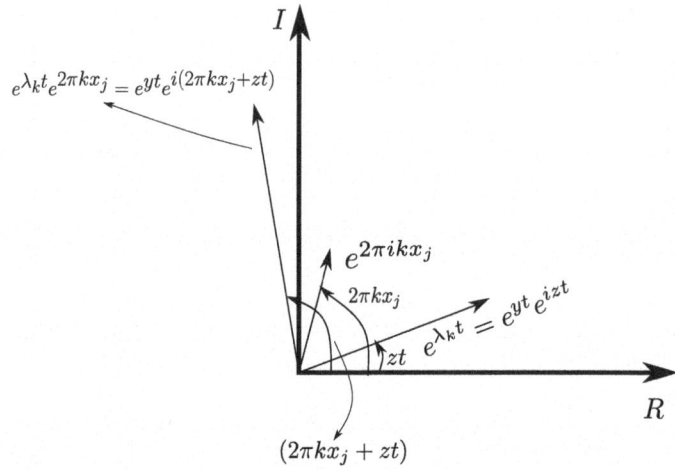

Fig. A.1 Phaser diagram depicting a single Fourier mode $e^{2\pi i k x_j}$, $e^{\lambda_k t}$. Moreover, $e^{\lambda_k t} e^{2\pi i k x_j}$, clearly demonstrates the change in the phase and the magnitude of the Fourier mode as a result of multiplication with $e^{\lambda_k t}$

and

$$e^{\lambda_k t} \psi_k(x_j) = e^{yt} e^{izt} e^{2\pi i k x_j} = e^{yt} e^{i(2\pi k x_j + zt)} \tag{A.100}$$

This equation implies that the real part of an eigenvalue merely stretches (or scales) the initial condition, while the imaginary part merely rotates the initial solution (Fig. A.1). Consequently, the initial modes are dampened if the real part is negative. For the left approximation, the real part is negative if $a > 0$. In other words, the left approximation is dissipative if it coincides with the upwind scheme. For $a < 0$, the left scheme has eigenvalues with positive real parts; thus, it is anti-dampening. Considering that the exact solution is merely shifted over time and does not grow in magnitude, the dampening approximation is stable, while the anti-dampening scheme is unstable. Hence, the left approximation is stable for $a > 0$, and unstable for $a < 0$.

The eigenvalue of the right approximation matrix (A.75) are

$$\lambda_k = \frac{a}{h} \left(e^{2\pi i k x_0} - e^{2\pi i k x_1} \right) = \frac{a}{h} \left[1 - \cos(2\pi k h) \right] - \frac{ia}{h} \sin(2\pi k h) \tag{A.101}$$

The right approximation is stable for $a < 0$ or when it coincides with an upwind scheme and unstable for $a > 0$ or when it coincides with a down-wind scheme.

Therefore, if the left or right approximation coincides with the upwind scheme or the direction of advection speed, it is stable; otherwise, it is unstable. Consequently, upwind schemes occupy an essential place in the numerical solutions of advection-type systems.

Finally, the eigenvalues of the central scheme (A.76) are

$$\lambda_k = \frac{a}{2h}\left(-e^{2\pi ikx_1} + e^{2\pi ikx_{m-1}}\right) = -\frac{ia}{h}\sin(2\pi kh) \quad (A.102)$$

Since the real part of the eigenvalue is zero, there is no stretching or shrinking of the initial mode. Thus, the central spatial scheme yields purely imaginary eigenvalues regardless of the direction of the wind, which is similar to the exact spatial differentiation scheme. For this reason, the central spatial scheme for the advection scheme yields a stable discretization.

Numerical dispersion and dissipation

Now, let us examine the impact of the above three spatial discretization schemes on the quality of the numerical solution. All three methods have the same imaginary part of the eigenvalue, $-\frac{a}{h}\sin(2\pi kh)$, and the approximate solution at each grid point, $w_{k,j}$, is

$$w_{k,j} = e^{\mathrm{Re}(\lambda_k t)} e^{2\pi ik(x_j - at\frac{\sin(2\pi kh)}{2\pi kh})} \quad (A.103)$$

Here, the first exponential, which differs from one scheme to the other, is real and responsible for stretching or shrinking the solution. The second exponential, on the other hand, contains the contribution of the imaginary part of the eigenvalues and reveals its impact on the phase of the Fourier mode, which results in deviations of the numerical advection speed from the exact speed. Comparing this with the exact solution

$$u_{k,j} = e^{2\pi ik(x_j - at))} \quad (A.104)$$

yields the numerical propagation speed in the form

$$a_k = a\frac{\sin(2\pi kh)}{2\pi kh} \quad (A.105)$$

which is a multiplicative factor of $\frac{\sin(2\pi kh)}{2\pi kh}$ different from the exact propagation speed a.

This deviation in the propagation speed of various modes is called dispersion error (numerical dispersion). On the other hand, the deviation in solution magnitude resulting from the real part of the eigenvalues is called a dissipation error (numerical dissipation). While the two stable schemes, the central and the upwind schemes, both suffer from numerical dispersion, only the upwind scheme suffers from numerical dissipation.

It is clear that since $\lim_{h\to 0}\frac{\sin(2\pi kh)}{2\pi kh} = 1$ (which can be shown using L'hospital rule), the second exponential in Eq. (A.103) becomes the same as the exact solution. For the vanishing grid spacing, the dispersion errors are vanishingly small. For a fixed h, however, higher wavenumber modes ($k \to \frac{m}{2}$) are contaminated with higher deviations in the propagation speed or higher dispersion errors, because

$$k \to \frac{m}{2} \implies \left[a \frac{\sin(2\pi kh)}{2\pi kh} \right] \to 0 \qquad (A.106)$$

Note that the Fourier modes are from $k = -m/2 + 1$ to $k = m/2$ if m is even, and from $k = -(m-1)/2 + 1$ to $k = (m-1)/2$ if m is odd.

Similarly, the numerical dissipation is higher for higher wavenumber modes because

$$k \to \frac{m}{2} \implies [\cos(2\pi kh) - 1] \to -2 \qquad (A.107)$$

However, in the upwinding schemes (the left and right schemes), the dispersion errors for increasing wavenumber modes are insignificant (or may not be observable) since the higher wavenumber modes are dampened quickly due to the numerical dissipation. On the other hand, due to the lack of any numerical dissipation, the solution of the central scheme can be polluted with significant numerical dispersion.

These pronounced dissipation or dispersion errors are inherent in all low-order methods. Overcoming these shortcomings requires using higher order methods, which are the main focus of the book and are discussed in Chaps. 2 through 7.

Nature of errors

Next, we explore the nature of the dissipative and dispersive behavior of the numerical scheme. An expression for the upwind approximation of the spatial derivative in the advection equation can be obtained using a Taylor series expansion around x and evaluated at $x - h$ as

$$\frac{1}{h}(u(t, x) - u(t, x - h)) = u_x - \frac{1}{2} h u_{xx}(t, x) - \mathcal{O}(h^2) \qquad (A.108)$$

Substituting u_x from above into the semi-discretized advection equation

$$\frac{\partial u(t, x)}{\partial t} + \frac{a}{h}(u(t, x) - u(t, x - h)) = 0 \qquad (A.109)$$

yields

$$\frac{\partial u(t, x)}{\partial t} + a u_x = \frac{1}{2} a h \, u_{xx}(t, x) + \mathcal{O}(h^2) \qquad (A.110)$$

This equation clearly shows that the solution to the advection equation obtained using the upwind scheme is closer to the solution of an advection-diffusion equation with a diffusion coefficient of ($\frac{1}{2} ah$). It reveals the diffusive or dissipative nature of the upwind scheme.

We can follow a similar analysis and obtain that the solution to the advection equation computed using the central scheme is closer to an advection-dispersion equation, i.e.,

$$\frac{\partial u(t,x)}{\partial t} + au_x = \frac{1}{2}ah^2 u_{xxx}(t,x) + \mathcal{O}(h^3) \quad \text{(A.111)}$$

Here, the dispersion arises from the existence of the third spatial derivative term.

Stability of the explicit Euler scheme for the upwind spatial method

Let us now address the stability of the explicit Euler scheme as applied to the system arising from the spatial discretization of the advection equation. Given the eigenvalues of the upwind scheme (A.93), the analysis of the θ-method for complex ODEs presented in Subsect. A.1.3 reveals the time step restriction for the stability of the explicit Euler scheme applied to the system arising from the first-order upwind scheme. In particular, Eq. (A.30) yields

$$\tau \leq \frac{2}{\max\left(\lambda_R + \frac{\lambda_I^2}{\lambda_R}\right)} \implies \tau \leq \frac{2h}{a \max_k \left[1 - \cos(2\pi kh) + \frac{\sin^2(2\pi kh)}{1 - \cos(2\pi kh)}\right]} \quad \text{(A.112)}$$

Since

$$\max_k \left[1 - \cos(2\pi kh) + \frac{\sin^2(2\pi kh)}{1 - \cos(2\pi kh)}\right] = 2 \quad \text{(A.113)}$$

the time step restriction becomes

$$\tau \leq \frac{h}{a} \quad \text{(A.114)}$$

Clearly, $\tau = \mathcal{O}(h)$. We may rewrite the stability condition in the form

$$\frac{\tau a}{h} \leq 1 \quad \text{(A.115)}$$

This condition is known as Courant-Friedrichs-Levy or CFL condition and $\frac{\tau a}{h}$ is referred as CFL number, or

$$\text{CFL} = \frac{\tau a}{h} \quad \text{(A.116)}$$

The stability condition based on the CFL number is thus

$$\text{CFL} = \frac{\tau a}{h} \leq 1 \quad \text{(A.117)}$$

For the central approximation, all eigenvalues are imaginary. Therefore, based on the analysis in Sect. A.1.2, the explicit Euler temporal integration for the system arising from the central spatial discretization of the advection equation is *unconditionally* unstable.

A.2.2 Diffusion equation

Consider the diffusion equation

$$\frac{\partial u}{\partial t} = d \frac{\partial^2 u}{\partial x^2} \qquad (A.118)$$

for $x \in [0, 1]$. Seeking an approximation of the spatial derivative, using the Taylor series, we may write

$$\frac{\partial^2 u(t, x)}{\partial x^2} = \frac{u_x(t, x + \frac{h}{2}) - u_x(t, x - \frac{h}{2})}{h} - \frac{h^2}{24} u_{xxxx}(t, x) + \cdots \qquad (A.119)$$

(This is obtained by combining two Taylor series, both about x, but one evaluated at $x + \frac{h}{2}$ and the other at $x - \frac{h}{2}$.) The grid spacing is $h = 2\pi/m$, and m is the number of grid points. Or on the discrete x_i with the use of Taylor series, we have

$$\frac{\partial^2 u(t, x_i)}{\partial x^2} = \frac{u_x(t, x_{i+\frac{1}{2}}) - u_x(t, x_{i-\frac{1}{2}})}{h} - \frac{h^2}{24} u_{xxxx}(t, x) + \cdots \qquad (A.120)$$

where $x_{i+\frac{1}{2}} = x_i + \frac{h}{2}$. Now, replacing the first derivative terms with their Taylor series expansion yields

$$\frac{\partial^2 u(t, x_i)}{\partial x^2} = \frac{\frac{u_{i+1} - u_i}{h} - \frac{h^2}{3} u_{xxx}(t, x_{i+\frac{1}{2}}) - \cdots - \frac{u_i - u_{i-1}}{h} + \frac{h^2}{3} u_{xxx}(t, x_{i-\frac{1}{2}}) + \cdots}{h} \qquad (A.121)$$

$$- \frac{h^3}{24} u_{xxx}(t, x) + \cdots \qquad (A.122)$$

where $u_i = u(t, x_i)$. Using

$$-\frac{h^2}{3} \left(u_{xxx}(t, x_{i+\frac{1}{2}}) - u_{xxx}(t, x_{i-\frac{1}{2}}) \right) = -\frac{h^2}{3} \left(h u_{xxxx}(t, x_i) + \mathcal{O}(h^3) \right) \qquad (A.123)$$

we have

$$\frac{\partial^2 u(t, x_i)}{\partial x^2} = \frac{u_{i-1} - 2u_i + u_{i+1}}{h^2} + \mathcal{O}(h^2) \qquad (A.124)$$

Appendix: Basic Schemes for Model Problems

We may now use the following approximation for the diffusion equation

$$w'_i(t) = \frac{d}{h^2}(w_{i-1} - 2w_i + w_{i+1}) \tag{A.125}$$

where $w_i(t)$ is a second-order spatial approximation to $u_i(t)$.

The eigenvalues of the associated matrix define the stability of the system. Since the matrix is circulant, Eq. (A.84) and $x_1 = h$ yields the eigenvalues

$$\lambda_k = \frac{d}{h^2}\left(e^{2\pi i k x_{m-1}} - 2e^{2\pi i k x_0} + e^{2\pi i k x_1}\right) \tag{A.126}$$

$$= \frac{2d}{h^2}(\cos(2\pi k h) - 1), \tag{A.127}$$

The eigenvalues have only real parts with negative values; therefore, they mimic the exact solution behavior, and hence, the discretization is stable. Using the explicit Euler scheme, the time step restriction for a stable solution is

$$\tau \leq \frac{2}{\max|\lambda|} = \frac{h^2}{2d} \tag{A.128}$$

where we have used

$$\max_k |\lambda_k| = \max_k \left|\frac{2d}{h^2}(\cos(2\pi k h) - 1)\right| = \frac{4d}{h^2} \tag{A.129}$$

with the maximum achieved for $k = m/2$ knowing that $h = 1/m$.

Equation (A.128) clearly shows $\tau = \mathcal{O}(h^2)$, which is an order of magnitude smaller than the allowable time step for the advection scheme.

Stiffness in numerical integration of the diffusion problem

The maximum eigenvalue of the circulant matrix for the discretized diffusion equation is given in (A.129) as

$$\lambda_{\max} = \max_k |\lambda_k| = \frac{4d}{h^2} \tag{A.130}$$

From (A.127), the minimum eigenvalue is obtained for $k = 1$. Substituting $k = 1$ in (A.127) and invoking $(\cos(2\pi k h) - 1) = -2\sin^2(\pi k h)$ yield the minimum eigenvalue as

$$\lambda_{\min} = \min_k |\lambda_k| = \frac{4d}{h^2}\sin^2(\pi h) \approx 4d\pi^2 \tag{A.131}$$

which we made use of $\sin^2(\pi h) \to (\pi h)^2$ for small h. Taking $N \propto 1/h$ as the number of grid points, the ratio of the maximum to the minimum eigenvalue is thus

$$\frac{\lambda_{\max}}{\lambda_{\min}} = \frac{\frac{4d}{h^2}}{4d\pi^2} = \frac{1}{\pi^2 h^2} = \frac{m^2}{\pi^2} \tag{A.132}$$

which indicates the spread of the time scales in the diffusion process and the degree of stiffness of the system. The stiffness grows with the number of grid points squared. The challenge is that the stiffness grows even more rapidly as the resolution increases. If one is interested in a steady solution, an explicit time integration scheme, however simple, requires prohibitively small or many times steps. Hence, implicit time integration schemes are more effective for diffusive systems.

A.3 Exercises

1. Write the recursion equation for the θ-method applied to a system of ODEs. Identify the stability function.
2. Describe three different ways of measuring error in a numerical solution of an ODE.
3. You have measured a particular numerical scheme's error vs. time step size. Verify the theoretical rate of convergence and the theoretical order of accuracy. In other words, how does one extract the convergence rate and effective order of accuracy?
4. For numerical integration of a system of ODES, the maximum errors for time step sizes $\tau = 0.001, 0.005, 0.025, 0.125$, and 0.625 are obtained as $|\epsilon| = 10^{-5}, 2.4 \times 10^{-4}, 5.5 \times 10^{-3}, 1.1 \times 10^{-1}$, and 2.0, respectively. Determine the effective rate of convergence and the effective order of accuracy of the method. Justify the ansatz made for the relation between the error and time step size.
5. Consider hemoglobin unbinding and binding with oxygen in the lung's alveoli as a two-way reaction. The reaction is symbolically shown as $W_1 \xrightarrow{k_1} W_2 \xrightarrow{k_2} W_1$ with reaction rates of k_1 and k_2, representing the hemoglobin-oxygen unbinding and binding. Hence, it is reasonable to assume that $k_2 \gg k_1$. Using mass action law, this two-way reaction can be modeled using the following system of ODEs

$$w_1'(t) = -k_1 w_1(t) + k_2 w_2(t) \tag{A.133a}$$
$$w_2'(2) = k_1 w_1(t) - k_2 w_2(t) \tag{A.133b}$$

$w_1(t)$ and $w_2(t)$ represent the bonded and unbonded hemoglobin and oxygen, respectively. **(a)** Write this system of ODEs in a matrix-vector form and identify the matrix entries. **(b)** Derive the exact solution to the problem using the eigendecomposition of the matrix, $A = U \Lambda U^{-1}$ and transforming the equations into the eigenspace with $v = U^{-1} w$. **(c)** Write out Euler's explicit method for time integration of the system. **(d)** Identify the eigenvalues of the system. **(e)** Determine the condition for a stable and accurate solution. **(f)** Write a computer code for the integration of the system using Euler's explicit method with $w_1(0) = 0.2$

Appendix: Basic Schemes for Model Problems

and $w_2(0) = 0.8$. For the final time of $T = 1$, obtain the approximate concentrations for $\tau = 1/50, 1/100, \cdots 1/5000$ and three pairs of the reaction rates $(k_1, k_2) = (1, 10), (1, 100)$ and $(1, 1000)$. Plot and discuss the data.

6. Consider the system of ODEs in the form

$$c_1'(t) = -0.1c_1(t) + 0.05c_2(t) \quad \text{(A.134)}$$
$$c_2'(t) = 0.1c_1(t) - 0.05c_2(t) \quad \text{(A.135)}$$

The explicit Euler scheme is proposed for numerical integration of this system. **(a)** Is there any stability condition? If yes, determine the condition. **(b)** Is this system considered stiff? Why?

7. The analysis in this chapter considers PDEs with spatial domains from 0 to 1. How does this analysis apply to PDEs on an arbitrary spatial domain such as $x \in [a, b]$. Discuss.

8. For the numerical solution of the advection (or diffusion) equation, derive an expression for the global error after applying both spatial and temporal discretizations. Discuss.

9. Consider the left approximation of the first-order spatial derivative and the forward Euler approximation of the temporal derivative applied to the advection equation. Find the stability condition of the overall scheme.

10. Consider the advection-diffusion-reaction equation in the form

$$\frac{\partial u}{\partial t} + a\frac{\partial u}{\partial x} = d\frac{\partial^2 u}{\partial x^2} - ku \quad \text{(A.136)}$$

where k is a positive real constant. Treating the advection and diffusion with first-order upwind and second-order central schemes, respectively, and the temporal derivative using the explicit Euler scheme, determine the stability condition of the overall scheme. .

11. Consider the advection equation in the form

$$\frac{\partial u}{\partial t} = 2\frac{\partial u}{\partial x} \quad \text{(A.137)}$$

(a) propose a second-order upwind scheme for the spatial discretization of this advection equation. Show the derivation. **(b)** Determine the stability condition for the numerical integration scheme arising from the second-order upwind scheme and the explicit Euler discretization of the time derivative.

12. Consider the advection equation in the form

$$\frac{\partial u}{\partial t} = 2\frac{\partial u}{\partial x} \quad \text{(A.138)}$$

Determine which of the following discretizations is a consistent discretization of the spatial derivative

$$\frac{\partial u_j}{\partial t} = \frac{1}{h}(-3u_j + 4u_{j+1} - u_{j+2}) \tag{A.139}$$

or

$$\frac{\partial u_j}{\partial t} = \frac{1}{h}\left(-\frac{3}{2}u_j + 2u_{j+1} - \frac{1}{2}u_{j+2}\right) \tag{A.140}$$

where u_j is an approximation of $u(t, x_j)$ over a uniform grid in x.

13. The concentration of a chemical (or nuclear substance) released at a specific time (initial time) is given as a normal distribution in the x-direction of space:

$$C(t = 0, x) = \frac{1 + \sin^3(2\pi x)}{2} \tag{A.141}$$

(a) Determine if this concentration profile is periodic over the spatial domain $[0, 1]$. (b) Decompose this initial profile into Fourier modes. (c) Rank the Fourier modes regarding their contribution to the concentration profile.

14. The concentration of a chemical (or nuclear substance) released in the air at a specific time (initial time) is given as a normal distribution in the x-direction of space:

$$C(t = 0, x) = \frac{1}{\sigma\sqrt{2\pi}} \exp\left(\frac{-(x - \bar{x})^2}{2\sigma^2}\right), \quad \sigma = 0.1, \, \bar{x} = 0.5 \tag{A.142}$$

(a) Determine if this concentration profile is periodic over the spatial domain $[0, 1]$. (b) Decompose this initial profile into the first 10 Fourier modes. (c) Rank the Fourier modes regarding their contribution to the concentration profile.

15. Derive a second-order upwind approximation for the advection equation. Discuss its stability.
16. Derive a third-order upwind approximation for the advection equation. Discuss its stability.
17. Derive a fourth-order central scheme for the advection scheme.
18. Consider the simple constant speed 1D advection equation with a periodic boundary condition and an initial condition with many modes such as $u(t = 0, x) = \sin(2\pi x)^{100}$. (a) Code the first-order upwind scheme, the first-order downwind scheme, and the second-order central scheme applied to the advection equation. . (b) For each spatial discretization scheme, what temporal scheme is the most effective one? Discuss. (c) Determine the stability of each pair of spatial-temporal schemes and compare with the theoretical results. Discuss. (d) Determine the global approximation order of the overall scheme using numerical experiments and exact solutions. (e) How would you investigate the impact of the three numerical schemes on the Fourier modes contained in the initial condition? Investigate the impact computationally and compare it with what the theory suggests.

Appendix: Basic Schemes for Model Problems

19. Consider the Burgers' equations

$$\frac{\partial u}{\partial t} + \frac{\partial u^2/2}{\partial x} = 0 \qquad (A.143)$$

Design a first-order upwind scheme for the spatial derivative. Hint. Split the flux into a positive and negative part in a form

$$\frac{u^2}{2} = \frac{1}{2}(\frac{u^2}{2} + \alpha u) + \frac{1}{2}(\frac{u^2}{2} - \alpha u) \qquad (A.144)$$

where

$$\alpha = \max_x |u(t,x)|$$

Moreover, apply an upwind approximation to each flux. Then, using these numerical fluxes, compute the derivative.

20. Give an upwind scheme for the general nonlinear conservation law

$$\frac{\partial u}{\partial t} + \frac{\partial f(u)}{\partial x} = 0 \qquad (A.145)$$

Hint: follow a similar splitting as the Burgers equation, i.e., splitting the flux into positive and negative.

21. Consider the viscous Burgers equation in the form

$$\frac{\partial u}{\partial t} + \frac{\partial u^2/2}{\partial x} = \mu \frac{\partial^2 u}{\partial x^2} \qquad (A.146)$$

Determine the stability condition assuming the explicit Euler scheme for the temporal integration, an upwind, and a central scheme for the first and second-order spatial derivatives.

22. For the advection equation in the form

$$\frac{\partial u}{\partial t} = \frac{\partial u}{\partial x} \qquad (A.147)$$

consider the following spatial discretization

$$\frac{\partial u_j}{\partial t} = \frac{1}{h}(-3u_j + 4u_{j-1} - u_{j-2}) \qquad (A.148)$$

where u_j is an approximation of $u(t, x_j)$ over a uniform grid in x. **(a)** Determine if this is an upwind, downwind, or central approximation. **(b)** Is this scheme stable? Why? Show details. (Useful trigonometric relation $\cos(2\theta) = 2\cos^2(\theta) - 1$.) **(c)** Determine if this is a consistent scheme. **(d)** Determine if this is a conservative scheme.

23. Using the matrices associated with the right and the central schemes for the advection equation and the eigenvalues of a circulant matrix, Eq. (A.84), verify Eqs. (A.101) and (A.102).

24. Consider a three-dimensional advection-diffusion scheme where the advection is treated using a first-order upwind scheme while the diffusion is treated using the second-order central scheme. An explicit Euler scheme is used for the temporal discretization. **(a)** What is the overall order of accuracy of the scheme? **(b)** The equation is marched in time to a final time of 10 with a spatial grid spacing of h, and the computation took 1 hr of the CPU time. What is the expected CPU time if the equation is marched in time with a fivefold finer spatial grid?

References

1. Hundsdorfer, W., Verwer, J.: Numerical Solution of Time-Dependent Advection-Diffusion-Reaction Equations, 1st edn. Springer, Berlin, Heidelberg (2003)
2. Strikwerda, J.C.: Finite Difference Schemes and Partial Differential Equations, 2nd edn. Society for Industrial and Applied Mathematics (2004)
3. Ahlfors, L.: An Introduction to The Theory of Analytic Functions of One Complex Variable, 3nd edn. McGraw-Hill (1997)
4. Dahlquist, G.: A special stability problem for linear multistep methods. BIT Numer. Math. **3**, 27–43 (1963)
5. Wanner, G.: Dahlquists talk in honour of g. dahlquist at the scicade05 meeting in nagoya, describes the two classical papers from 1956 and 1963 of dahlquist and their enormous impact on the research of decades to come; it also allows the author to present a personal testimony of his never ending admiration for the scientific and personal qualities of this great man (2006)
6. Schechter, E.: Handbook of Analysis and Its Foundations, 1st edn. Elsevier (1996)
7. Dahlquist, G., Bjork, A.: Numerical Methods. Dover, Mineola, New York (1974)

Index

A
Advection or transport equation
  characteristic curve, 8, 51
  characteristic speed, 8, 13, 16, 18, 23, 64, 170
  exact solution, 22, 231, 233, 235–238, 241, 246, 248, 249, 253
  initial and boundary conditions, 243
  time rate of change following a fluid element or blob, 2
A-stability, 107, 108, 236

B
Blast wave, 148, 159, 196, 197
Boundary intercept, 85
Boundary normal, 66, 68, 85
Brent method, 145
Burgers equation, 16, 23, 124, 147, 157, 257

C
Cartesian domain, 63, 64, 66, 68, 73, 75, 84
CFL condition, 94, 140, 143, 167, 178, 195, 197, 251
Characteristic boundary treatment
  adiabatic no-slip wall, 75
  Dirichlet boundary condition, 78, 106, 109, 123
  incoming characteristics, 64, 67
  inviscid wall, 73
  Neumann boundary conditions, 83
  outgoing characteristics, 89
  periodic boundary condition, 76, 78, 245
  reflecting boundary condition, 76–78
  subsonic inflow, 72, 73
  subsonic outflow, 71
  supersonic inflow, 70
  supersonic outflow, 71
Chemical potential, 203–205, 207, 208, 215, 218, 223
Circulant matrix, 60, 246, 247, 253, 258
Concavity, 142, 143, 157, 178, 179, 199
Consistency, 42, 50, 64, 131, 132, 160, 163, 177, 178, 189, 231, 233, 244
Convex, 142, 143, 177–179, 195
CPU time, 93, 94, 140, 141, 153, 172, 177, 179, 189, 197, 258

D
Decay rate, 226, 238
Defect correction
  original problem, 113
  simplified problem, 113
Deferred correction, 95, 112, 114, 121, 228
Diffuse interface, 223
Diffusive effects
  multiscale nature of diffusion, 231
  stiffness due to diffusion, 119, 231
Diffusive fluxes, 67, 119, 128, 191, 194–196, 228
Discontinuities and large gradients
  advection of fluid interfaces and parallel characteristics, 15
  fluid interfaces, 4, 15, 160
  shock waves, 5, 63, 132
Dispersion, 132, 249–251

Double Mach reflection, 200

E

Equations of compressible fluid dynamics (or transport phenomena)
   differential form, 1, 4, 18, 22, 210
   energy of a fluid element, 2
   integral form, 1, 5, 20
   linear momentum of a fluid element, 2
   mass of a fluid element, 1
   Reynolds Transport Theorem, 2
Equilibrium
   chemical, 203, 205, 206
   mechanical, 203, 205, 207
   thermal, 192, 193, 197, 200, 201, 205
Essentially Non-Oscillatory (ENO), 35, 46, 48, 51, 52
Euler system
   characteristic curves, 8, 16
   characteristic speed. *see also* speed of sound, 10, 12, 64, 127, 128
   converging, diverging, and parallel characteristics, 14
   diffusive effects: viscous effects and heat conduction, 5, 6
   eigen-decomposition, 13
   eigen-values, 12, 13, 19, 23, 67, 86, 87, 107, 138, 170, 177, 223, 236–240, 246, 248, 249, 251–253, 258
   entropy generation due to diffusive effects, 18
   quasilinear form, 11, 20, 51, 67, 69, 70, 139, 223
Euler thermodynamics equation, 203, 207

F

Finite difference scheme
   conservative form, 43, 134
   derivative approximation, 31, 32, 76, 90, 134, 139
   discontinuity, 25, 30, 132
   interior domain, 82, 88
   mid-point approximation, 32, 35, 37, 60, 194
   near the boundary, 25, 29, 64, 75, 78, 81, 82
   on the boundary, 66, 67, 70, 71, 84

G

Gauss-Lobatto quadrature, 99, 115, 117, 118

Ghost data, 86
Ghost points, 64, 78, 84, 85, 87, 89, 90
Gibbs property relation, 5, 161, 204

H

Helium bubble, 94, 173, 175

I

Incremental stencil interpolation, 56
Interpolation
   error, 25, 27, 29, 30
   Gibbs phenomenon, 25
   Hermit, 85, 87
   high-order, 29, 56, 76
   Lagrange, 31, 35, 45, 90, 96
   linear, 26, 30, 48, 58, 79
   nonlinear, 31, 35, 52
   Runge phenomenon, 25, 29, 30, 99
Inverse Lax-Wendroff (ILW), 64, 66, 67, 85, 87, 89
Isentropic vortex, 154–156

L

Lax-Friedrichs (LF) splitting/method/scheme

   dissipation, 131
   negative flux, 129, 139
   positive flux, 129, 139
   stability, 157
Legendre transformation, 208, 210
L-stability, 107, 111, 241

M

Maximum principle preserving, 135, 136, 148
Modified pressure, 177–179, 184, 189
Monotonicity, 128, 130, 131, 178, 200
Multi-implicit SDC
   algorithm, 120

N

Non-Cartesian domain, 63, 64, 66–68, 84

O

Oblique shock wave, 200

P

Phaser diagram, 248

Positivity-preserving, 127, 151, 154, 180, 184, 191, 197, 202
Pressure relaxation time, 227

R

Relative velocity, 3
Runge-Kutta (RK) method
 construction, 96, 112
 diagonally implicit, 96, 99, 108, 123
 error, 103, 106
 explicit, 95, 96, 100, 106, 108, 109, 111, 112, 120, 122
 Gauss-Lobatto nodes, 120
 implicit, 95, 96, 98, 99, 106–108, 111, 120

 Neumann expansion, 105
 order reduction, 99, 106, 109
 stiff, 95, 99, 106, 109, 111, 112, 121
 total variational diminishing (TVD), 109

S

Shock-interface, 154
Shock waves
 conservative form, 13, 19–22, 54, 130, 134

 quasi-conservation form and error in shock in speed, 129
 shock speed, 19, 54, 160, 184
 shock waves in nonlinear scalar conservation laws, 233
 time to shock formation, 140, 147, 157
Specific heat at constant volume, 211, 214
Specific heat ratio, 148, 160, 161, 170, 189, 197, 214

Spectral Deferred Correction (SDC), 95, 112, 114–117, 120, 121, 228
Speed of sound
 definition, 211
 ideal gas, 15, 210
 stiffened-gas, 160, 197, 203
Strong stencil separation, 76, 78, 79

T

Targeted Essentially Non-Oscillatory (TENO), 43, 56–58, 79, 83, 127, 133, 135, 138, 139, 164, 169
Thermal conductivity, 18, 192, 193, 197, 200, 201, 225, 226
Thermal diffusivity, 18
Thermodynamic cyclic relation, 11
Two-fluid, 128, 133, 135
Two-phase, 1, 118, 180, 210, 217, 228

U

Ultrasound pulse, 139, 141

W

Weighted Essentially Non-Oscillatory (WENO)
 algorithm, 44, 48
 characteristic variables, 50, 51, 138
 optimal weights, 56
Wood speed of sound, 15, 70, 71, 142, 161, 177, 199, 203, 210, 211, 215, 221, 223, 227

The manufacturer's authorised representative in the EU is Springer Nature Customer Service Centre GmbH, Europaplatz 3, 69115 Heidelberg, Germany. If you have any concerns regarding our products, please contact ProductSafety@springernature.com

Printed and bound by CPI Group (UK) Ltd, Croydon, CR0 4YY

26/03/2026

02078965-0002